W9-AYI-228

Urban
Remote
Sensing

Urban Remote Sensing

Edited by
Qihao Weng
Dale A. Quattrochi

CRC Press
Taylor & Francis Group
Boca Raton London New York

CRC Press is an imprint of the
Taylor & Francis Group, an informa business

CRC Press
Taylor & Francis Group
6000 Broken Sound Parkway NW, Suite 300
Boca Raton, FL 33487-2742

© 2007 by Taylor & Francis Group, LLC
CRC Press is an imprint of Taylor & Francis Group, an Informa business

No claim to original U.S. Government works
Printed in the United States of America on acid-free paper
10 9 8 7 6 5 4 3 2 1

International Standard Book Number-10: 0-8493-9199-7 (Hardcover)
International Standard Book Number-13: 978-0-8493-9199-6 (Hardcover)

Library of Congress Cataloging-in-Publication Data

Weng, Qihao.
 Urban remote sensing / Qihao Weng and Dale A. Quattrochi.
  p. cm.
 Includes bibliographical references and index.
 ISBN 0-8493-9199-7 (978-0-8493-9199-6)
  1. City planning--Remote sensing. 2. Land use, Urban--Remote sensing. 3. Urban geography--Remote sensing. I. Quattrochi, Dale A. II. Title.

HT166.W46 2006
307.1'216--dc22   2006013706

Visit the Taylor & Francis Web site at
http://www.taylorandfrancis.com

and the CRC Press Web site at
http://www.crcpress.com

Editors

Qihao Weng was born in Fuzhou, China in 1964. He received a B.A. in geography from Minjiang University in 1984, an M.S. in physical geography from South China Normal University in 1990, an M.A. in geography from the University of Arizona in 1996, and a Ph.D. in geography from the University of Georgia in 1999. He is currently an associate professor of geography and director of the Center for Urban and Environmental Change at Indiana State University. His research focuses on remote sensing and GIS analysis of urban ecological and environmental systems, land-use and land-cover change, urbanization impacts, and human–environment interactions.

Dr. Weng is the author of more than 50 peer-reviewed journal articles and book chapters. He has been the recipient of the Robert E. Altenhofen Memorial Scholarship Award by the American Society for Photogrammetry and Remote Sensing (1999) and the Best Student-Authored Paper Award from the International Geographic Information Foundation (1998). In 2006, he received the Theodore Dreiser Distinguished Research Award by Indiana State University, the university's highest research honor bestowed to faculty. He has worked extensively with optical and thermal remote sensing data, primarily for urban heat island study, land-cover and impervious surface mapping, urban growth detection, spectral mixture analysis, and socioeconomic characteristics derivation.

Dale Quattrochi is a senior research scientist with the NASA Marshall Space Flight Center in Huntsville, Alabama, and has over 26 years of experience in the field of Earth science remote sensing research and applications. Dr. Quattrochi's research interests focus on application of thermal remote sensing data for analysis of heating and cooling patterns across the diverse urban landscape as they impact overall local and regional environments. He is also conducting research on applications of geospatial statistical techniques, such as fractal analysis and multiscale remote sensing data.

Dr. Quattrochi is the recipient of numerous awards, including the NASA Exceptional Scientific Achievement Medal, NASA's highest science award, which he received for his research on urban heat islands and remote sensing. He is also a recipient of the Ohio University College of Arts and Sciences Distinguished Alumni Award. Dr. Quattrochi is the coeditor of two books: *Scale in Remote Sensing and GIS* (with Michael Goodchild), published in 1997 by CRC/Lewis Publishers, and *Thermal Remote Sensing in Land Surface Processes* (with Jeffrey Luvall), published in 2004 by CRC Press. He received his Ph.D. degree from the University of Utah, his M.S. degree from the University of Tennessee, and his B.S. degree from Ohio University, all in geography.

Contributors

Sharolyn Anderson
University of Denver
Denver, Colorado

Keith C. Clarke
University of California, Santa Barbara
Santa Barbara, California

Daniel Comarazamy
University of Puerto Rico
Mayagüez, Puerto Rico

Fabio Dell'Acqua
University of Pavia
Pavia, Italy

Manfred Ehlers
University of Osnabrück
Osnabrück, Germany

Christopher D. Elvidge
National Geophysical Data Center
of NOAA
Boulder, Colorado

Paolo Gamba
University of Pavia
Pavia, Italy

Jorge E. González
Santa Clara University
Santa Clara, California

Jeff Hemphill
University of California, Santa Barbara
Santa Barbara, California

Geoffrey M. Henebry
South Dakota State University
Brookings, South Dakota

Martin Herold
Friedrich Schiller University
Jena, Germany

Patrick Hostert
Humboldt University of Berlin
Berlin, Germany

John A. Kelmelis
U.S. Department of State
Washington, DC

Guiying M. Li
Indiana State University
Terra Haute, Indiana

Weiguo Liu
The University of Toledo
Toledo, Ohio

Xiaohang Liu
San Francisco State University
San Francisco, California

Dengsheng Lu
Indiana University
Bloomington, Indiana

Jeffrey C. Luvall
NASA, Marshall Space Flight Center
Huntsville, Alabama

Soe W. Myint
Arizona State University
Tempe, Arizona

Janet Elizabeth Nichol
The Hong Kong Polytechnic University
Kowloon, Hong Kong

Ana J. Picón
University of Puerto Rico
Mayagüez, Puerto Rico

Douglas L. Rickman
NASA, Marshall Space
 Flight Center
Huntsville, Alabama

Dar A. Roberts
University of California, Santa Barbara
Santa Barbara, California

Aparajithan Sampath
Purdue University
West Lafayette, Indiana

Sebastian Schiefer
Humboldt University of Berlin
Berlin, Germany

Karen C. Seto
Stanford University
Stanford, California

Jie Shan
Purdue University
West Layfayette, Indiana

Uwe Soergel
University of Hanover
Munich, Germany

Conghe Song
University of North Carolina at
 Chapel Hill
Chapel Hill, North Carolina

Uwe Stilla
Technical University of Munich
Munich, Germany

Zhanli Sun
University of Illinois at
 Urbana–Champaign
Champaign, Illinois

Paul C. Sutton
University of Denver
Denver, Colorado

Matthew J. Taylor
University of Denver
Denver, Colorado

Yong Tian
University of Massachusetts
Boston, Massachusetts

Qihao Weng
Indiana State University
Terre Haute, Indiana

Man Sing Wong Charles
The Hong Kong Polytechnic University
Kowloon, Hong Kong

George Xian
U.S. Geological Survey Center
Sioux Falls, South Dakota

Guoqing Zhou
Old Dominion University
Norfolk, Virginia

Acknowledgments

We wish to extend our most sincere thanks to all the contributors to this book for making this endeavor possible. Moreover, we offer our deepest appreciation to all the reviewers, who have taken precious time from their busy schedules to review the chapters in this book: Toby Carlson, Giles Foody, Tung Fung, Paolo Gamba, Ayman Habib, Jack Harvey, Geoffrey Henebry, Martin Herold, Guiying Li, Weiguo Liu, Xiaohang Liu, Dengsheng Lu, David Martin, Soe W. Myint, Janet Nichol, Ruiliang Pu, Fang Qiu, Jeffery Shan, Uwe Soergel, Conghe Song, Uwe Stilla, David Streutker, Paul Sutton, Changshan Wu, Fulong Wu, George Xian, Yichun Xie, and Guoqing Zhou. Finally, we are indebted to our families for their enduring love and support. It is our hope that the publication of this book will provide stimulation to students and researchers to perform more in-depth work and analysis on the applications of remote sensing to urban and suburban areas.

Table of Contents

An Introduction to Urban Remote Sensing

Qihao Weng and Dale A. Quattrochi

The twenty-first century is the first "urban century," according to the United Nations Development Program. The focus on cities reflects awareness of the growing percentage of the world's population that lives in urban areas. In environmental terms, as has been pointed out at the U.N. Conference on Human Settlement, cities and towns are the original producers of many of the global problems related to waste disposal and air and water pollution. The need for technologies that will enable monitoring the world's natural resources and urban assets and managing exposure to natural and man-made risks is growing rapidly.

This need is driven by continued urbanization. In 2000, about 3 billion people, representing about 40% of the world's population, lived in urban areas. Urban population will continue to rise substantially over the next several decades according to the United Nations, and most of this growth will be in developing countries. The number of megacities (i.e., cities with populations of more than 10 million) will increase to 100 by 2025. Thus, there is critical need to understand urban areas to help improve and foster environmental and human sustainability of cities around the world.

Over the past decades, the majority of remote sensing work has been focused on natural environments. Applying remote sensing technology to urban areas is relatively new. With the advent of high-resolution imagery and more capable techniques, urban remote sensing is rapidly gaining interest within the remote sensing community. Driven by advances in technology and societal needs, biannual international symposia on remote sensing of urban areas (since 1997) and remote sensing and data fusion (since 2001) have been very successful. Recently, several journals have published special issues on remote sensing of urban areas, including: *Remote Sensing of Environment*, 2003, vol. 86, issue 3; *IEEE Transactions on Geoscience and Remote Sensing*, 2003, vol. 41, issue 9; *Photogrammetric Engineering and Remote Sensing*, 2003, vol. 69, issue 9; and the *Remote Sensing of Environment* special issue on urban thermal remote sensing published in 2006. It appears that increasing numbers of universities in the United States and other countries have started offering courses on remote sensing of urban and suburban areas.

To meet the growing interest in applications of remote sensing technology to urban and suburban areas, we have assembled a team of experts to write a book on urban remote sensing. For the first time, this book systematically examines all aspects of the field. Each chapter follows a similar prototype, including such elements as literature review of basic concepts and methodologies, case studies, methods for applying up-to-date techniques, and analysis of results. This book may be used as a textbook for upper-division undergraduate and graduate students; however, it can also serve as a reference book for researchers or individuals in academia and governmental and commercial sectors who are interested in remote sensing of cities.

This book consists of five parts. Part I focuses on data, sensors, and system considerations and algorithms for urban feature extraction. Part II analyzes urban landscapes in terms of composition and structure, using subpixel analysis techniques particularly. Part III presents methods for monitoring, analyzing, and modeling urban growth. Part IV illustrates various approaches to urban planning and socioeconomic applications of urban remote sensing. Part V assesses progress made to date, identifies existing problems and challenges, and demonstrates new developments and trends in urban remote sensing.

The three chapters in Part I are concerned with extracting urban buildings and other features. These researchers utilize an electro-optical sensor and two range sensors — LIDAR and interferometric SAR, respectively. Chapter 1 describes algorithms and methods for large-scale urban orthoimage generation. The experiment conducted by these contributors in Denver, Colorado, demonstrates that buildings and bridges can be placed with accurate upright, planimetric locations and that sidewalks and roads can be completely visible. LIDAR (light detection and ranging) technology provides a unique and promising solution to extracting urban features (Ackermann, 1999). Chapter 2 presents an approach to building extraction from nonground LIDAR points, with three sequential steps: building segmentation, boundary tracing, and regulation. The approach was tested with success in urban areas in Baltimore, Maryland, Osaka, Japan, and Toronto, Canada. Chapter 3 focuses on acquisition and segmentation of interferometric SAR data for reconstruction of buildings with a model-based approach.

One of the most widely used applications of remote sensing technology in urban areas focuses on the characterization, identification, classification, and quantification of urban construction materials, composition, and structure. Chapter 4 applies the vegetation–impervious surface–soil concept (Ridd, 1995) and spectral mixture analysis technique for a subpixel analysis of the urban landscape structure and dynamics in Indianapolis, Indiana. The potentials and limitations of spectral mixture analysis for characterizing urban landscapes are also examined. Chapter 5 introduces a new approach, the Bayesian spectral mixture analysis, in which endmember spectral signatures are no longer assumed as constants. Instead, they are represented by probability density functions and thus can incorporate the natural variability of endmember spectral signatures. Because of the complexity of urban landscapes, lack of spatial consideration in traditional per-pixel classifiers, and inconsistencies between scale of observation (i.e., pixel resolution) and spatial characteristics of the target (Mather, 1999), traditional image classification approaches such as the maximum-likelihood classifier are ineffective in classifying urban land use and land cover. Chapter 6 examines how various geospatial approaches can be used to extract textures of land-use or land-cover classes to improve classification accuracy. Urban areas are characterized by a large diversity of materials, such as impervious surfaces, vegetation, soils, water, and so on. Chapter 7 applies imaging spectrometry to urban areas, especially for characterization of artificial and man-made surfaces. The authors provide a summary of the current state of knowledge of imaging spectrometry and, through case studies, also show how this technology can support urban applications.

Part III focuses on urban land-change detection, growth monitoring, modeling, and prediction. Chapter 8 applies a neural network-based spatiotemporal data mining method to simulate and predict urban expansion in St. Louis, Missouri. Chapter 9 uses subpixel impervious surfaces derived from satellite remote sensing data in conjunction with digital orthophotography to analyze urban expansion in the Las Vegas, Nevada, metropolitan area from 1984 to 2002 and in the Tampa Bay, Florida, area from 1991 to 2002. Subpixel impervious surfaces were found to be capable of providing quantitative measurements of the spatial extents, development densities, and temporal changes of urban land. Chapter 10 examines the potential of remote sensing as it may contribute to urban growth theory and modeling. By citing examples of urban dynamics analysis, this chapter has outlined a general framework for urban growth and developed the basis for combining remotely sensed data and spatial measurements (metrics) to aid in development and validation of new urban growth theory assessments.

Remote sensing data and research results have been applied to many environmental and socioeconomic applications, such as urban heat islands (Quattrochi et al., 2000; Weng, 2001), urban environmental quality (Nichol and Wong, 2005), and estimation of demographic and socioeconomic variables (Lo and Faber, 1997; Thomson and Hardin, 2000; Li and Weng, 2005). The five chapters in Part IV illustrate the current state of these applications. Chapter 11 investigates the impact of urbanization on land-surface temperatures and urban heat island phenomenon in San Juan, Puerto Rico, using remote sensing, *in situ* field measurement, and numerical modeling techniques. Chapter 12 investigates integration of environmental data sets derived from remotely sensed images with other environmental variables for assessment of urban environmental quality in Hong Kong. Urban environmental quality index maps were generated at the levels of pixels and administrative regions, with principal component analysis and GIS overlay as the methods of data integration. Chapter 13 examines various methods for population estimation and interpolation and illustrates them with specific examples. The examples in this chapter particularly highlight the use of recently available high spatial-resolution satellite data to study intraurban population characteristics. Chapter 14 provides a summary of ways in which nighttime imagery has been used to study socioeconomic variables and urban environments and suggests potential improvements on these methods if finer resolution sensors become available. In particular, the Defense Meteorological Satellite Program's Operational Linescan System (DMSP OLS) data products are explored for use in understanding urban and exurban areas. Chapter 15 develops a methodology for assessing urban quality of life based on integration of Landsat Enhanced Thematic Mapper Plus (ETM+) imagery and Census 2000 data within a GIS framework. The model developed for Marion County, Indiana, was applied to Monroe and Vigo Counties in the same state for validation.

The last part of the book is concerned with the current state of urban remote sensing, problems encountered in the past, and trends for future development.

Remote sensing of urban areas has relied primarily on three spectral regions: visible through near infrared, thermal infrared, and microwave. Chapter 16 explores strengths and weaknesses of using the middle infrared (3 to 5 μm) spectral region for characterization of urban and suburban environments and makes suggestions for future direction of development. Chapter 17 discusses recent development of very high- and ultrahigh-resolution satellite, digital airborne, and LIDAR sensors and their impacts on processing techniques. The final chapter, Chapter 18, compares the capacities and trade-offs of very high spatial resolution and very high spectral resolution sensors for urban mapping. A case study of land-cover classification around the area of the castle of Pavia, Italy suggests that when high-spectral and high-spatial resolutions are not available at the same time, the former seems to be more valuable than the latter, provided that some minimum requirements are met for both.

REFERENCES

Ackermann, F., 1999. Airborne laser scanning — present status and future expectations, *ISPRS J. Photogrammetry Remote Sensing*, 54(2–3), July, 64–67.

Li, G. and Weng, Q., 2005. Using Landsat ETM+ imagery to measure population density in Indianapolis, Indiana, *Photogrammetric Eng. Remote Sensing*, 71, 947–958.

Lo, C.P. and Faber, B.J., 1997. Integration of Landsat Thematic Mapper and census data for quality-of-life assessment, *Remote Sensing Environ.*, 62, 143–157.

Mather, P., 1999. Land-cover classification revisited, in Atkinson, P.M. and N.J. Tate (Eds.), *Advances in Remote Sensing and GIS Analysis*, John Wiley & Sons, New York, 7–16.

Nichol, J. and Wong, M.S., 2005. Modeling urban environmental quality in a tropical city, *Landscape Urban Plann.*, 73, 49–58.

Quattrochi, D.A., Luvall, J.C., Rickman, D.L., Estes, M.G., Laymon, C.A., and Howell, B.F., 2000. A decision support information system for urban landscape management using thermal infrared data, *Photogrammetric Eng. Remote Sensing*, 66, 1195–1207.

Ridd, M.K., 1995. Exploring a V-I-S (vegetation–impervious surface–soil) model for urban ecosystem analysis through remote sensing: comparative anatomy for cities, *Int. J. Remote Sensing*, 16, 2165–2185.

Thomson, C.N. and Hardin, P., 2000. Remote sensing/GIS integration to identify potential low-income housing sites, *Cities*, 17, 97–109.

Weng, Q., 2001. A remote-sensing GIS evaluation of urban expansion and its impact on surface temperature in the Zhujiang Delta, China, *Int. J. Remote Sensing*, 22, 1999–2014.

Part I

Urban Feature Extraction

1 True Orthoimage Generation for Urban Areas with Very High Buildings

Guoqing Zhou and John A. Kelmelis

CONTENTS

1.1 INTRODUCTION

Digital orthoimages are a critical component of the national spatial data infrastructure (NSDI) and *The National Map* (Federal Geographic Data Committee, 1997; Kelmelis et al., 2003; Kelmelis, 2003; Maitra, 1998). Digital orthophotos contain the image characteristics of a photograph and the geometric properties of a map and thus can (1) serve as a geospatial foundation upon which to add detail and attach attribute information; (2) provide a base on which to register and compile other themes of data accurately; and (3) orient and link the results of an application to the landscape (Federal Geographic Data Committee, 1995).

Digital orthophotos are also a necessary part of the strategy to update other layers of *The National Map* (http://nationalmap.usgs.gov/) so that the features mapped on the U.S. Geological Survey (USGS) national geospatial database and derived maps are current. The highly detailed images will serve as a backdrop to other layers and as source material for locating some of the features to be mapped or used independently for analysis or as backdrops for other maps. Thus, the highly versatile orthoimages can serve many purposes for the public.

The National Digital Orthophoto Program (NDOP) was first proposed in 1990 by the U.S. Department of Agriculture's (USDA) Natural Resources Conservation Service, Farm Service Agency, and the USGS (USGS, 1998). The primary goal of this program is to ensure public domain availability of digital orthophoto quadrangle (DOQ) data for the nation. The USGS began to produce DOQs in 1991 and currently has nearly 50,000 available for distribution (USGS, 1996, 1998). When digital orthophoto quarter quadrangles are considered, there are more than 220,000 available for distribution. Once-over DOQ coverage of the conterminous United States is complete and revisions have been completed for some locations.

The original program plan was that the DOQs would be updated on a 5- or 10-year cycle in areas where land use change is most rapid. The success of this effort can be seen in the large number and variety of digital orthophoto products and services available today (USGS, 1996). The NDOP was based on quarter-quadrangle-centered aerial photographs (3.75 minutes of longitude and latitude in geographic extent) obtained at a nominal flying height of 20,000 ft above mean terrain with a 6-in. focal-length camera (photo scale = 1:40,000). The photograph was acquired under the cooperatively funded National Aerial Photography Program (NAPP). The Standard for Digital Orthophotos formulated by USGS (1996) did not consider the requirements of large-scale city orthophoto generation. Additionally, early procedures and algorithms for digital orthophoto generation were based on earlier USGS mapping operations, such as field control, aerotriangulation (using photogrammetric equations derived in the early 1920s), and 2.5-dimensional digital elevation models.

The procedures and algorithms used in the 1990s are not appropriate for large-scale city orthorectification when generating large-scale urban orthophotos. Using the existing algorithms and methods available in commercial software can create problems such as incomplete orthorectification, occlusion, ghost image, shadow, etc. A comprehensive discussion regarding these problems has been given by Zhou et al. (2003, 2004, 2005). The existence of these problems indicates that the conventional orthorectification methods (procedures and algorithms) used in the 1990s are not able to orthorectify the objects into their correct and upright positions and remove sufficient radiometric differences for large-scale urban aerial images. As a result, the usefulness of digital orthoimages in industry, government, and elsewhere is significantly reduced.

Because errors in these incompletely rectified large-scale city orthoimage maps cannot be tolerated when used for updating and planning urban tasks, the generation of so-called *true* orthoimages has been of increasing interest in the past several years. For example:

Skarlatos (1999) and Joshua (2001) demonstrated that building occlusions significantly influenced image quality and accuracy of orthoimages.

Amhar et al. (1998) and Schickler and Thorpe (1998) considered the hidden effects introduced by abrupt changes of surface height (e.g., buildings and bridges).

Schickler and Thorpe (1998) and Mayr (2002) considered seamless mosaicking around fill-in areas to reduce gray-value discontinuities.

Rau et al. (2002) treated enhancements of image radiometry, demonstrating a suitable enhancement technique to restore information within building shadow areas.

Jauregui et al. (2002) presented a procedure for orthorectifying aerial photographs to produce and update terrain surface maps.

Vassilopoulou et al. (2002) used IKONOS images to generate orthoimages for monitoring volcanic hazards on Nisyros Island, Greece, and Siachalou (2004) used IKONOS images to generate the urban orthoimage.

Cameron et al. (2000) analyzed orthorectified aerial photographs to measure changes of native pinewood of Scotland, and Passini and Jacobsen (2004) analyzed the accuracy of orthoimages from very high resolution imagery.

Biason et al. (2004) further explored the automatic generation of true orthoimages.

Despite these great efforts, the issues of true orthorectification in urban areas are not thoroughly addressed. This chapter presents our comprehensive study and in-depth understanding of urban orthophoto generation, including algorithms and data-processing procedures. The study was undertaken jointly by Old Dominion University, USGS, and the private sector for future large-scale orthophoto generation and deployment in cities. Our research results on true three-dimensional orthophoto generation, including algorithms and data processing procedures for future national large-scale city orthophoto generation deployment, are presented in this chapter.

1.2 PRINCIPLE OF URBAN TRUE ORTHOIMAGE GENERATION

As described in the previous section, the digital surface model (DSM)-based orthorectification method for urban large-scale aerial images cannot make vertical features in the orthorectified image appear in their correct and upright positions (Zhou et al., 2003, 2005). Therefore, true orthoimage generation methods usually divide the orthorectification into three basic steps: (1) digital terrain model (DTM)-based orthoimage generation; (2) digital building model (DBM)-based orthoimage generation; and (3) their merging (Figure 1.1). Here, the DTM is an elevation model that describes the surface of the terrain without buildings and vegetation; the DBM is defined to describe the surface of man-made object details, for which a real three-dimensional representation

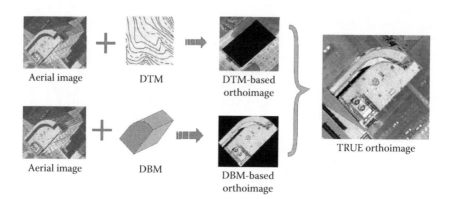

Aerial image   DTM   DTM-based orthoimage

Aerial image   DBM   DBM-based orthoimage

TRUE orthoimage

FIGURE 1.1 The procedure of true orthophoto generation.

leads to a much more complex data structure than is generally used for DTM; and DSM is defined as the representation of the entire surface of the observed region. Therefore, the DSM is a combination of DTM and one or multiple DBMs.

1.2.1 DTM-Based Orthoimage Generation

DTM-based orthoimage generation for correcting relief displacement caused by terrain can be completed by the traditional orthorectification method. This method has been described in many textbooks and in the literature, such as Chen and Rau (1993) and Zhou et al. (2002). The basic mathematical model of this algorithm is the photogrammetric collinearity equation, which requires a DTM and the interior and exterior orientation elements of an aerial image. The collinearity equations are used to calculate the relation of features on the image to their corresponding ground position.

1.2.2 DBM-Based Orthoimage Generation and Occlusion Detection

DBM-based orthoimage generation only orthorectifies the displacement caused by buildings. In this process, the displacement caused by terrain is ignored; the occluded buildings need to be detected and the occlusion is compensated for. At present, an effective and commonly used method for occlusion detection is to apply the DBM to calculate the distance between object surfaces to projection center. This distance is called *Z distance*, and the method is called *Z*-buffer algorithm (Amhar et al., 1998).

Based on the DBM, the Z-buffer is generated by projecting each DBM surface polygon onto the first image plane by using collinearity equations. The projected polygon is rasterized and filled with the Z distance and the polygon's identification (ID) code. The identification code is used to distinguish between walls and roofs. The process of filling by means of resampling original image pixels will first check the ID codes and then rectify only pixels from the "roof" — the wall will not be orthorectified.

Because occluded buildings cannot be orthorectified, this step leaves holes in the orthoimage. These are filled with brightness value 0 so that holes can be compensated for by using another orthoimage in which the building is visible. This orthoimage is

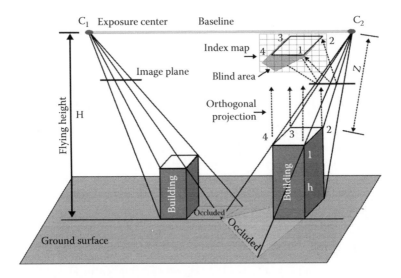

FIGURE 1.2 Occlusion analysis and DBM-based orthoimage generation.

called a *slave orthoimage* (Rau et al., 2002; Amhar et al., 1998; Zhou et al., 2005). The generated orthoimage contains buildings only and is thus called a *DBM-based orthoimage*. Figure 1.2 shows the relationship of the ground surface, buildings, occluded areas, and exposure centers and the various distances that must be measured or calculated to conduct occlusion analysis and other relevant photogrammetric analyses.

1.2.3 NEAR-TRUE ORTHOIMAGE GENERATION

A true orthoimage is created by merging the DTM-based orthoimage and DBM-based orthoimage. Because the background is black (i.e., the brightness value 0), Amhar et al. (1998) and Zhou et al. (2003, 2005) employed a very simple operation, called logical <OR> operation, to carry out this merging operation. This operation is very effective and fast in computation. Since this step still cannot compensate all occluded areas, the resulting orthoimage contains black patches. This intermediate orthoimage is thus called near-true orthoimage (NTOI).

1.2.4 OCCLUSION COMPENSATION

An NTOI still contains black patches that need to be refilled from neighboring slave orthoimages with a considerable overlap. For this purpose, an algorithm for optimum selection of slave orthoimage patches has been developed based on the fact that the relief displacement at image nadir is zero. Thus, this method first calculates the distance of each patch to its nadir and then chooses the patch with the shortest distance.

In addition, when the black patches in the NTOI are refilled, the seam lines might fall on ground features not modeled in the DBM, such as trees, cars, or shadows, that probably have different brightness values between master NTOI and slave NTOI images. Therefore, the least-squares correlation algorithm has been employed to minimize radiometric difference between seam lines to obtain a seamless mosaic when compensating for occluded areas. This algorithm is used to calculate the similarity of the image contents between "master" and "slave" orthoimages. The pull range and search space are determined on the basis of the image texture and quality.

1.3 THREE-DIMENSIONAL URBAN MODEL GENERATION

As mentioned above, DBM-based orthorectification requires an exact DBM, which describes the building structure, three-dimensional coordinates, topologic relationship, etc. Therefore, a three-dimensional urban model must be developed. This can be done using photogrammetric means, building digital models from engineering drawings, or, much more efficiently, using airborne light detection and ranging (LIDSR) technology.

1.3.1 DATA STRUCTURE

Although many models have been proposed (e.g., Grün and Wang, 1998; Zhou et al., 2000; Zlatanova, 2000), the effective organization of a three-dimensional urban (building) model is a rather challenging task because different applications (e.g., city planning, communication design, tourism, pollution distribution, military security operation, etc.) require different data types and manipulation functions (Zhou et al., 2005; Zlatanova, 2000). The geospatial data needed for true orthoimage generation include (1) digital terrain data; (2) images; and (3) spatial objects (e.g., buildings). The vector triangular irregular network (TIN) data structure is capable of precisely representing the ground surface even though the urban terrain is complex (Schickler and Thorpe, 1998). The digital images, including original images, generated orthoimages, and generated DTM or DBM images, are usually represented in grid format with rows and columns. These data structures are used in this chapter.

Spatial objects are abstractly understood as the four types of geometric objects (Zhou et al., 2000): point objects, line objects, surface objects, and body objects. They represent zero-, one-, two-, and three-dimensional objects, respectively. Of these four types of objects, "point" is the basic geometric element. For example, a point can present a point object or the starting or ending point of an edge. An edge is a line segment that is an ordered connection between two points: beginning point and ending point. A facet is completely described by the ordered edges that define its border. A polyhedron is described by the ordered facets. Thematic data can be attached to a point, line, facet, or polyhedron as appropriate. Image data can be attached to a facet and each facet is related to an image patch through a corresponding link (Figure 1.3).

These defined object types are taken as the basic classes of primary geographic entities from which other geographic entities are derived. In addition, in the proposed data model, each object is identified by a defined attribute data, such as type identifier (TI), and thus the four types of objects are referred to as PI (point identifier), LI (line identifier), FI (facet identifier), and PHI (polyhedron identifier), respectively. The other attribute data (e.g., thematic data and geometric data) can be attached to each type of object.

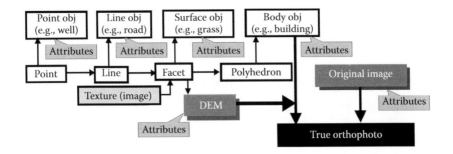

FIGURE 1.3 Data structure of the proposed three-dimensional urban model for true orthoimage generation.

1.3.2 Implementation via Relational Database

The data structure described in Figure 1.3 can be implemented by relational database technology in which each type of object is defined as a table. For example, a table describing a point object includes three terms: point identification (PID), attribute linkage (AL), and XYZ coordinates. The PID is for identification of a point object and the AL is a linkage to the attribute table describing point properties. The XYZ coordinates are for geometric description of a point. Similarly, other types of objects, such as lines, can be referenced.

On the basis of the relational structure, the query of a geometrical description for a type of object is easily realized, and the topological relationships between geometrical elements are implicitly defined by the data structure and explicitly stored in a relational database model (Figure 1.4). For example, a point object is represented by a distinct point element. The line object is described by ordered edges. Similarly, the body object is described by a polyhedron that is described by the ordered facets. Thus, the topological relationships between point and edge, edge and facet, and facet

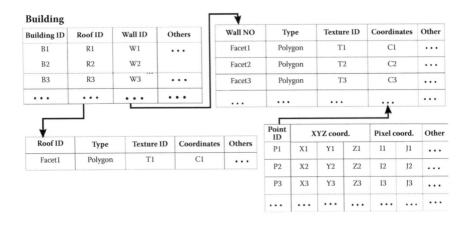

FIGURE 1.4 Relational data model applied for true orthoimage generation.

and polyhedron are registered by links between the geometrical elements. Zhou et al. (2003) offer a detailed discussion for a relational database model of three-dimensional city data for true orthoimage generation.

1.4 CITY STREET VISIBILITY ANALYSIS

Traditional flight planning usually requires approximately a 65% endlap for two adjacent photographs along a flight strip and 30% sidelap across flight strips. It has been demonstrated that the early specifications in urban areas with high buildings encountered challenges as a result of impacts of the occlusions, deep discontinuities, and shadows (Zhou et al., 2003, 2005). To minimize their impact on urban true orthophoto generation, we investigated and analyzed the relationship between occlusion and building height, street width, flying height, and AFOV (angular field of view) of the camera. They are modeled by Zhou et al. (2005):

$$B*h/H \leq W \qquad (1.1)$$

where B represents the length of the air base, H is flying height, h denotes a building height, and W is street width.

Using the model derived in Figure 1.4, we simulated three experiments to investigate the relationship among the four factors: flying height, air base, building height, and street width. Some significant results are drawn up. For example, if a street width is 20 m and air base is 654 m, a flying height of at least 3270 m must be achieved to ensure complete visibility of the street.

1.5 EXPERIMENTS AND ANALYSES

1.5.1 DATA SET

The experimental field is located in downtown Denver, Colorado. The six original aerial images from two flight strips were acquired using an RC30 aerial camera at a focal length of 153.022 mm on April 17, 2000. The flying height was 1650 m above the mean ground elevation of the imaged area. Figure 1.5 illustrates the configuration of the photogrammetric flight mission and the available DSM. Two of the six aerial images (DV1119 and DV1120) covered the downtown area. The endlap of the images is about 65% and the sidelap is approximately 30%. A DSM obtained via Z/I photogrammetric workstation with computer-interactive operation in the central part of downtown is available. The accuracy of planimetric and vertical coordinates in the DSM is approximately 0.1 and 0.2 m, respectively. The horizontal datum is GRS 1980 and the vertical datum is NAD83. The visualization of the DSM is depicted in the bottom right-hand section of Figure 1.5.

1.5.2 URBAN THREE-DIMENSIONAL MODEL

With the DSM data provided by Analytic Survey Inc. at Colorado Springs, Colorado, each building vector datum is extracted using the ERDAS/Imagine software in a manner of computer–human interaction, and all building vector data are managed

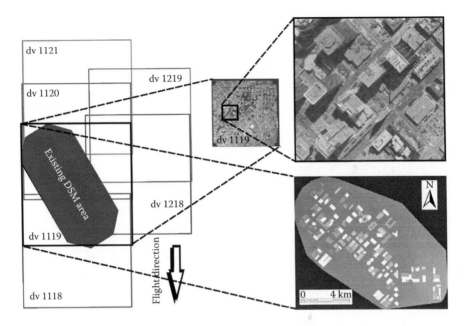

FIGURE 1.5 Six aerial images from two strips and digital surface model covering downtown Denver, Colorado. The brightness in the DSM represents surface (ground or buildings) height.

using the developed data structure described in Section 1.3. Figure 1.6 is the visualization of the DBM model at a ground resolution of about 25.4 cm for each pixel. The DTM is generated by removing the digital buildings from the DSM.

1.5.3 SHADOW DETECTION AND RESTORATION

Another important concern in urban large-scale orthoimage generation is detection and removal of shadow. Many attempts in this research field have been made in the past decades (e.g., Liow and Pavlidis, 1990; Noronha and Nevatia, 2001; Jaynes et al., 2004). Most did not consider the self-shadow and typically focused on the umbra, considering the penumbra as a particular case of umbra. Moreover, they only took advantage of the image gray information and two-dimensional object geometric shapes when detecting the shadowed areas.

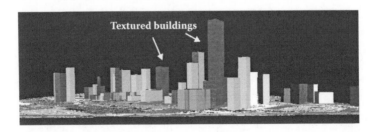

FIGURE 1.6 The city model of Denver created by our system.

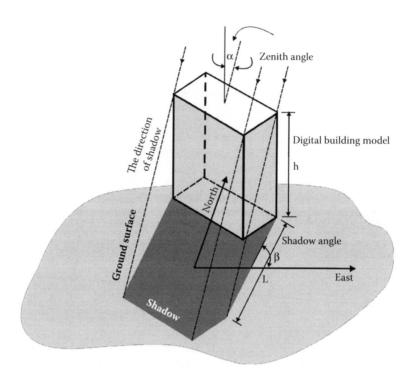

FIGURE 1.7 Shadow detection using the DBM.

This chapter presents an innovative method that makes full use of DBM data for building shadow detection. However, it does not focus on detecting the self-shadow because walls of buildings will be invisible in the true orthoimage. This proposed method is based on the following facts (Figure 1.7):

The length of shadow depends on building height and the zenith of the light source (sun); the length of shadow of each building can be measured using a building model if the sun's zenith is known. The direction of shadow depends on the position of the sun relative to the buildings.

The zenith and the position of the light source relative to buildings at the time of imaging remain a constant. This means that the direction of the shadow for all buildings within an aerial image is the same.

Based on the method outlined here, the relationship between the length of the shadow and the building height is established by

$$\tan(\alpha) = L/h \qquad (1.2)$$

where h is building height and α is the zenith of the light source (sun).

The solved zenith from Equation 1.2 can be used to calculate the length of other buildings' shadows because the zenith is constant within an aerial image. Furthermore, the coordinates of shadow boundary in a defined local coordinate system can be calculated by

$$x = \text{L} \cos (\beta)$$

$$y = \text{L} \sin (\beta) \tag{1.3}$$

where x and y are coordinates of the shadow boundary and β is the direction of shadow in a given local coordinate system. Because the local coordinates of the shadow boundary can be transformed to image coordinates, the shadowed areas can be detected.

We first manually determined the relationship between the length and the direction of one building shadow and the three-dimensional building model. With the established shadow relationship, the other building shadows can be detected using the DBM, which is stored in a DBM database.

After the occluded areas are detected, the following work is to restore the shadowed areas. Schott et al. (1988) used a histogram adjustment algorithm. We applied this approach to change the mean and variance of the shadowed areas by matching the mean and variance of an image with no shadows. Figure 1.8 shows the experimental results before and after the shadow treatment. As can be seen, the shadowed objects caused by tall buildings in Figure 1.8 have been substantially restored and the image quality of these shadowed areas is now comparable to the ones around them.

1.5.4 OCCLUSION DETECTION AND COMPENSATION

Occlusion detection in true orthorectification applies a visibility analysis technology, called Z-buffer, that was presented in Section 1.2.2 (also see Amhar et al., 1998, and Rau et al., 2002). The method in our study consists of two matrixes. One stores the distance from the projection center to the surface point, whose corresponding position in the image plane is first calculated with the imaging system parameters (interior and exterior orientations) using the collinearity condition. In the other, the ground surface plane is used as a binary index map with the same resolution, dimension, and projection properties as the DBM. When an area is occluded, several surface points have the same coordinates in the image plane. Under these circumstances, we will select the shorter distance to store and mask the index map at the corresponding location, where the distance to the projection center is longer.

This method allows us to identify all the hidden areas in the orthophoto (Figure 1.8a). Using the index map (Figure 1.9) generated by the Z-buffer algorithm, we can mask occluded areas. When the actual orthorectification begins, we need only find the brightness values for the surface points that are not masked in the index map. All masked areas can therefore be set as blank or other background values to indicate that they are invisible or occluded in the resulting orthophoto (Figure 1.8b).

Occlusion compensation is implemented by refilling the occluded areas from neighbor slave orthoimages. This process involves mosaicking, for which the automatic selection of seam lines from slave orthoimages has been described in Section 1.2.3.

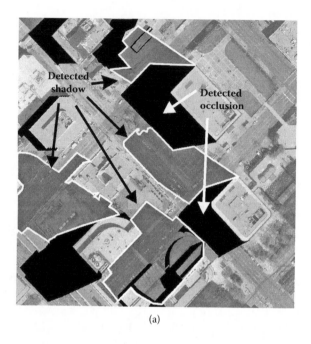

(a)

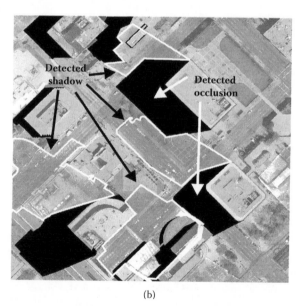

(b)

FIGURE 1.8 Comparison between (a) before shadow treatment and (b) after shadow treatment.

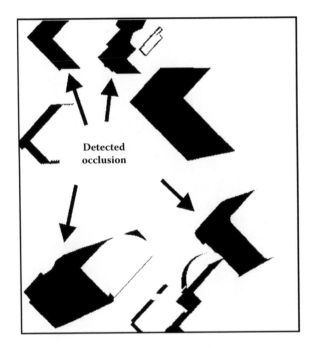

FIGURE 1.9 Index map generated using the Z-buffer algorithm that the occlusion has identified by using the orthophoto, as depicted in Figure 1.8b.

Before mosaicking, radiometric balances and blending between images must be conducted because significant scene-to-scene radiometric variations are obvious in urban aerial images. To perform the radiometric balancing, we developed an Arc/Info-based tool in combination with a number of available GIS functions, including line and polygon buffering, distance calculation, and grid algebraic operations. This tool allows us to calculate the weights for blending individual scenes along the specified buffer zone using the cubic Hermite function (Zhou et al. (2002), gave a detailed description).

1.5.5 TRUE ORTHOIMAGE GENERATION

After the above procedures are implemented, a true orthoimage can be created. Figure 1.10b depicts a true orthoimage generated by using our method. For comparison purposes, Figure 1.10a is an orthoimage generated by using a conventional differential model. As displayed in this figure, "doubling" of building roofs and occlusions caused by buildings still exist in the resulting orthoimage. In Figure 1.10b, the two phenomena have been completely removed. The buildings are correctly orthorectified in their true planimetric positions, and the shadows of buildings have been removed. Therefore, the results in Figure 1.10b demonstrate that the proposed methodology is capable of generating true orthoimages from large-scale images in urban areas. However, there is room for improvement (e.g., identifying conjugate areas of multiple slave images and compensating for the spectral difference between refilled and surrounding areas).

(a)

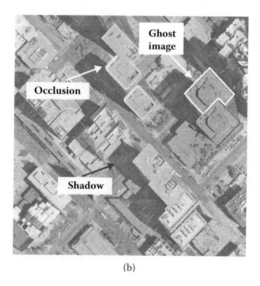

(b)

FIGURE 1.10 Comparison of orthorectified large-scale aerial images: (a) original image; (b) using traditional differential method; and (c) using proposed method in this chapter.

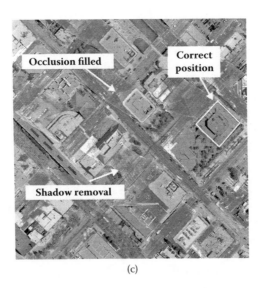

(c)

FIGURE 1.10 (Continued).

1.6 CONCLUSIONS

This chapter has described the theories, algorithms, and methods of large-scale urban orthoimage generation. Its main contributions include:

The data structure to model urban buildings and terrains accurately (i.e., DBM and DTM)

Algorithms and methods for urban true orthoimage generation, including DTM- and DBM-based orthoimage generation and their merging

Algorithms for detection and compensation of shadowed and occluded areas

The model of the relationship of air base, flying height, building height, and street width

We used large-scale aerial images of downtown Denver, Colorado, to validate the applicability of our methodology. Our experimental results demonstrated that (1) buildings and bridges can be placed in their proper and accurate upright, planimetric locations; (2) sidewalks and roads can be completely visible and building walls invisible; and (3) the seam lines of mosaicked images can be eliminated. These results demonstrate that our method can effectively and correctly orthorectify displacements caused by terrain and buildings in large-scale aerial images.

Although not all algorithms developed in this chapter are fully automatic, the preceding four essential contributions to urban large-scale orthoimage generation will be helpful to government agencies, private industry, and academia to produce large-scale digital orthoimages with clear and unobstructed representations of the highly complex built-up urban landscape. This chapter has demonstrated a method to produce orthoimages of urban areas with a visibility far superior to standard orthoimages. The improved visibility expands the orthoimage's value for safety and security applications as well as urban planning, traffic management, and many other uses. Because of these advantages and those identified in the chapter, this improved orthoimage can provide a baseline for standards development. To improve the economic viability and production efficiency of true orthoimage production, a more highly automated production system should be developed. The National Science Foundation has provided funding to initiate these next steps.

ACKNOWLEDGMENTS

This project was funded by the U.S. National Science Foundation (NSF) under contract number NSF 0131893. The authors sincerely thank Wolfgang Schickler for providing aerial images, building data, and the DSM of Denver, Colorado. We also would like to thank the project administrators of the city and county of Denver for granting permission to use their data. Postdoctoral researchers Dr. Qin and Dr. Chen participated in the project. The Ph.D. students Rajkiran Gottumukkal and C. Song, and M.S student Z. Tan also participated in this project. We thank Mark Demulder and Susan Benjamin, who provided useful suggestions for this chapter and USGS editorial staff for their assistance.

Disclaimer: Any use of trade, product, or firm names in this document is for descriptive purposes only and does not imply endorsement by the U.S. government.

REFERENCES

Amhar, F., J. Josef, and C. Ries. (1998). The generation of true orthophotos using a 3D building model in conjunction with a conventional DTM, *Int. Arch. Photogrammetry Remote Sensing*, 32, Part 4, 16–22.

Biason A., S. Dequal, and A. Lingua. (2004). A new procedure for the automatic production of true orthophotos, *Int. Arch. Photogrammetry, Remote Sensing Spatial Inf. Sci.*, Orthan Altan, Ed., Vol. XXXV, July 12–23, 2004, DVD.

Cameron, A.D., D.R. Miller, F. Ramsay, I. Nikolaou, and G.C. Clarke. (2000). Temporal measurement of the loss of native pinewood in Scotland through the analysis of orthorectified aerial photographs, *J. Environ. Manage.*, 58, 33–43.

Chen, L.C. and J.Y. Rau. (1993). A unified solution for digital terrain model and orthoimage generation from SPOT stereopairs, *IEEE Trans. Geosci. Remote Sensing*, 31(6), 1243–1252.

Federal Geographic Data Committee. (1995). Development of a national digital geospatial data framework, April 1995, http://www.fgdc.gov/framework/framdev.html.

Federal Geographic Data Committee. (1997). Fact sheet: National Digital Geospatial Data Framework: A Status Report, Federal Geographic Data Committee, July 1997, 37 p. http://www.fgdc.gov /framework/framdev.html, July 1997.

Grün, A. and X. Wang. (1998). CC-modeler: a topology generator for 3-D city models, *ISPRS J. Photogrammetry Remote Sensing*, 53, 286–295.

Jauregui, M., J. Vílchez, and L. Chacón. (2002). A procedure for map updating using digital mono-plotting, *Computers Geosci.*, 28(4), 513–523.

Jaynes, C., S. Webb, and R. Steele. (2004). Camera-based detection and removal of shadows from interactive multiprojector displays, *IEEE Trans. Visualization Computer Graphics*, 10(3), 290–301.

Joshua, G. (2001). Evaluating the accuracy of digital orthophotos quadrangles (DOQ) in the context of parcel-based GIS, *Photogrammetrlc Eng. Remote Sensing*, 67(2), 199–205.

Kelmelis, J.A. (2003). To the national map and beyond, *Cartography Geogr. Inf. Sci.*, 30(2), 185–198.

Kelmelis, J.A., M. Demulder, C. Ogrosky, N. VanDriel, and B. Ryan. (2003). The National Map, from geography to mapping and back again, *Photogrammetric Eng. Remote Sensing*, 69(10), 1109–1118.

Liow, Y.-T. and T. Pavlidis. (1990). Use of shadows for extracting buildings in aerial images, *Computer Vision, Graphics, Image Process.*, 49(2), 242–277.

Maitra, J.B. (1998). The national spatial data infrastructure in the United States: standards, metadata, clearinghouse, and data access, Federal Geographic Data Committee, USGS, http://www.gisdevelopment.net/policy/gii/gii0002.htm.

Mayr, W. (2002). True orthoimages, *GIM Int.*, 37, April, 37–39.

Noronha, S. and R. Nevatia. (2001). Detection and modeling of buildings from multiple aerial images, *IEEE Trans. Pattern Anal. Mach. Intelligence*, 23(5), 501–518.

Passini, R. and K. Jacobsen. (2004). Accuracy analysis of digital orthophotos from very high resolution imagery, *Int. Arch. Photogrammetry, Remote Sensing Spatial Inf. Sci.*, Orthan Altan, Ed., vol. XXXV, July 12–23 (DVD).

Rau, J.Y., N.Y. Chen, and L.C. Chen. (2002). True orthophoto generation of built-up areas using multiview images, *Photogrammetric Eng. Remote Sensing*, 68(6), June, 581–588.

Schickler, W. and A. Thorpe. (1998). Operational procedure for automatic true orthophoto generation, *Int. Arch. Photogrammetry Remote Sensing*, 32, Part 4, 527–532.

Schott, J.R., C. Salvaggio, and W.J. Volchok. (1988). Radiometric scene normalization using pseudoinvariant features, *Remote Sensing Environ.*, 26, 1–16.

Siachalou, S. (2004). Urban orthoimage analysis generated from IKONOS data, *Int. Arch. Photogrammetry, Remote Sensing Spatial Inf. Sci.*, Orthan Altan, Ed., Vol. XXXV, July 12–23, 2004 (DVD).

Skarlatos, D. (1999). Orthophotograph production in urban areas, *Photogrammetric Rec.*, 16(94), 643–650.

USGS. (1996). Digital orthophoto standards, *National Mapping Program — Technical Instructions*, Part I, general; part II, specifications. U.S. Department of the Interior, U.S. Geological Survey, National Mapping Division. http://rmmcweb.cr.usgs.gov/public/nmpstds/doqstds.html, December 1996.

USGS. (1998). Digital Orthophoto Program, http://mapping.usgs.goc/www/ndop/index.html, U.S. Department of the Interior, U.S. Geological Survey.

Vassilopoulou, S., L. Hurni, V. Dietrich, E. Baltsavias, M. Pateraki, E. Lagios, and I. Parcharidis. (2002). Orthophoto generation using IKONOS imagery and high-resolution DEM: a case study on volcanic hazard monitoring of Nisyros Island (Greece), *ISPRS J. Photogrammetry Remote Sensing*, 57(1–2), 24–38.

Zhou G., M. Xie, and J. Gong. (2000). Design and implementation of attribute database management system for GIS (GeoStar), *Int. J. Geogr. Inf. Sci.*, 6(2), 170–180.

Zhou, G., K. Jezek, W. Wright, J. Rand, and J. Granger. (2002). Orthorectifying 1960s declassified intelligence satellite photography (DISP) of Greenland, *IEEE Trans. Geosci. Remote Sensing*, 40(6), 1247–1259.

Zhou, G., Z. Qin, S. Benjamin, and W. Schickler. (2003). Technical problems of deploying national urban large-scale true orthoimage generation, *2nd Digital Gov. Conf.*, Boston, May 18–21, 2003, pp. 383–387.

Zhou, G. and W. Schickler. (2004). True orthoimage generation in extremely tall building urban areas, *Int. J. Remote Sensing*, January 2004, 25(22), 5161–5178.

Zhou, G., W. Chen, and J. Kelmelis. (2005). A comprehensive study on urban aerial image orthorectification for national mapping program, *IEEE Geosci. Remote Sensing*, 43(9), 2138–2147.

Zlatanova, S. (2000). 3D GIS for urban development, ISBN: 90-6164-178-0, ITC dissertation number 69. Doktor der technischen Wissenschaften an der Technischen Universitate Graz.

2 Urban Terrain and Building Extraction from Airborne LIDAR Data

Jie Shan and Aparajithan Sampath

CONTENTS

2.1 INTRODUCTION

One of the primary tasks in urban remote sensing is to extract terrain and building information from various data sources. Such information is useful for urban run-off and flooding analysis in hydrology and hydraulics studies. Terrain and buildings are also primary input data for urban development planning, forecast, and simulation. Urban environmental monitoring and modeling need such information to model the distribution of heat, pollution, and population. Besides, they recently find increasing applications in the real estate industry, disaster management, homeland security, and realistic visualization.

LIDAR (*light* *d*etection *a*nd *r*anging) technology provides a unique and promising solution to extracting such urban information (Ackermann, 1999; Baltsavias, 1999a). Compared to other conventional optical air- and space-borne technologies, LIDAR is an active remote sensing technology and thus data can be collected during the night. This can be a beneficial feature for certain applications and especially in large municipalities. LIDAR is also a three-dimensional remote sensing technology with which planimetric and vertical positions can be obtained as direct measurements. As a result, no image matching, which is often difficult and problematic in urban areas, is needed as in conventional optical remote sensing. In terms of quality, present LIDAR data can have submeter resolution and centimeter position accuracy. This property is superior to almost all other remote sensing data and therefore greatly extends the capabilities and potentials of urban remote sensing.

The introduction of LIDAR technology challenges innovative theory and methodology in data processing. Conventional image processing methods may not be directly used for LIDAR data analysis. This is basically because LIDAR data primarily is a collection of three-dimensional geometric points (object coordinates), while remote sensing image records the spectral properties of the objects. Because distinctions in spectrometry may not exist in geometry and vice versa, LIDAR data processing will primarily be carried out in spatial domain rather than in spectral domain.

This chapter will first briefly describe the principles and properties of LIDAR technology. Characteristics of LIDAR data will then be illustrated with a number of examples. The bulk of the chapter is two sections: one focuses on filtering algorithms for terrain extraction and the other on segmentation and regularization for building extraction. Detailed algorithmic steps are presented, and the extraction results are assessed visually and quantitatively. To facilitate further studies, a comprehensive literature review is included in these two sections.

2.2 PRINCIPLES AND PROPERTIES OF LIDAR

LIDAR, as its name conveys, is basically a remote sensing technology measuring the distances from a light source to targets. For surveying and mapping applications, the targets in general can be water or land. The bathymetric LIDAR usually uses a light source at the blue/green portion of the electromagnetic spectrum to penetrate the water and obtain two reflections: one from the water surface and the other from the water floor (Guenther, 2001). The light source of the topographic LIDAR typically works at the infrared portion of the electromagnetic spectrum, which is mostly reflected by land other than water (Fowler, 2001). Depending on the targets to be measured, LIDAR equipment can be mounted on a tripod, ground vehicle, airplane, or even a satellite. For urban remote sensing, airborne topographic LIDAR is the most popular one and hence will be the topic of this chapter.

LIDAR instruments must work with other instruments as a calibrated system to provide accurate georeferenced three-dimensional measurements. Typically, the entire system consists of three key subsystems or units (Fowler, 2001; Wehr and Lohr, 1999). The LIDAR unit transmits laser pulses to the target and receives the

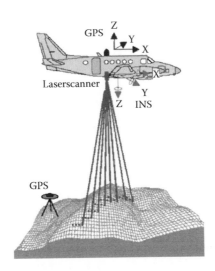

FIGURE 2.1 Units and principles of LIDAR system. (From the World Wide Web.)

returned pulses. An accurate clock records the round-trip travel time between the transmitted and received pulses. Dividing this travel time by two and multiplying the result with the speed of light will then yield the distance between the LIDAR unit and the target. The onboard inertial measurement unit (IMU) measures the orientation of the LIDAR scanner, and the GPS (global positioning system) receiver (unit) determines the position of the airplane. Postprocessing the measurements from the three units will determine the three-dimensional geospatial coordinates of every LIDAR return (Figure 2.1).

Compared to conventional frame cameras, LIDAR has certain unique distinctions. It works in a scanning mode — that is, it transmits laser pulses while scanning a swath of terrain across the flight direction. The scanning can be 40 swaths per second (40-Hz scan rate), with each swath 40° relative to the LIDAR sensor in the vertical plane. As many as 80,000 pulses can be transmitted per second (i.e., the pulse rate is 80 kHz). The distance between the footprints of two adjacent laser pulses is called ground spacing. This is an important property of LIDAR data and will be illustrated with examples in the next section.

Depending on the reflective property of the target, a transmitted laser pulse may be returned multiple times to the receiver because the pulse is distended when it travels through the air and reaches the target. As a result, if part of the pulse hits the roof edge, the rest of the pulse may continue to travel to reach a tree or ground, or both. Such a scenario will cause multiple returns of a pulse: in this example, the first return is from the roof, the second from the tree, and the third from the ground. At present, some LIDAR receivers are able to record as many as dozens of or even over a hundred returns.

Another distinction of LIDAR technology is that the receiver can also record the intensity — the energy of the returned pulses. The intensity is related to the reflective properties of the targets. For example, roads paved with asphalt usually

TABLE 2.1
Properties of a Typical LIDAR System

Specification	Typical Values
Laser wavelength	1.064 μm
Pulse repetition rate	5–33 kHz (50 kHz max)
Pulse energy	100s μJ
Pulse width	10 ns
Beam divergence	0.25–2 mrad
Scan angle (full angle)	40° (75° max)
Scan rate	25–40 Hz
Scan pattern	Zig-zag; parallel; elliptical; sinusoidal
GPS frequency	1–2 Hz
INS frequency	50 Hz (200 Hz max)
Operating altitude	100–1000 m (6000 m max)
Footprint	0.25–2 m (at altitude 1000 m)
Multiple elevation capture	1–5
Ground spacing	0.5–2 m
Vertical accuracy	~15 cm
Horizontal accuracy	10–100 cm

Source: Fowler, R., 2001, in *Digital Elevation Model Technologies and Applications: The DEM Users Manual*, David F. Maune, Ed. The American Society for Photogrammetry and Remote Sensing, Bethesda, MD, 539 pp.

appear darker than concrete roofs, and water is often the darkest image because it absorbs the infrared laser beam. Compared to passive or optical remote sensing technologies, where the gray values of the image are from reflected sunlight, the intensity image in LIDAR comes from reflected laser light. Multiple returns are often used to determine heights of trees and vegetation. However, the intensity image is not yet fully exploited, though it can be useful for recognizing different types of features.

As a summary, Table 2.1 lists the major properties of a LIDAR system. It should be noted that such values may vary from system to system and will certainly change as technology advances. For information on LIDAR systems and their accuracy, see Huising and Gomes Pereira (1998) and Baltsavias (1999b).

2.3 LIDAR DATA AND ITS PROCESSING OVERVIEW

This section will first describe the five LIDAR data sets used in this chapter and then present an overview of LIDAR data processing. Suburban Baltimore, Maryland, mostly has trees, vehicles, and long buildings with flat roofs. Downtown Baltimore is covered densely by tall buildings with complex roofs. Its nonground features include bridges, vehicles, and trees. In addition, this data set has a harbor where only boats and decks in the water returned LIDAR pulses. As an example, its

FIGURE 2.2 LIDAR elevation map and selected ground truth for assessment, downtown Baltimore. Dark polygons: buildings; light polygons: ground.

color-coded elevation map is shown in Figure 2.2. The Osaka, Japan, data set consists of trees and mostly long and flat buildings. The Toronto, Canada, data set is full of tall and complex buildings, and the Purdue data set covers the university campus at West Lafayette, Indiana.

Table 2.2 summarizes the properties of the first four data sets. The total number of LIDAR points used in the study is listed. These points are a portion of the entire data sets. The dimension of the study area is given by the number of points across track (scanning direction) and along track (flight direction). To characterize the topographic relief of the study areas, the 5, 50 (median), and 95 percentiles of the ground slopes are listed. The Toronto area has very complex topography with a ground slope range of 44.3°, whereas the other three areas are relatively moderate with a maximum ground slope range of 18.5° (Baltimore suburb). The point density stands for the number of LIDAR points per 10 m². It is calculated by the total number of LIDAR points divided by the entire area covered by the LIDAR data. As shown in Table 2.2, the point density varies from a minimum of 1.3 points (downtown Baltimore) to a maximum of 7.2 points (Toronto) per 10 m².

It should be noted that the distribution of LIDAR points is not even. As shown in Table 2.2, the ground spacing across track can be twice as small (Baltimore suburb) as the one along track. To illustrate and further examine this, Figure 2.3

TABLE 2.2
Properties of Four Test LIDAR Data Sets

Site Name	Baltimore Suburb	Downtown Baltimore	Osaka	Toronto
Total points #	30,000	50,000	50,000	50,000
Dimension (across*along track, in square meters)	~548*312	~1,216*405 (minus 101,660 water)	~500*400 (minus 32,958 no data area)	~167*420
Percentile ground slope (5%, 50%, and 95% in deg.)	0.2; 2.7;18.7	0.2; 1.6; 15.8	0; 1.6; 10.7	0; 2.7; 44.3
Ground spacing (across/ along track, in meters.)	2.3/4.5	2.5/4.0	1.6/1.7	1.0/1.5
Point density (points/10 m²)	1.8	1.3	3.0	7.2

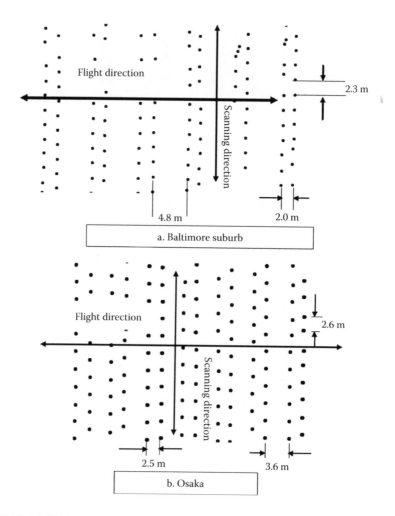

FIGURE 2.3 LIDAR data distribution. a: Baltimore suburb; b: Osaka.

compares the distribution of sample points from suburban Baltimore and Osaka data sets. For the Baltimore data, the ground spacing across track is an average of about 2.3 m, whereas the distance between two neighboring LIDAR profiles alternates with two different spacing intervals. The smaller spacing along track is at about 2.0 m, whereas the larger one is about 4.8 m. The Osaka data set presents a similar pattern, though it is not as apparent as in the Baltimore data set.

This uneven spacing can be explained with the principles described by Wehr and Lohr (1999). The ground spacing across track depends on the laser pulse repetition frequency and the scan angle off the nadir. Therefore, the spacing across track can be treated as a constant in a small area (as shown in Figure 2.3) and becomes larger at the swath boarder. Due to acceleration or slow-down of the scan mechanism, the points at the swath borders exhibit unwanted characteristics and are sometimes removed from the raw data set (Wehr and Lohr, 1999). Ground spacing along track is determined by airplane speed and the period of one scan. The bidirectional oscillating scanning causes a zigzag pattern with uneven distances along the flight direction. The ground spacing is smaller where the oscillating mirror finishes one scan line and changes its direction to start a new scan line; it is larger between the start of one scan line and the end of the next scan line. This pattern repeats, as shown in Figure 2.3.

As described earlier, LIDAR data is actually clouds of points from which the LIDAR receiver recorded the returned pulses. Because the point clouds are mixed with all reflected terrain targets, the tasks in LIDAR data processing are therefore threefold: separate different categories of feature classes, such as trees, buildings, and ground; separate each individual feature, such as a single tree and building, from the same feature class; and model the individual feature, i.e., use certain mathematical surface or parameters to describe the feature. Such parameters can be simply a sequence of structured points for roads, the height and crown size of trees, and dimensions of buildings.

Although different terminologies can be used to describe the tasks of LIDAR data processing, a convention is followed in most of the literature:

Classification means the process to categorize the data into feature classes, such as trees, buildings, and ground. As a special case of classification, the term *filtering* is specifically used for separating ground from the rest of the LIDAR points. This is often the first step in LIDAR data processing.

Segmentation is a process to determine the LIDAR points that belong to one homogenous region. It can be applied to classified data or to the original data. In the latter case, classification is needed to further separate individual features in the segmented results. Some processing approaches tend to combine the classification and segmentation steps to determine individual features.

Reconstruction is the final step that uses classified and segmented data to generate digital terrain model (DTM) from ground points, or to determine the building or tree parameters from the LIDAR points of a building or a tree, respectively.

The discussion in this chapter addresses the classification (filtering) for terrain extraction, segmentation, and reconstruction for building extraction. Building reconstruction is formulated as a regularization problem in two dimensions for building boundary extraction. Complete three-dimensional extraction is addressed in the literature review because it is still an active research subject.

2.4 TERRAIN EXTRACTION

This section describes a filtering approach to urban terrain extraction and assesses its performance visually and quantitatively. A literature review on related studies is presented at the end of the section.

2.4.1 METHODOLOGY

Filtering is a special case of classification in which the primary objective is to classify the LIDAR data into two classes: ground points and nonground points (Vosselman, 2000). The former will be used to generate DTM, while the latter will be used for building extraction. Shan and Sampath (2005) designed a filter primarily for urban areas. It detects the nonground points along the LIDAR scan line in two opposite directions and thus is called a one-dimensional bidirectional filter.

Two criteria are used to determine if a point is ground or nonground. First, slopes at the border between building and ground are usually significantly larger than the ground slope calculated at common LIDAR point density. Second, ground points have lower elevations than nonground points in the neighborhood. The slope criterion detects the edge points of nonground objects, such as buildings, bridges, vehicles, and trees; the elevation criterion considers the situation in which the building roofs are not flat. Because complex roofs have many local changes in slope, using the slope criterion alone may lead to erroneous results. Mathematically, the filter is described as:

$$\forall P_i : \begin{cases} if\,(S_v > S_T \ and \ Z_i > Z_T) & non-ground\ point \\ otherwise & ground\ point \end{cases} \tag{2.1}$$

That is, for any LIDAR point, P_i, if the slope S_v at its vicinity is greater than a given threshold S_T and its elevation Z_i is greater than an adaptive threshold Z_T, then the point P_i is classified as a nonground point; otherwise, it is classified as a ground point. The check on the vicinity slope S_v essentially detects the presence of a nonground object, and the elevation condition determines whether the LIDAR point belongs to the object or ground. In Equation (2.1), the slope S_i is calculated with two consecutive points along the LIDAR (scanning) profile using the following equation

$$S_i = \arctan\left(\frac{Z_i - Z_{i-1}}{\sqrt{(X_i - X_{i-1})^2 + (Y_i - Y_{i-1})^2}}\right) \quad S_i \in \left[-\frac{\pi}{2}, \frac{\pi}{2}\right] \tag{2.2}$$

where $(X_{i-1}, Y_{i-1}, Z_{i-1})$ and (X_i, Y_i, Z_i) are the coordinates of two consecutive points in a LIDAR profile, with the Z coordinate standing for elevation.

Note that the slope in Equation (2.2) can take positive and negative signs. Point sequence entering a building from the ground will get a positive (ascending) slope at the ground–building border, while point sequence leaving a building will get a negative (descending) slope at the building–ground border. Initially, one may assume the point sequence entering a building from the ground. Therefore, once a LIDAR point gets a slope greater than the given slope threshold S_T, the LIDAR profile has reached a building edge at this point. It will then be classified as a nonground point. The elevation and slope of the last detected ground point will be noted for prediction use.

The algorithm continues classifying the subsequent points as building points until a negative slope is encountered. This negative-slope (descending) point may possibly be a ground point and needs to be examined. Its elevation is compared with the elevation of the last ground point where the LIDAR profile reached the building. If the two elevations are close within a tolerance, the algorithm classifies the descending point as ground, which means the LIDAR profile leaves the building and returns to ground at this point. This method works well when the topography is relatively flat. For hilly regions, elevation prediction is needed. The slope of the last ground point in a profile is used to predict the elevation of the descending point. If the elevation of the descending point is higher than the predicted one within a tolerance, it will still be classified as a building; otherwise, the LIDAR profile leaves the building and returns to ground.

This process continues until the last LIDAR point is in the data set. This is regarded as the forward process. To consider the situation when the first LIDAR point in the file is a nonground point, the preceding process is repeated in the reverse direction by starting the process from the last point in the data set — namely, the backward process. A final regression along the LIDAR profile can be implemented to further smooth the results. Points classified as ground in both forward and backward processes are regarded as ground points. Figure 2.4 illustrates the filtering process with LIDAR points overlaid atop the original LIDAR surface model in suburban Baltimore.

2.4.2 ASSESSMENT

Filtering results may contain two types of mistakes. Type I error, or false positive, refers to the mistake that classifies nonground points as ground points, and type II error, or false negative, stands for the mistake that classifies a ground point as a nonground point. Filtering results should be evaluated against these two mistakes. At present, using ground truth or visual check in comparison to available orthoimages is the most common and reliable method.

A number of homogenous ground and building regions (polygons) are traced manually over the LIDAR data color-coded by elevation. In Figure 2.2, building and ground polygons are shown in dark (black) and light (yellow), respectively. The number of raw LIDAR points inside the traced building and ground polygons is used as ground truth. Similarly, the number of ground and building points in the filtering results is also counted for each traced building and ground polygon. Rows in Table 2.3 under "filtering results" contain the number of points and their percentage (within parentheses) that are correctly or wrongly classified for each study site. As a general assessment, the overall filtering results for the four study sites are listed in the last row of the table, which essentially is an average performance measure. A comparison of the number of points counted from filtering results and the ground truth can be used for evaluation.

Table 2.3 shows that, for the ground regions, an average 97.3% of the total points is correctly classified, while 2.7% are wrongly (type II error) classified as building points. As for building regions, the average correct classification rate is 97.4%, with an average type I error rate of 2.6%. The similar error rates (2.7 vs. 2.6%) of the two possible mistakes suggest that a good balance is achieved, which is desirable for an unbiased filtering algorithm. Table 2.3 also suggests that the performance of the filtering approach varies with the topographic complexity. The study sites of

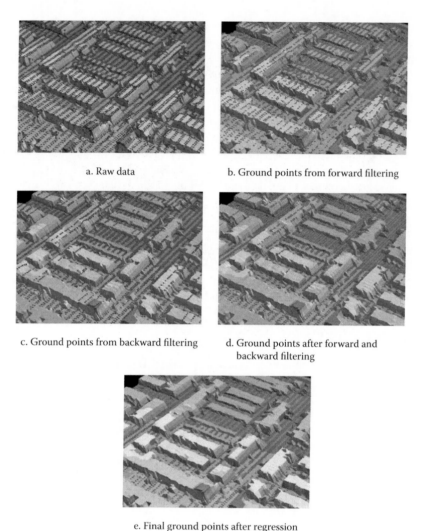

a. Raw data   b. Ground points from forward filtering

c. Ground points from backward filtering   d. Ground points after forward and
                                                    backward filtering

e. Final ground points after regression

FIGURE 2.4 (Color Figure 2.4 follows page 240.) Stepwise three-dimensional illustration of the filtering process (Baltimore suburb). a: Raw data; b: ground; c: ground points from backward filtering points from forward filtering; d: ground points after forward and backward filtering; e: final ground points after regression.

Baltimore suburbs and downtown and of Osaka are relatively flat. Their filtering error rate ranges from 0.5% (Baltimore suburb, false positive) to 2.1% (Baltimore urban, false negative). For the Toronto site, cliffs or steep ramps exist, thus causing higher error rates up to 4.9% (false positive) and 4.5% (false negative).

Figure 2.5 is several selected closer examinations on handling typical urban features. Shown in this figure are complex buildings, bridges, trees, and a water

TABLE 2.3
Quality of Filtering Results

		Ground Truth		Filtering Results	
Site Name	Point Type	No. of Regions	No. of Points	Ground (no. and %)	Building (no. and %)
Baltimore	Ground	18	1692	1645 (97.2)	47 (2.8)
suburb	Building	22	4012	21 (0.5)	3991 (99.5)
Downtown	Ground	19	1784	1747 (97.9)	37 (2.1)
Baltimore	Building	24	8081	122 (1.5)	7959 (98.5)
Osaka	Ground	15	3069	3036 (98.9)	33 (1.1)
	Building	14	4668	72 (1.5)	4596 (98.5)
Toronto	Ground	14	3387	3234 (95.5)	153 (4.5)
	Building	15	9282	455 (4.9)	8827 (95.1)
Overall	Ground	66	9932	9662 (97.3)	270 (2.7)
(Average)	Building	75	26043	670 (2.6)	25373 (97.4)

body (harbor) along with the orthoimages from the Baltimore downtown data. Notice that the two small concave parts in the middle left and right of Figure 2.5a are not identical in height. The left is ground, while the right is part of the building. This is successfully detected by the filter and then verified by manually measuring the elevations in the vicinity. Figure 2.5b shows the filtering of a possible ring-type building. The filtering process identifies the inner part as a building rather than ground, which is further verified by manual check. Notice that the black object on the roof does not produce any return. The LIDAR data is void in this place.

Figure 2.5c attempts to demonstrate the capability of classifying bridges as nonground. As is shown, all three bridges are correctly classified. In Figure 2.5d, almost all vehicles in the surface parking place in the middle of the figure are correctly classified. However, some ground points on the horizontal street at the bottom are mistakenly classified as nonground, which is a trade-off that many vehicles on that street are identified correctly. The result of the water body is shown in Figure 2.5e. The returns from water surface or its objects are very sparse. Because of its flat surface, water and the deck are classified as ground. To ultimately determine their topographic nature, spectral information from LIDAR intensity or images is needed.

2.4.3 LITERATURE REVIEW

Many filtering algorithms have been developed. For the convenience of review, we summarize them into two general categories: classification approach and adjustment approach. A classification approach determines the ground points using certain operators that are often related to slopes calculated from the LIDAR data. As one of the early efforts, Lindenberger (1993) introduced mathematical morphologic operators for this task. In this approach, an opening operator with horizontal structure element was first used to detect possible ground points. Points within

a. Building 1

b. Building 2

c. Bridges, trees

d. Vehicles

e. Water

FIGURE 2.5 (Color Figure 2.5 follows page 240.) Filtering results of urban features.
a: Building 1; b: building 2; c: bridges, trees; d: vehicles; e: water.

a certain vertical offset from the estimated local average elevation were classified as ground points. A regression process was then applied to refine the initial results obtained from the morphologic operation. Noticing that this algorithm was vulnerable to the size of the structure element, Kilian et al. (1996) used a series of morphological operators with different sizes to discover the ground points.

To consider local relief, Vosselman (2000) proposed a slope-based filter, which was shown to be closely related to the erosion operator in mathematic morphology. Training may be needed to determine the parameters when implementing this filtering process. Zhang et al. (2003) utilized the classical morphological opening and gradually increased the size of the structure element to classify the LIDAR data into ground and nonground points. Arefi and Hahn (2005) separated ground and nonground points using a morphological reconstruction algorithm that was designed based on geodesic distance dilation (Lantuejoul and Maisonneuve, 1984). Compared to classical morphological algorithms, this approach does not need to decide the size of the structure elements. In addition to the morphology-based or -related operators, terrain slope (Axelsson, 1999; Sithole, 2001; Yoon and Shan, 2002) and local elevation difference (Wang et al., 2001) have also been used as criteria in filtering operations. Zhang et al. (2004) first detected candidate ground points by using partial derivatives. A local surface fitting was then followed to remove nonground points.

Recently, segmentation-based filtering approaches have been reported. They detect nonground objects in one dimension (Shan and Sampath, 2005) or look for continuous ground in two dimensions (Sithole and Vosselman, 2005). Ground points can also be identified through an adjustment process. A mathematical function, usually selected as a two-dimensional polynomial surface, can be used to approximate the ground. In this approach, ground is essentially represented as a continuous or at least piecewise continuous surface. Points within a certain vertical offset above the surface are treated as ground points. Least squares adjustment is used to detect the nonground points as if they were blunders by reducing their weights in each iteration calculation. Typical methods of this kind have been proposed by Kraus and Pfeifer (1998, 2001) and Schickler and Thorpe (2001), based on classical surface modeling, and by Elmqvist (2002), based on active shape modeling. Recently, such a method was combined with the segmentation process for urban terrain extraction (Tóvári and Pfeifer, 2005). It should be noted that the adjustment approach can be used together with the classification approach. As a matter of fact, many of the preceding classification methods apply surface fitting after the classification operation.

The performance of filters has been studied recently. Primarily for coastal applications, Zhang and Whitman (2005) compared the performance of three filters and showed that progressive morphology yields the best results and the topographic slope is the most sensitive factor. Sithole and Vosselman (2004) reported the results of a test on filter performance organized by the International Society for Photogrammetry and Remote Sensing.

2.5 BUILDING EXTRACTION

This section presents a solution to building extraction from nonground LIDAR points. It consists of three sequential steps (Sampath and Shan, 2004). After the filtering operation, segmentation is first applied to the nonground class. Regions larger than a certain area will be regarded as buildings. Next, a modified convex hull-formation algorithm is applied to find the building boundary points and connect them to form the boundary. The final step is to model or regularize the building boundary by determining its parametric equations. A literature review on related studies is included at the end of the section.

2.5.1 BUILDING SEGMENTATION

Once the nonground class is obtained via filtering operations, LIDAR points belonging to individual buildings can be determined. This is accomplished through building segmentation. The raw LIDAR points usually have a rather uniform spatial distribution. In the nonground data set, however, only the cluster of points that belong to one building will still have the same spatial distribution. This uniform distribution of points within one cluster and nonuniform distribution of points among clusters is used to map each point to an individual building. The solution is based on a region-growing algorithm by successively collecting points of the same building. This algorithm consists of the following steps:

1. Start from a building point P_0.
2. Center a window at the point and collect all the points $A = \{P_1, P_2, ..., P_k\}$ within the window.
3. Move the window center to P_1.
4. Collect the points within the window and store them in a temporary array, $T = \{tP_1, tP_2, ..., tP_r\}$.
5. Move the window center to point P_2. Append the newly collected points to the array T, and in this process make sure that no two points are identical.
6. Continue the process until the window has been placed over all the points in the set A.
7. Merge points in A and points in T and store them in B, (i.e., $B = \{B \cup A \cup T\}$). Initially B is a null set.
8. Replace points in A with points in T so that the newly populated set A is equivalent to $\{T \not\subset A\}$.
9. Go back to step 3.
10. Stop when no new point is added to the set B.

At the end of these steps, the set B has the points that belong to the same building. The set of points in B is removed from the data set and the algorithm repeats for the rest of the points until all the points are mapped to a building. The window described in step 2 is oriented along and perpendicular to the scan directions. The dimensions of the window are set as slightly larger than two times the point spacing,

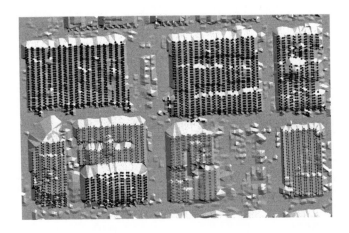

FIGURE 2.6 Segmented buildings labeled with symbols of different sizes and shapes.

which is usually different in the across-tract and along-track scan directions. Segmented clusters containing less than a certain number of points are rejected because they may be trees or vehicles that are not successfully removed in the filtering process. Figure 2.6 shows segmented buildings labeled with symbols of different sizes and shapes.

2.5.2 BOUNDARY TRACING

Once all points of a building are found, the next step is to determine the building boundary. Jarvis (1977) showed that a modified convex hull algorithm can be used to determine the shape of a set of points. Such modification restricts the searching space of the convex hull formation algorithm to a neighborhood. For a given set of points, the convex hull is the smallest convex boundary containing all the points (de Berg et al., 2000). It can be understood as a rubber-band wrapped around the "outside" points. Figure 2.7 illustrates the principal steps of the modified convex hull approach to tracing the boundary for a set of points.

As shown in the first row, the algorithm starts with selecting the left-most point (shown by an empty circle) as the boundary point. All points (shown by gray circles) within a neighborhood (shown by a larger circle) are selected. The convex hull algorithm is then used to determine the next point on the boundary by only considering the points within this neighborhood. After that, the algorithm will proceed to this newly determined point and repeat the same procedure until the boundary is determined (fourth row). As can be expected, the performance of the algorithm depends on the neighborhood used in the tracing process. Because the point spacing in along-track and across-track scan directions is usually different, a rectangular neighborhood is used whose dimensions are slightly larger than twice the point spacing in the along- and across-scan directions. In this way, only immediately adjacent points at about one ground spacing are considered for the convex hull algorithm so that a compact boundary is found.

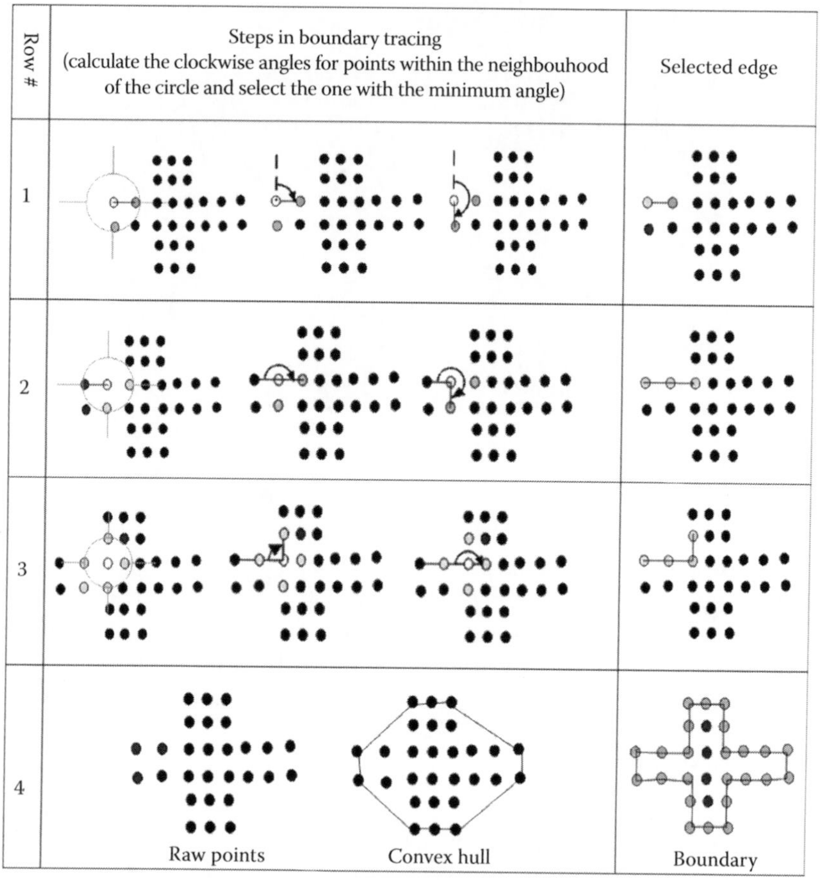

Row #	Steps in boundary tracing (calculate the clockwise angles for points within the neighbouhood of the circle and select the one with the minimum angle)	Selected edge
1		
2		
3		
4	Raw points Convex hull	Boundary

FIGURE 2.7 Modified convex hull algorithm for building boundary tracing.

The boundary tracing approach is designed as follows. Let B be the set of points belonging to one building:

1. Start from point P, which is necessarily a boundary point (e.g., the left-most point).
2. Select a set of points, $Pts = \{P_1, P_2, \ldots P_m\} \subset B$, so that all points in Pts lie within the neighborhood of the point P.
3. Using the convex hull approach, determine the next boundary point, P_k, from $P_k \subset Pts$. The point is chosen so that the line segment $\overline{PP_k}$ does not intersect any existing line segments.
4. Choose P_k as the next current point and repeat steps 2, 3, and 4.
5. Continue these steps until the point P_k corresponds with the point P selected in step 1.

Demonstrated in Figure 2.8 are three building boundaries obtained with this approach. The first row presents the building points and the second shows the generated

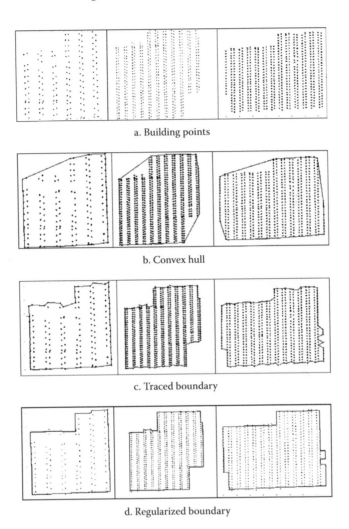

a. Building points

b. Convex hull

c. Traced boundary

d. Regularized boundary

FIGURE 2.8 Building extraction steps: building points (a), convex hull (b), traced boundary (c), and regularized boundary (d).

building convex hull. The third row is the boundary traced by the modified convex hull approach where the searching space is restricted to a rectangular neighborhood.

2.5.3 BOUNDARY REGULARIZATION

Boundary reconstruction is formulated as a regularization problem. Because LIDAR points are randomly collected, the traced boundary cannot be directly used as the final building boundary because of its irregular shape and possible artifacts introduced in the previous steps. Further refinement must be carried out before the traced boundary can be input into a geospatial database. For this objective, building models must be introduced. As a rather generic and yet realistic assumption, many buildings should have mutually perpendicular directions. Under such an assumption, the traced boundary points can be used to determine the parameters of the building boundary. This process is called regularization; a hierarchical least squares solution is presented next.

The regularization algorithm is designed based on the observation that longer edges of a building are more likely to represent its dominant directions and form the basic frame of the building boundary. Therefore, a hierarchical approach is designed. Initially, relatively long line segments are extracted from the building boundary and their least squares solution is determined, assuming that these long line segments lie in two mutually perpendicular directions. After that, all the line segments are included to determine the least squares solution, using the slopes of the long line segments as weighted approximations.

Four steps are involved in this process. The first step is to extract the points that lie on long line segments. This is done by sequentially following the boundary points and looking for positions where the slopes of two consecutive edges are significantly different. Points on consecutive edges with similar slopes are grouped to one line segment. In the second step, each of the long line segments is modeled by line equation $A_i x + B_i y + 1 = 0$. For each line segment (A_i, B_i), the slope $M_i = -A_i/B_i$ is obtained. Line segments that are parallel within a given tolerance are sorted into one group. In this way, the long line segments are grouped to "horizontal" and "vertical" groups based on their slopes. The third step is to determine the least squares solution to these long line segments, with the constraints that the slopes of these segments are equal (parallel lines) or their product is equal to -1 (perpendicular lines), depending on whether they belong to the same category or different categories. The solutions consist of a set of parameters that describe each of the long line segments.

The preceding process can be formulated as shown here. For each line segment, the following equation is formed:

$$A_i x_j + B_i y_j + 1 = 0 \qquad i = 1, 2...n;$$
$$j = j(i) = 1, 2, ...m_i \tag{2.3}$$

where n is the number of line segments and m_i is the number of points on the line segment i. Let M_u and M_v be the slopes that define the two mutually perpendicular directions of a building; for each line segment we have

$$\frac{A_i}{B_i} + M_s = 0 \quad i = 1, 2, ..., n \tag{2.4}$$

where M_s is the slope for the "vertical" or "horizontal" line segment groups and takes M_u or M_v. In addition, the following orthogonal constraint equation is included:

$$M_u M_v + 1 = 0 \tag{2.5}$$

The least squares criterion is used to solve the equation system of Equation (2.3), Equation (2.4), and Equation (2.5). The unknowns include all the line segment parameters A_i and B_i ($i = 1, 2, \ldots, n$), and the dominant slopes M_u and M_v.

The final regularization step is to include all (long and short) line segments into the least squares solution. The slopes, M_u and M_v, obtained from the previous step are used as approximate values. The slope parameters for long line segments are given high weights in the regularization adjustment. Therefore, no explicit constraint (Equation 2.5) is enforced in this final step. In this way, the line segments of a building can be properly constrained to its dominant directions and have the flexibility to fit to the LIDAR boundary points. Figure 2.8 (fourth row) presents the determined parametric line segments for the building boundaries.

2.5.4 ASSESSMENT

Shown in Figure 2.9 are representative regularized buildings and their orthoimages in downtown Baltimore. The LIDAR points and regularized building boundary are overlaid atop the LIDAR surface model; d is the maximum distance of LIDAR points off the corresponding parametric line segments and σ is the standard deviation of the least squares adjustment-based regularization. Several observations can be made based on the results in Figure 2.9.

It is first seen that almost all building edges are very well determined. The regularized boundary fits to the LIDAR points and reflects the building's shape. The building outline provides an authentic appearance compared to the orthoimages and the LIDAR surface model. Second, many minor rectilinear features (e.g., the short right-angle edges labeled as "A" in a circle) are determined correctly through the regularization process. This forms the fine details in the determined building boundary that can possibly be inferred from the LIDAR data. The effect of LIDAR data resolution is the third observation. The regularized building boundary may miss details or introduce artifacts due to the limited resolution of the LIDAR data. As shown by the "B" labels, right-angle corners formed by short edges may not be observed in the regularized building boundary. Similarly, the "C" labels demonstrate the right-angle corners introduced as artifacts. For either of the two situations, the regularization process produces slightly distorted and shifted building boundary segments.

Finally, it is observed that very low places of a building may be identified as ground, e.g., the "D" labels in the figure. This in turn will cause missing parts in the final building boundary so that the regularization result is similar to the roof print rather than to the footprint. As a quantitative evaluation, the average of the maximum distances and the average of distance variances of ten buildings from each study site are calculated. The maximum distance of the regularized building boundary off the LIDAR points is about the same as the LIDAR point spacing: 2.4 vs. 2.7 m for downtown Baltimore, 1.1 vs. 1.2 m for Toronto, and 1.2 vs. 1.0 m for Purdue. The standard deviations of the regularized boundary off the LIDAR points are at 18 to 21% of the LIDAR point spacing. Such a relationship varies little among different data sets, suggesting a trend of linear relationship between building extraction uncertainty and LIDAR point spacing.

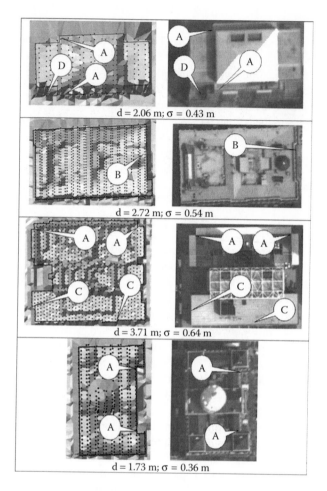

FIGURE 2.9 Regularized building, its orthoimage, and quality (downtown Baltimore).

Finally, it should be noted that the preceding regularization process can be extended to three-dimensional building reconstruction (Sampath and Shan, 2004). Buildings with flat roofs or multiple levels of flat roofs are first vertically sliced into different levels according to their elevation. Each level is individually regularized in the same manner described here. For ridge buildings, slopes of roof points are calculated based on the triangulation of roof points. Points with similar slopes are then grouped and used to determine a slant roof plane. A similar regularization process can then be applied to determine the planar roof boundary.

2.5.5 LITERATURE REVIEW

Many attempts to build extraction from LIDAR points or digital surface models (DSMs) generated from stereo images have been made. Weidner and Förstner (1995), Brunn and Weidner (1997), and Ameri (2000) used the difference between DSM and

DTM to determine building outlines. Haala et al. (1998), Brenner (2000), and Vosselman and Dijkman (2001) used building plan maps, and Sohn and Dowman (2003) used Ikonos imagery to facilitate detection and reconstruction of buildings from LIDAR points. Maas and Vosselman (1999) presented two direct solutions to the parameters of buildings with gable roofs by using invariant moments. Masaharu and Hasegawa (2000) segmented building polygons from neighboring non-building regions and used boundary-tracing methods to segment individual buildings. Wang and Schenk (2000) generated the TIN (triangulated irregular network) model from the LIDAR point clouds. Triangles were then grouped based on orientation and position to form larger planar segments. The intersection of such planar segments resulted in building corners or edges. Al-Harthy and Bethel (2002) determined building footprints by subtracting DTM from DSM. The building polygon outline was then obtained by using a rotating template to determine the angle of the highest cross-correlation, which suggested the dominant directions of the building. Morgan and Habib (2002) first determined the breaklines in a raw LIDAR data set and formed the TIN model. Through a connected component analysis on the TIN model, individual buildings were segmented. The final building boundary was formed by performing the Hough transform to the centers of the boundary triangles in the TIN model. Rottensteiner and Briese (2002) used hierarchical robust interpolation (Kraus and Pfeifer, 1998) with a skew error distribution function to separate building and ground points. After applying morphological filters to the candidate building points, an initial building mask was obtained, which was then used to determine polyhedral building patches with a curvature-based segmentation process. The final individual building regions were found by a connected component analysis. Schwalbe et al. (2005) used two-dimensional GIS data to segment buildings and determine their orientation. Building points were then projected to X–Z and Y–Z planes so that planar roofs could be detected as straight lines in these planes. Rottensteiner et al. (2005) focused on roof plane detection and grouping in building reconstruction. Roof planes were initially extracted by segmentation based on local homogeneity of surface normals. Domain-specific information was used to determine intersection lines and step edges among the roof planes. A polyhedral model was then formed through a consistency check and regularization adjustment. Sampath and Shan (2006) first detected breaklines based on eigenvector and eigenvalue analysis. The planar roofs were then found by clustering the normal vectors of a number of planar patches. Figure 2.10 presents color-coded planar roofs extracted through the clustering process.

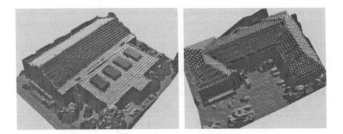

FIGURE 2.10 (Color Figure 2.10 follows page 240.) Color-coded planar roof points via clustering (Purdue campus).

For further literature review, see Vosselman et al. (2004), who summarize several techniques for segmenting aerial and terrestrial LIDAR point clouds into various classes and extracting different types of surfaces. To a similar extent, Brenner (2005) reviews several building reconstruction techniques from images and LIDAR data. Kaartinen et al. (2005) report the accuracy of building reconstruction tests from 11 participants. They show that laser scanning is more suitable in deriving building heights and extracting planar roof faces and ridges of the roof, whereas aerial images are more suitable in building outline and length determination.

2.6 OUTLOOK

Despite decade-long, abundant studies on urban terrain and building extraction from LIDAR data, the industry and academia still face challenges. Filtering processing for terrain extraction has become a routine functionality offered by many LIDAR service providers. However, visual quality assessment and manual editing on the filtered terrain are needed to clean up remaining nonground objects. Such a task becomes a dominant factor in terms of time and workforce planning for postflight processing. Also, terrain products in complex topography and complex urban environments are still not satisfactory.

As for building extraction, research has progressed in determining simple planar building faces and roofs. Such functionality can be found in some commercial products; however, it is mostly implemented in an interactive manner. Segmenting complex roofs and roof components is premature and deserves more effort. Furthermore, parameterization and regularization of three-dimensional building models need to be formulated; the results should be represented in three-dimensional data structure and integrated with GIS and CAD systems.

ACKNOWLEDGMENTS

The two Baltimore data sets were provided by EarthData Technology, Maryland. The Osaka and Toronto data sets were provided by Optech, Inc., Toronto, Canada. This chapter was written when Prof. Shan was on sabbatical leave with the Institute for Photogrammetry and GeoInformation (IPI), University of Hannover, Germany, under the support of the Alexander von Humboldt–Stiftung.

REFERENCES

Ackermann, F., 1999. Airborne laser scanning — present status and future expectations, *ISPRS J. Photogrammetry Remote Sensing*, 54(2–3), 64–67.

Al-Harthy, A. and J. Bethel, 2002. Heuristic filtering and 3D feature extraction, ISPRS Commission III Symposium, Graz, Austria, CD-ROM.

Ameri, B., 2000. Automatic recognition and 3D reconstruction of buildings from digital imagery, Deutsche Geodaetische Kommission, Series C, No. 526.

Arefi, H. and M. Hahn, 2005. A morphological reconstruction algorithm for separating off-terrain points from terrain points in laser scanning data, ISPRS WG III/3, III/4, V/3 Workshop "Laserscanning 2005," Enschede, The Netherlands, September 12–14.

Axelsson, P., 1999. Processing of laser scanner data — algorithms and applications, *ISPRS J. Photogrammetry Remote Sensing*, 54(2–3), 138–147.

Baltsavias, E., 1999a. A comparison between photogrammetry and laser scanning, *ISPRS J. Photogrammetry Remote Sensing*, 54(2–3), 83–94.

Baltsavias, E., 1999b. Airborne laser scanning: existing systems and firms and other resources, *ISPRS J. Photogrammetry Remote Sensing*, 54(2–3), 164–198.

Brenner, C., 2000. Dreidimensionale Gebäuderekonstrutkion aus digitalen Oberflächenmodellen und Grundrissen, Deutsche Geodaetische Kommission, Series C, No. 530.

Brenner C., 2005. Building reconstruction from images and laser scanning, *Int. J. Appl. Earth Obs. Geoinf.*, 6, 187–198.

Brunn, A. and U. Weidner, 1997. Extracting buildings from digital surface models, *Int. Arch. Photogrammetry Remote Sensing*, 32(3–4W2), 27–34.

de Berg, M., O. Schwarzkopf, M. van Kreveld, and M. Overmars, 2000. *Computational Geometry: Algorithms and Applications*, 2nd ed., Springer–Verlag, New York, 367 pp.

Elmqvist, M., 2002. Ground surface estimation from airborne laser scanner data using active shape models, *Int. Arch. Photogrammetry Remote Sensing Spatial Inf. Sci.*, XXXIV, Part 3A, Commission III, September 9–13, Graz, Austria, 115–118.

Fowler, R., 2001. Topographic LIDAR, in *Digital Elevation Model Technologies and Applications: The DEM Users Manual*, David F. Maune, Ed., The American Society for Photogrammetry and Remote Sensing, Bethesda, MD, 539 pp.

Guenther, G.C., 2001. Airborne LIDAR bathymetry, in *Digital Elevation Model Technologies and Applications: The DEM Users Manual*, David F. Maune, Ed., The American Society for Photogrammetry and Remote Sensing, Bethesda, MD, 539 pp.

Haala, N., C. Brenner, and K.-H. Anders, 1998. Urban GIS from laser altimeter and 2D map data, *Int. Arch. Photogrammetry Remote Sensing*, 32(3/1), T. Schenk and A. Habib, Eds., *ISPRS Commission III Symp. Object Recognition Scene Classification Multispectral Multisensor Pixels*, Columbus, OH, 339–346.

Huising, E.J. and L.M. Gomes Pereira, 1998. Errors and accuracy estimates of laser data acquired by various laser scanning systems for topographic applications, *ISPRS J. Photogrammetry Remote Sensing*, 53(5), 245–261.

Jarvis, R.A., 1977. Computing the shape hull of points in the plane, *Proc. IEEE Computer Soc. Conf. Pattern Recognition Image Process*, 231–241.

Kaartinen, H., J. Hyyppä, E. Gülch, G. Vosselman, H. Hyyppä, L. Matikainen, A.D. Hofmann, U. Mäder, Å. Persson, U. Söderman, M. Elmqvist, A. Ruiz, M. Dragoja, D. Flamanc, G. Maillet, T. Kersten, J. Carl, R. Hau, E. Wild, L. Frederiksen, J. Holmgaard, and K. Vester, 2005. Accuracy of 3D city models: EuroSDR comparison, ISPRS WG III/3, III/4, V/3 Workshop "Laserscanning 2005," Enschede, the Netherlands, September 12–14.

Kilian, J., N. Haala, and M. English, 1996. Capture and evaluation of airborne laser scanner data, *Int. Arch. Photogrammetry Remote Sensing*, XXXI, Part B3, Vienna, 383–388.

Kraus, K. and N. Pfeifer, 1998. Determination of terrain models in wooded areas with airborne laser scanner data, *ISPRS J. Photogrammetry Remote Sensing*, 53(4), 193–203.

Kraus, K. and N. Pfiefer, 2001. Advanced DEM generation from LIDAR data, in *Proc. ISPRS Workshop Land Surface Mapping Characterization Using Laser Altimetry*, Hofton, Michelle A., Ed., Annapolis, MD, *Int. Arch. Photogrammetry, Remote Sensing Spatial Inf. Sci.*, XXXIV, part 3/W4, Commission III.

Lantuejoul, Ch. and F. Maisonneuve, 1984. Geodesic methods in image analysis, *Pattern Recognition*, 17(2), 177–187.

Lindenberger, J., 1993. Laser-Profilmessungen zur topographischen Gelaedeaufnahme, Deutsche Geodaetische Kommission, Series C, No. 400, Munich.

Maas, H.-G. and G. Vosselman, 1999. Two algorithms for extracting building models from raw laser altimetry data, *ISPRS J. Photogrammetry Remote Sensing*, 54(2–3), 153–163.

Masaharu, H. and H. Hasegawa, 2000. Three-dimensional city modeling from laser scanner data by extracting building polygons using region segmentation method, *Int. Arch. Photogrammetry Remote Sensing*, 33, Part B3, Amsterdam, The Netherlands, CD-ROM.

Morgan, M. and A. Habib, 2002. Interpolation of LIDAR data and automatic building extraction, *ACSM-ASPRS Annu. Conf. Proc.*, CD-ROM.

Rottensteiner, F. and C. Briese, 2002. A new method for building extraction in urban areas from high-resolution LIDAR data, ISPRS Commission III Symposium, Graz, Austria, CD-ROM.

Rottensteiner, F., J. Trinder, S. Clode, and K. Kubik, 2005. Automated delineation of roof planes from LIDAR data, ISPRS WG III/3, III/4, V/3 Workshop "Laserscanning 2005," Enschede, The Netherlands, September 12–14.

Sampath, A. and J. Shan, 2004. Urban modeling based on segmentation and regularization of airborne LIDAR point clouds, *ISPRS Congr.*, Commission III, Istanbul, Turkey, CD-ROM.

Sampath, A. and J. Shan, 2006. Clustering-based planar roof extraction from LIDAR data, *Proc. ASPRS Annual Conf.*, May 1–5, Reno, NV, CD-ROM.

Schickler, W. and A. Thorpe, 2001. Surface estimation based on LIDAR, *Proc. ASPRS Annu. Conf.*, April 23–27, St. Louis, MO, CD-ROM.

Schwalbe, E., H.-G. Maas, and F. Seidel, 2005. 3D building model generation from airborne laser scanner data using 2D GIS data and orthogonal point cloud projections, ISPRS WG III/3, III/4, V/3 Workshop "Laserscanning 2005," Enschede, The Netherlands, September 12–14.

Shan, J. and A. Sampath, 2005. Urban DEM generation from raw LIDAR data: a labeling algorithm and its performance, *Photogrammetric Eng. Remote Sensing*, 71(2), 217–226.

Sithole, G., 2001. Filtering of laser altimetry data using a slope adaptive filter, in Hofton, *Proc. ISPRS Workshop Land Surface Mapping Characterization Using Laser Altimetry*, Michelle A., Ed., Annapolis, MD, *Int. Arch. Photogrammetry, Remote Sensing Spatial Inf. Sci.*, XXXIV, Part 3/W4 Commission III, 203–210.

Sithole, G. and G. Vosselman, 2004. Experimental comparison of filter algorithms for bare earth extraction from airborne laser scanning point clouds, *ISPRS J. Photogrammetry Remote Sensing*, 59(1–2), 85–101.

Sithole, G. and G. Vosselman, 2005. Filtering of airborne laser scanner data based on segmented point clouds, ISPRS WG III/3, III/4, V/3 Workshop "Laserscanning 2005," Enschede, the Netherlands, September 12–14.

Sohn, G. and I. Dowman, 2003. Building extraction using LiDAR DEMs and IKONOS images, *Int. Arch. Photogrammetry, Remote Sensing*, 34(3/W13), WG III/3 Workshop on 3-D reconstruction from airborne laserscanner and InSAR data, Dresden, Germany, October 8–10. CD-ROM.

Tóvári, D. and N. Pfeifer, 2005. Segmentation-based robust interpolation — a new approach to laser data filtering, ISPRS WG III/3, III/4, V/3 Workshop "Laserscanning 2005," Enschede, the Netherlands, September 12–14.

Vosselman, G., 2000. Slope-based filtering of laser altimetry data, *Int. Arch. Photogrammetry Remote Sensing*, 33, Part B3/2, Amsterdam, the Netherlands, 935–942.

Vosselman, G. and S. Dijkman, 2001. 3D building model reconstruction from point clouds and ground plans, *Int. Arch. Photogrammetry Remote Sensing*, 34(3W4), 37–43.

Vosselman, G., B.G.H. Gorte, G. Sithole, and T. Rabbani, 2004. Recognizing structure in laser scanner point clouds, *Int. Arch. Photogrammetry Remote Sensing*, 46(8/W2), 33–38.

Wang, Y., B. Mercer, C. Tao, J. Sharma, and S. Crawford, 2001. Automatic generation of bald earth digital elevation models from digital surface models created using airborne IFSAR, *Proc. ASPRS Conf.*, April 23–27, St. Louis, MO, CD-ROM.

Wang, Z. and T. Schenk, 2000. Building extraction and reconstruction from LIDAR data, *Int. Arch. Photogrammetry Remote Sensing*, 33, Part B3, Amsterdam, The Netherlands, CD-ROM.

Wehr, A. and U. Lohr, 1999. Airborne laser scanning — an introduction and overview, *ISPRS J. Photogrammetry Remote Sensing*, 54(2–3), 68–82.

Weidner, U. and W. Förstner, 1995. Towards automatic building extraction from high-resolution digital elevation models, *ISPRS J. Photogrammetry Remote Sensing*, 50(4), 38–49.

Yoon, J.-S. and J. Shan, 2002. Urban DEM generation from raw airborne LIDAR data, *Proc. Annu. ASPRS Conf.*, Washington D.C., April 22–26, CD-ROM.

Zhang, K., S.C. Chen, D. Whitman, M.L. Shyu, J. Yan, and C. Zhang, 2003. A progressive morphological filter for removing nonground measurements from airborne LIDAR data, *IEEE Trans. Geosci. Remote Sensing*, 41(4), 872–882.

Zhang, K. and D. Whitman, 2005. Comparison of three algorithms for filtering airborne LIDAR data, *Photogrammetric Eng. Remote Sensing*, 71(3), 313–324.

Zhang, Y., C.V. Tao, and J.B. Mercer, 2004. An initial study on automatic reconstruction of ground DEMs from airborne IfSAR DSMs, *Photogrammetric Eng. Remote Sensing*, 70(4), 427–438.

3 Reconstruction of Buildings in SAR Imagery of Urban Areas

Uwe Stilla and Uwe Soergel

CONTENTS

3.1 INTRODUCTION

Three-dimensional city models are of great interest for visualization, simulation, and monitoring purposes in different fields (Stilla et al., 2005). A typical application is the visualization of the influence of a planned building to the surrounding townscape in city and regional planning. City models are used as a basis for simulation — for example, in the fields of environmental engineering for microclimate investigations or telecommunications for transmitter placement. Additionally, demand for such models is growing in civil and military mission planning. Furthermore, three-dimensional information can be used for monitoring (e.g., damage assessment after an earth-quake). All the mentioned tasks require knowledge about the three-dimensional structure of buildings.

In photogrammetry, the distance from the camera to a spatial surface is classically obtained by triangulation based on corresponding image points from two or more pictures of the surface. The required tie-points are chosen and matched manually or automatically by image analysis. In addition to this indirect approach, depending on passive acquisition of the surface reflectance of natural illumination, active range sensors such as LASER scanner systems or synthetic aperture radar (SAR) systems derive object distances directly based on time-of-flight measurements independently from sun illumination.

For topographic mapping, data acquisition in nadir view is advantageous, especially in dense urban scenes with elevated objects. The *light detection and ranging* (LIDAR) principle allows airborne applications in oblique (e.g., obstacle warning systems) and nadir (e.g., building reconstruction [Stilla and Jurkiewicz, 1999]) views as well. In contrast to electro-optical sensors or LIDAR, the SAR principle requires a side-looking illumination. Due to the large signal wavelength (e.g., 3 cm), radar shows almost no sensitivity to weather influence in contrast to the other mentioned important remote sensing spectra.

Three-dimensional scene structure can be obtained by SAR interferometry (InSAR), which requires at least two SAR images. Another advantage of SAR is the opportunity to record large areas in a short time and from a large distance. The latter feature enables space-borne Earth observation by SAR; this has been demonstrated successfully in several satellite and shuttle missions, such as ERS1/2 and SRTM, respectively. Upcoming satellite systems (e.g., TerraSAR-X, scheduled 2006 [Roth, 2003], and Radarsat 2) will achieve significant improvement of geometric resolution that is expected to be in the order of a few meters and modern airborne systems even provide spatial sampling with decimeter spacing (Ender and Brenner, 2003).

The increasing resolution of SAR sensors opens the possibility to utilize such data for a variety of new applications. For example, Slatton et al. (2000) proposed fusion techniques of LIDAR and InSAR data to improve characterization of vegetated areas. On the other hand, mapping and monitoring of urban areas by radar have become major fields of remote-sensing research and applications in recent years. Examples in this context are road network extraction (Tupin et al., 1998) and disaster management (Shinozuka et al., 2000; Takeuchi et al., 2000) based on SAR data. The focus of this chapter is on the reconstruction of buildings using InSAR data. Approaches to three-dimensional building recognition from SAR and InSAR data have been proposed for rural areas (Bolter, 2001), industrial plants (Soergel et al., 2001), and inner-city areas with tall buildings (Gamba et al., 2000; Gamba and Houshmand, 2000; Stilla et al., 2003).

Different SAR-specific phenomena (Schreier, 1993), like foreshortening, layover, shadow, and multipath propagation, burden the scene interpretation or make it impossible. These phenomena arise from the side-looking scene illumination of SAR sensors. The mentioned methods from the literature achieved good results for rural areas or large and detached buildings. Especially in high-density areas with tall buildings, large portions of the data can be interfered with by illumination effects.

However, even in a dense urban environment, building recognition is feasible if the typical appearance of buildings in the data is properly modeled. It is possible to determine building height and roof structure from length and size of the occluded shadow cast from a building on the ground behind it. Other hints to buildings' existence are layover areas and bright double-bounce scatterers at building footprints. This will be demonstrated with a novel approach to detect and reconstruct buildings. The model-based iterative approach exploits context knowledge — for example, neighborhood relations between buildings.

First, the basic principles of SAR and InSAR techniques are introduced briefly in Section 3.2. The sensing of buildings by SAR is discussed in Section 3.3. Significant problem areas with respect to building recognition are caused by occlusion or layover. Geometric relations for the existence and size of such areas of disturbed data are derived. The impact of illumination phenomena is demonstrated for an InSAR test data set covering a part of the inner city of Karlsruhe, Germany, in Section 3.4. The model-based approach for the detection and reconstruction of buildings is proposed in Section 3.5.

3.2 SAR PRINCIPLE

Side-looking SAR sensors are mounted on satellites or airplanes. The basic sensor principle is to illuminate large areas on the ground with the radar signal and to sample the backscatter (Figure 3.1a). The signal of the electromagnetic waves travels with velocity of light, c. From the different time of flight of the incoming signal, the range between the sensor and the scene objects is obtained. The smaller the pulse duration, τ, becomes, the finer the slant range resolution, δ_{sr}, gets:

$$\delta_{sr} = \frac{c \cdot \tau}{2} \qquad (3.1)$$

The pulse length, τ, is inversely proportional to the bandwidth, B, of the signal. Hence, it is advantageous to choose the bandwidth as high as possible. In contrast to the slant range resolution, δ_{sr}, that is constant over the image, the range resolution, δ_{gr}, on the ground (ground range) depends on the viewing angle θ:

$$\delta_{gr} = \frac{c \cdot \tau}{2 \cdot \sin(\theta)} \qquad (3.2)$$

The resolution in the other direction, called azimuth, is given by the width of the sensor footprint of the ground for radar with real aperture. The real aperture depends on the ratio of the signal wavelength, λ, and the antenna length, D. Because the size of the footprint increases with the range r, the azimuth resolution, δ_{a_real}, is usually poor:

$$\delta_{a\_real} = \frac{\lambda \cdot r}{D} \qquad (3.3)$$

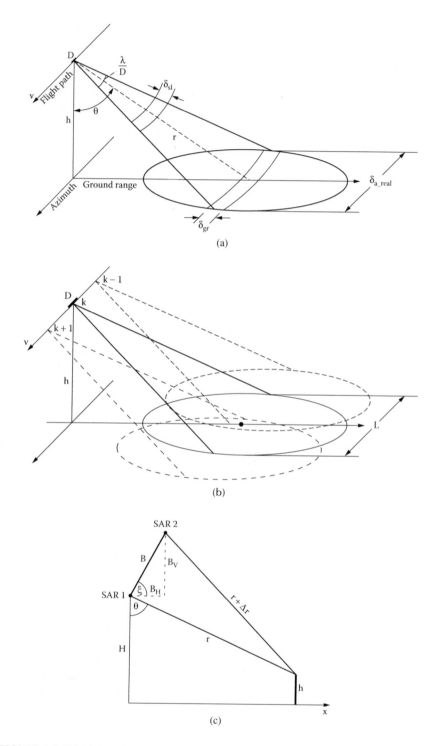

FIGURE 3.1 Principles of a) side-looking radar; b) SAR; c) InSAR.

The synthetic aperture radar principle overcomes this limitation. The measurement is repeated many times along the flight track of the carrier (Figure 3.1b). Due to the relative motion between sensor and ground, a frequency shift of the received signal occurs (Doppler frequency). The Doppler frequency curve depends on range and azimuth of a given point target. This effect is exploited to enhance the azimuth resolution: by a combined frequency analysis of these overlapping one-dimensional signals, a virtual large antenna is generated. In this manner, the backscatter contributions of different objects are integrated into the related position in the SAR image's two-dimensional grid. The aperture of this synthetic antenna matches the footprint width on the ground. The SAR azimuth resolution δ_{a_SAR} is about the half-length D of the real antenna:

$$\delta_{a\_SAR} = \frac{D}{2} \tag{3.4}$$

The grid step size can differ in range and azimuth. Geocoding is usually performed by forward projection on planes (flat plane or ellipsoid) or by backward projection based on digital elevation models (DEMs). The latter method is more accurate (Meier et al., 1993) because the effects of topography undulation can be corrected.

3.2.1 INSAR TECHNIQUE

The complex SAR signal is commonly expressed in polar coordinates with amplitude a and phase φ. The analysis of single SAR images is usually restricted to signal amplitude. Complete complex information is required for interferometric processing of at least two SAR images. The interferometric SAR (InSAR) technique offers an opportunity to gather information about the third dimension as well. Digital elevation models of large areas of the globe have been collected using InSAR.

Figure 3.1c illustrates the basic principle of InSAR. In the following section, a single-pass measurement is discussed; the SAR images are acquired simultaneously from a single carrier. Two radar antennas are mounted perpendicular to the carrier path with geometric displacement B. One of the antennas illuminates the scene and both antennas receive the backscattered complex signals, s_1 and s_2, at the same time:

$$s_1 = a_1 \cdot e^{j\varphi_1} \quad \text{and} \quad s_2 = a_2 \cdot e^{j\varphi_2}. \tag{3.5}$$

The two SAR images are processed independently from each other. Due to the baseline B, the distances from the antennas to the scene differ by Δr. Therefore, the two SAR images are shifted in range direction and coregistration is a prerequisite for interferometric processing; the slave image, s_2, is resampled to the grid of the master image, s_1. The interferogram S is then calculated by a pixel-by-pixel complex multiplication of image s_1 with the complex conjugated and shifted image s_2. The geometric displacement during data acquisition results in a phase difference in the interferogram:

$$S = s_1 \cdot s_2^* = a_1 \cdot a_2 \cdot e^{j\Delta\varphi}. \tag{3.6}$$

The product of the amplitudes is called *intensity*. The phase difference, $\Delta\varphi$, is caused by the range difference, Δr:

$$\Delta\varphi = -\frac{2\pi}{\lambda} \cdot \Delta r \tag{3.7}$$

The phase difference, $\Delta\varphi$, is unambiguous in the range $[-\pi,\pi]$ only. Thus, phase unwrapping is often required before the heights are calculated. From the illumination geometry, the range difference, Δr, can be expressed in another way:

$$\Delta r = \sqrt{r^2 + B^2 + 2 \cdot B \cdot r \cdot \sin\left(\xi - \theta\right)} - r \quad with \quad \xi = \arctan\left(\frac{B_v}{B_H}\right). \tag{3.8}$$

B_V and B_H are the vertical and horizontal components of baseline B, respectively. Equation 3.4 is now solved for the unknown viewing angle θ (off-nadir angle) to the object. Finally, the object height, h, and position, x, are obtained by

$$h = H - r \cdot \cos(\theta) \quad and \quad x = r \cdot \sin(\theta), \tag{3.9}$$

with carrier height, H.

The accuracy of DEM produced by the InSAR technique varies locally depending on the signal-to-noise ratio (SNR). The so-called coherence, γ, is a measure of the local SNR. The coherence is usually estimated from the data by a window-based computation of the magnitude of the complex cross-correlation coefficient of the SAR images. The noise sensitivity of SAR interferometry results often in data holes or competing elevation values after geocoding with the forward transformation (Equation 3.9). Hence, the InSAR height data must be further processed before the geocoding step is carried out.

Figure 3.2 illustrates the InSAR data set of the test site in ground range projection and a LIDAR DEM for comparison. The data were recorded by the airborne AER-II experimental multichannel SAR system (Ender, 1998) of FGAN. This system has been tested in several flight campaigns onboard a Transall airplane and is equipped with a phased array antenna and several receiver channels. Center frequency of this X-band system is 10 GHz ($\lambda = 3$ cm) with bandwidth of 160 MHz. The ground range data have an approximate resolution of 1 m × 1 m. Range direction is top down. Assuming a constant noise power, it is evident that in areas with low backscatter power the SNR is poor. This results in low coherence (Figure 3.2c) and distorted height data (Figure 3.2b).

3.2.2 Buildings in SAR Images

3.2.2.1 Phenomena Caused by Side-Looking Illumination

Figure 3.3 illustrates typical effects in SAR images in the vicinity of buildings. The so-called layover phenomenon occurs at locations with steep elevation gradients facing towards the sensor, such as vertical building walls (A and B). Because object areas

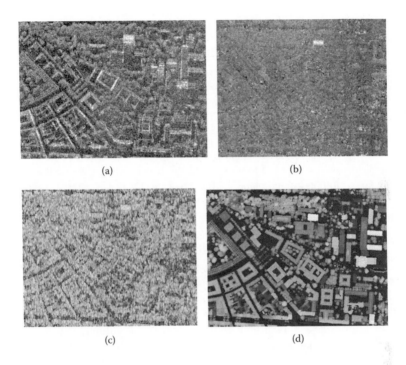

(a)   (b)

(c)   (d)

FIGURE 3.2 a) Intensity; b) height; c) coherence (bright = good); d) LIDAR DEM (ground truth).

located at different positions (roofs, walls, and the ground in front of buildings) have the same distance to the sensor, the backscatter is integrated to the same range cell. This signal integration leads to the bright appearance of layover areas in SAR images. Points A and B differ in height but have the same x–y coordinates in the world coordinate system. However, in the slant range image, I_s, the elevated point B is projected to

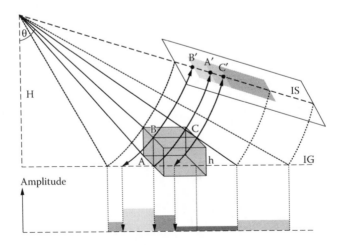

FIGURE 3.3 SAR phenomena at a flat-roofed building.

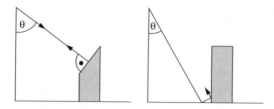

FIGURE 3.4 Single scattering (left) and double-bounce scattering (right) at building.

position B′, which is closer to the sensor than A′. If geocoding is carried out without DEM by projection on ground plane, I_G, this reversed point order cannot be corrected.

Perpendicular alignment of buildings to the sensor leads to strong signal responses by double-bounce scattering at the dihedral corner reflector between the ground and the building wall (Figure 3.4, right). This results in a line of bright scattering in the azimuth direction at the building footprint. At the opposite building side, the ground is partly occluded from the building shadow. This region appears dark in the SAR image because no signal returns into the related range bins. The sketch in the lower part of Figure 3.3 illustrates the typical appearance of buildings in SAR amplitude images: a bright layover area that may be terminated by a very bright linear double-bounce signal located at building walls (A and B), followed by roof signal, depending on material and roughness, and finally a dark-shadow area.

In addition to material and roughness influence, depending on the illumination aspect the roof structure, geometry may lead to strong signal response. The gabled building roof sketched in Figure 3.4 (left) is oriented perpendicular to the sensor. Because the entire power is mirrored back to the sensor, this reflection leads to a line of dominant scattering in the azimuth direction, similar to the corner reflector. This bright line caused from the roof appears closer to the sensor in the SAR image compared to the corner reflector signal. In addition to the offset in range direction, both effects can be discriminated by their polarimetric properties (single bounce and double bounce, respectively).

The mentioned effects can be studied in Figure 3.5, which compares the section of an aerial image and a SAR image showing the castle of Karlsruhe, Germany. The SAR image is superimposed with the building footprints from a map. The scene was illuminated from the top. The signal from a corner reflector at the castle's main building is located at the building footprint (pos. 1). The bright signal from the gabled roof is projected on the terrace in front of the castle (pos. 2). These two lines enclose the layover area. Layover can be observed as well at the castle wing (pos. 3) and at the tower (pos. 4). At the wing, no line of bright scattering appears because the double-bounce signal is reflected away from the sensor. Another double-bounce event happens at a little wall at the border of the terrace (pos. 5).

3.2.2.2 Geometric Constraints

In the following section, the phenomena of layover and shadow are discussed in more detail. The sizes of the layover areas, l_g, and shadow areas, s_g, on the ground

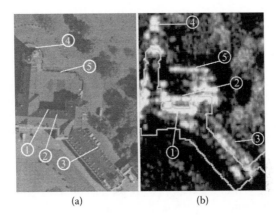

(a)   (b)

FIGURE 3.5 Karlsruhe Castle: a) aerial image; b) SAR image overlaid with building foot-prints and pointers to SAR phenomena at (1) main building wall, (2) main building roof, (3) wing, (4) tower, and (5) terrace wall (SAR illumination bottom-up).

in range direction depend on the viewing angle, θ, and the building height, h. The layover area (see Figure 3.6a) is given by:

$$l_g = h \cdot \cot(\theta). \qquad (3.10)$$

For building analysis, the roof area, l_{rt}, is of interest; this is influenced by layover. In extreme cases, layover might be present at the entire building. Only at the far building side may part of the roof not be interfered with by layover. For a given

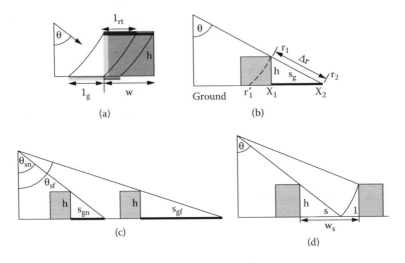

FIGURE 3.6 a) Layover in front and on a flat-roof building; b) shadow behind a building; c) angle dependency of shadow length; d) shadow and layover from buildings displaced in range direction.

building of width w, such an area of undisturbed data exists if the inequation is fulfilled:

$$h < w \cdot \tan(\theta). \tag{3.11}$$

In the case of shadow, geometric relations can be obtained too (Figure 3.6b). The slant range shadow length, Δr, is the hypotenuse of the rectangular triangle with the two sides, h and s_g. Hence, the building elevation, h, is given by:

$$h = \Delta r \cdot \cos(\theta). \tag{3.12}$$

A simple projection of the slant range SAR data on a flat ground plane (ground range), ignoring the building elevation, leads to a wrong mapping of the roof's edge, r_1, to point r_1'. Starting from point r_2, the true position, x_1, of the building wall can be determined:

$$x_2 = r_2 \cdot \sin(\theta) \tag{3.13}$$

$$x_1 = x_2 - \Delta r \cdot \sin(\theta) \tag{3.14}$$

$$s_g = x_2 - x_1 = h \cdot \tan(\theta). \tag{3.15}$$

However, the shadow analysis is reliable only in the case of flat ground behind the building and if no signal of other elevated objects interferes with the shadow area (e.g., caused by a neighboring building). It is obvious that, at building locations, a steep viewing angle θ leads to large layover areas on the ground and the roofs, but to small shadow areas and vice versa. Therefore, the viewing angle must be chosen carefully to maximize the portion of useful SAR data.

The viewing angle increases in range direction over the swath. In Figure 3.6c, such a situation is depicted (shadow length s_{gn}, s_{gf}). Assuming a viewing angle range between 40° and 60°, the shadow length of a certain building is more than doubled from near to far range. This means that the shadow of the building might occlude other objects or not, depending only on its distance to the sensor or on its location in the signal swath.

In the worst case, a road between two building rows is oriented parallel to the sensor trajectory (Figure 3.6d). The street is partly occluded (shadow area) and partly covered with layover. An object on the road can only be sensed undisturbed if a condition for the road width, w_s, holds:

$$w_s > s_{gn} + l_g = h \cdot (\tan(\theta_{sn}) + \cot(\theta_l)). \tag{3.16}$$

Assuming the same viewing angle of 45° and height of 20 m for both buildings, the road width w_s would be 40 m according to Equation 3.16. Consequently, applications

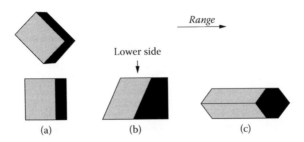

FIGURE 3.7 The appearance of different types of buildings in SAR amplitude or intensity images: a) flat roof; b) pent roof; c) gable roof.

such as vehicle detection seem hardly feasible from a single SAR image alone and require, in general, multiaspect SAR data.

Different types of roof structures of buildings lead to special shapes of the shadow cast on the ground. The sketch in Figure 3.7 illustrates the appearances of buildings with common roof types in SAR amplitude or intensity images, assuming that the scene is illuminated from left to right. The shadow shape depends on the aspect. Flat-roofed buildings usually cast a stripe-like or L-shaped shadow. A building with pent-roof structure (sloped roof) may cause a trapezoidal shadow. A gabled-roof building may cast a hexangle-shaped shadow. However, both of the latter hexangle-shaped building types may cast a stripe-like or L-shaped shadow as well (e.g., if the range direction is top–down or bottom–up as in Figure 3.7).

3.3 SIMULATION OF LAYOVER, SHADOW, AND DOMINANT SCATTERING AT BUILDINGS

Based on ground truth DEM, it is possible to simulate layover and shadow areas (Meier et al., 1993). Such a simulation was carried out for the test site according to the given parameters of the real SAR data. The result is illustrated in Figure 3.8a. A detailed visibility analysis (Soergel et al., 2005) reveals that only 20% of the road area and 43% of the roof area can be sensed properly with this SAR measurement. For the rest, layover, shadow, or both interfere. A fusion of complementing SAR data acquired from different aspects offers the opportunity to fill occluded areas and to correct layover artifacts. By combining four carefully chosen SAR acquisitions, the portion of undisturbed visible areas (at least from one view) rises to 62% for streets and over 85% for roads.

Figure 3.8b shows a detail of the test data with strong signals at building locations mainly caused by specular and double-bounce reflections at facade and roof elements. Based on a three-dimensional model and given sensor parameters, scattering events can be predicted by simulation. Here, the locations of possible strong scattering (black lines in Figure 3.8c) were detected using three-dimensional vector data derived from the LIDAR DEM (Stilla and Jurkiewicz, 1999). This was done by scanning the DEM in range direction for planes facing perpendicularly towards the sensor and dihedral plane couples pointing to the sensor. Particularly interesting is

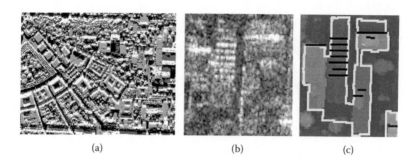

(a)   (b)   (c)

FIGURE 3.8 a) Simulation with the given SAR parameters (range from top to down): layover (white), shadow (black), layover and shadow (dark gray), and reliable data (bright); b) SAR image of scene detail on the center right; c) DEM with building footprints (white) and possible corner structures (black).

the rippled roof structure on the left, which is causing strong signal response from specular reflection. Similar SAR simulation techniques are incorporated in the iterative building detection and reconstruction approach presented in the next section.

3.4   APPROACH FOR BUILDING RECOGNITION

3.4.1   INTRODUCTION

In this section, an approach for building detection and recognition from InSAR data is described. Based on the imagery depicted in Figure 3.9, the concept for assessment of results will be discussed first. Figure 3.9a shows an aerial image acquired during the SAR measurement. The related InSAR intensity and height are illustrated in Figure 3.9b and Figure 3.9c. For better comparison with the photo, the SAR imagery was rotated by 180°. The range direction is now bottom–up. The InSAR data are still in slant range geometry, with better resolution in the azimuth direction (horizontal coordinate). Figure 3.9e shows again the intensity image overlaid with the building footprints drawn by a human expert without any context information of the scene. This information will be called "sensed truth." The comparison with the LIDAR DEM (Figure 3.9f) and the aerial image in nadir view (Figure 3.9d) reveals that even a human expert cannot spot every building in the scene. Especially small buildings are sometimes hardly visible due to layover caused by objects such as high trees or neighboring buildings.

Scene interpretation from remote sensing imagery is a demanding task. The human ability to interpret even complex scenarios is usually not accomplished by automatic machine vision systems. Hence, the sensed truth represents a best effort result of any automatic approach. The assessment of building detection is carried out in a twofold manner. The performance of the proposed approach will be assessed with respect to the sensed truth. Furthermore, it is interesting to compare the results with the "real ground truth" (Figure 3.9f). The latter gives insight into the feasibility of building detection and recognition with this approach and from InSAR data of this quality.

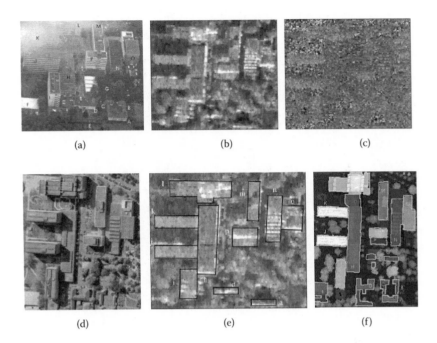

(a)   (b)   (c)

(d)   (e)   (f)

FIGURE 3.9 a) Aerial image shot during InSAR measurement; b) InSAR intensity in slant range; c) InSAR height in slant range; d) aerial image in nadir view; e) "sensed truth"; f) "real ground truth."

3.4.2 ALGORITHM OVERVIEW

An overview of the approach is given in Figure 3.10. The building recognition is performed iteratively. The workflow consists of separated modules for the tasks of detection and reconstruction of buildings. First, preprocessing (e.g., multilooking, smoothing of the phase, and speckle reduction [Desnos and Matteini, 1993]) is required to prepare the InSAR data for the subsequent segmentation of primitive objects. This object segmentation is carried out in the original slant range data of phase and magnitude. This is advantageous so as to avoid artifacts due to geocoding, such as the distorted appearance of building edges in the ground range projection. From this point, the analysis is restricted to the symbolic object level.

From the object primitives, more complex building hypotheses are assembled in the detection module. After a projection of these building candidates coordinates from the slant range into the world coordinate system, the building recognition step follows. In the reconstruction procedure, model knowledge (e.g., rectangular shape of buildings or preferred parallel alignment of buildings along roads) is exploited. If several SAR or InSAR data sets acquired from different directions are available, the results are fused. Intermediate results are used for a simulation of layover, shadow, dihedral corner reflectors, and an InSAR DEM. The simulation results are reprojected into the SAR geometry and compared with the real data.

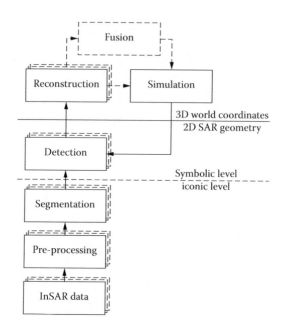

FIGURE 3.10 Workflow of building recognition.

Differences between the simulation and the real data steer the update of the process: new building hypotheses are generated and false ones eliminated. Hence, the resulting scene description is expected to converge to the real three-dimensional objects in the scene with increasing numbers of steps. The iteration stops if the RMSE between the simulated and the real InSAR DEM falls below a threshold or after a given number of cycles.

3.4.3 SEGMENTATION OF PRIMITIVE OBJECTS

Primitive objects are segmented in the intensity and height data. In the intensity data, edge and line structures are detected. Three types of structures are of interest:

- Salient bright lines caused by layover or double-bounce reflection will be referred to as objects STRONGE_SCATTER_LINES.
- Edge structures at the border of a dark region are potentially caused by building shadow. Two sets of objects are distinguished. The first set builds those border edges of the dark region that face the sensor (object NEAR_ SHADOW_EDGE). The other set consists of edges at the far side of dark regions (object far from shadow edge).
- The remaining edges build the set of other object BUILDING_EDGES. Those may coincide with building edges oriented in range direction.

In the height data, segmentation of objects with significant elevation above ground is required. For this purpose, a normalized DEM (NDEM) is derived from the InSAR height data that represents the elevation of objects over ground. First, a rank filtering of the height data is carried out. The filter window size is chosen larger than the expected maximum building area. Height values coinciding with poor coherence or low intensity are not considered in this step in order to exclude blunders. This filtering results in a digital terrain model (DTM) representing the topography without elevated objects such as trees or buildings. The NDEM is derived from the difference of the original InSAR height from the DTM. With an elevation threshold, elevated objects are separated from the ground in the NDEM. Connected regions of elevated pixels build the set of objects' elevated region.

3.4.4 Detection of Building Candidates in the Slant Range

From the line- and edge-structured primitive objects more complex object QUAD-RANGLES are assembled by a production system:

- At least one edge of the object QUADRANGLE facing the sensor must be derived from an object strong scatter line.
- Analogously, an object NEAR_SHADOW_EDGE is required at the far edges of the object QUADRANGLE.

Usually, many object QUADRANGLES are produced. The object QUADRANGLES from the test data are shown in Figure 3.11a. Only a subset of those actually coincides with buildings. A variety of different object configurations is a source of false hints, such as fences or car rows that may cause strong scattering as well.

For discrimination of buildings from the objects, ELEVATED_REGIONS are used. The intersection area of each object QUADRANGLE with all objects' ELEVATED_REGIONS is determined. The ratio of the intersection area to the quadrangle area is used as the object feature "elevated area ratio." Only object quadrangles with an elevated area ratio larger than 0.7 are considered as object BUILDING_CANDIDATES. The number of object BUILDING_CANDIDATES is

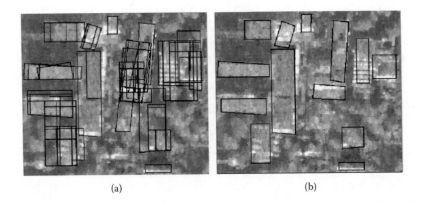

(a)   (b)

FIGURE 3.11 a) The set of all assembled object QUADRANGLES; b) object BUILDING_ CANDIDATES: the elevated and best assessed subset of (a).

further reduced: from mutual intersecting candidates, only the best assessed one is considered for the reconstruction step in this iteration. The assessment value depends on:

* The overlap and parallelism of the primitive objects
* The feature elevated area ratio

The remaining object BUILDING_CANDIDATES of the first iteration are illustrated in Figure 3.11b.

3.4.5 BUILDING RECONSTRUCTION

Building reconstruction is based on the following scene model:

* The footprints of buildings have right-angled shapes.
* Buildings are elevated objects with different roof structures. Three types of parametric building models are considered: flat-roof buildings, gabled-roof buildings, and pent-roof buildings. A common feature of the parametric building models is the rectangular footprint.
* A generic building model addresses complex building structures consisting of several parts such as wings. These parts may have different height, which is assumed constant for each part. The footprint of a generic building is modeled as a right-angled polygon. The number of parts of a generic object is not determined a priori.
* Neighboring buildings and the parts of complex buildings often have the same orientation because they are aligned parallel to roads.

For the reconstruction step the polygon coordinates of the object's BUILDING_CANDIDATE are transformed from SAR geometry into the world coordinate system. This step requires knowledge of object height. A pixel-by-pixel transformation according to Equation 3.9 leads to distorted object edges because of the noise sensitivity of interferometric measurement. Averaging the height values inside the borders of each BUILDING_CANDIDATE reduces the noise impact. In the case of flat-roofed buildings, the average is the maximum likelihood estimate of building height. To consider the different reliability of the individual height pixel values, the related coherence value is used as weight in the averaging procedure (Soergel et al., 2001).

Because of tolerances in the building candidate assembly, the building footprint reconstructed so far is usually not perfectly right angled after geocoding. Therefore, the footprint is approximated by a rectangular or right-angled polygon. The roof structure is analyzed with two different methods. The first is restricted to the height data. After the reprojection of the now right-angled footprints of the building hypotheses into the slant range, planes are fitted to the data. The second method analyzes size and shape of the shadow areas (objects NEAR_SHADOW_EDGE and FAR_SHADOW_EDGE). Both results are combined to determine the roof shape and to improve the footprint location. The corrected footprint and roof structure are features of new produced objects BUILDING.

3.4.6 ITERATIVE IMPROVEMENT OF RESULTS

The first iteration is restricted to reconstruction of object BUILDINGS with rectangular footprints. In subsequent cycles, the generic building model is considered as well. If several InSAR data sets are available, they are analyzed in parallel and the results are fused (Soergel et al., 2003). In the case of competing results, only the object BUILDINGS with the best assessment is accepted for the fused result. Shadow and layover areas in one view can be filled by information from images of different aspects. Due to parameter tolerances, the reconstructed orientations of neighboring objects' buildings might differ slightly. Hence, the orientations are corrected according to the parallelism constraint of the model.

The assessment of the object BUILDINGS is used as weight for this adjustment step. Based on intermediate results, simulations of layover, shadow, dihedral corner reflectors, and the InSAR DEM are carried out with respect to parameters of the real data. The InSAR DEM height value can be determined from the range difference derived from given sensor positions according to Equation 3.7. Differences between the simulation and the real data are hints to inconsistencies. These govern update of the process: building hypotheses are accepted and modified, new ones are generated, and false ones are eliminated. Then the reconstruction step is repeated.

3.5 RESULTS

The result of the first iteration reprojected into the slant range after the reconstruction step is illustrated in Figure 3.12a (dark rectangles). The shadow analysis did not yield good results for this built-up test site, due to the proximity of the buildings and the presence of many tall trees. Therefore, the calculation of the building height was based mainly on the InSAR height data. Thirteen buildings were detected.

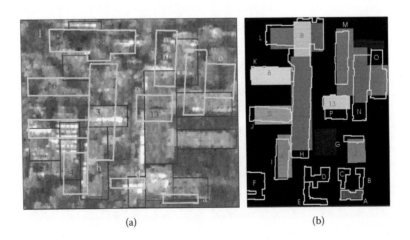

(a)   (b)

FIGURE 3.12 a) Result of first iteration reprojected into the slant range after the reconstruction step (red) and sensed truth; b) final result after five iterations (gray value coding of building height) superimposed with real ground truth.

The result is assessed with respect to the two sets of ground truth data (Figure 3.9e, f) described earlier.

The comparison with the sensed truth (bright in Figure 3.12a) gives: nine buildings are correct, one is missing (*d*), and one is oversegmented (*l*). The oversegmentation was caused by superstructures on the rooftop with significantly different heights. One false building is present (object 2) at the location of some large trees. However, the major parts of those buildings were detected and labeled manually.

In Figure 3.12b, the result after five iterations is illustrated. The gray level corresponds with the reconstructed building height. The real ground truth is superimposed (bright lines). With respect to this ground truth, an additional five buildings at the bottom of the scene are missing. The reason is often occlusion or layover caused by high trees. Building *G*, for example, is not visible at all, even in the aerial image shown in Figure 3.9a. However, mainly small buildings were not detected. In particular, the height of tall buildings was underestimated. The buildings *J* and *K* on the left-hand side are about 40 m high; however, their estimated height was 7 m lower. The reason for this is probably layover causing a mixture of elevation values from the rooftop, the wall, and the ground in the InSAR DEM. A wrong height estimate leads to an erroneous position of the footprint after the forward transformation into the world coordinate system.

The buildings appear shifted towards the sensor. The main structures of the two large building complexes were detected. The recognition of the gabled roof of the small buildings *A* and *D* failed: they were recognized to be flat-roofed buildings. Probably, these roofs would not be reconstructed correctly from this InSAR data set even if the buildings were detached. The reason is the orientation of the two buildings in the azimuth direction, which is disadvantageous for the shadow analysis. For an illumination from the right, better results could be expected.

A three-dimensional view of the result is given in Figure 3.13a. The same result is shown again in Figure 3.13b together with the reference LIDAR DEM (darker gray). The visual comparison reveals that the algorithm was able to recognize the major building structures in this built-up scene.

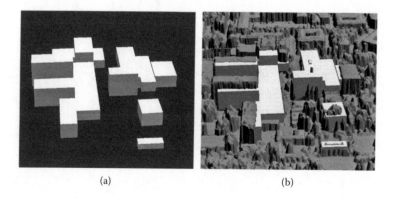

(a)   (b)

FIGURE 3.13 a) Three-dimensional view of the reconstructed prismatic building models; b) three-dimensional building models together with laser elevation data.

3.6 CONCLUSION

In general, parts of the urban scene cannot be recognized from single SAR measurements. This is a consequence of the inherent side-looking illumination by SAR causing unfavorable phenomena such as layover, occlusion, and multipath signal propagation. Geometric relations discussed here can predict the size of certain types of problem areas. The geometric constraints show that the greater the heights of buildings become, the larger the size of the related area becomes. Similarly, too close proximity of buildings is disadvantageous. These tendencies were confirmed by results of the proposed approach for building detection and recognition.

In addition to tall buildings, the presence of tall trees turned out to be problematic. In built-up areas, the analysis of shadow length and size, which shows good results for detached buildings and suburban environments, seems not to be suitable because the shadow is often superimposed by layover from neighboring objects. Again, due to the layover, the InSAR elevation values at the roof parts close to the sensor are disturbed. If the entire roof area is affected, the building height is underestimated. However, the main buildings have been detected and reconstructed by the proposed method, even though only a single InSAR data set of the test site was available.

The accuracy of results obtained from InSAR until now cannot compete with three-dimensional reconstructions derived from LIDAR data. A potential for improvement of results offers the fusion of several SAR and InSAR data. This must be further investigated with focus on high-density areas. In the case of multiaspect analysis, an iterative approach is particularly suitable because hints from one image may initiate a refined analysis at related locations in other images. The conclusion of this chapter is that a detailed scene reconstruction of inner-city areas from InSAR data alone seems to be feasible only by an iterative analysis of multiaspect InSAR data.

In this chapter, urban three-dimensional extraction from SAR data alone was discussed. If additional data are available, a combined analysis offers the opportunity for further result enhancement by exploiting complementary information. With respect to high-resolution SAR data, such fusion strategies have been proposed using aerial imagery (Tupin and Roux, 2003) or GIS data (Soergel et al., 2005). In this manner, deeper insight into the appearance of urban objects in remote sensing data of different natures can be gained. Such knowledge will be valuable with respect to change-detection tasks such as disaster monitoring — for example, by comparing an actual SAR image with other imagery or GIS data.

ACKNOWLEDGMENT

We thank Dr. Ender and Dr. Brenner at FGAN-FHR Research Institute for High Frequency Physics and Radar Techniques for providing the InSAR image data. The data were recorded by the AER II experimental system of FGAN.

REFERENCES

Bolter, R., 2001. Buildings from SAR: detection and reconstruction of buildings from multiple view high resolution interferometric SAR data, Ph.D. thesis, University of Graz, Austria.

Desnos, Y.L. and Matteini, V., 1993. Review on structure detection and speckle filtering on ERS-1 images, *EARSeL Adv. Remote Sensing*, 2(2), 52–65.

Ender, J.H.G., 1998. Experimental results achieved with the airborne multi-channel SAR system AER-II, *Proc. Eur. Conf. Synthetic Aperture Radar* (EUSAR'98), pp. 315–318.

Ender, J.H.G. and Brenner, A.R., 2003. PAMIR — a wideband phased array SAR/MTI system, *IEEE Proc. Radar, Sonar, Navigation*, 150(3), 165–172.

Gamba, P. and Houshmand, B., 2000. Digital surface models and building extraction: a comparison of IF-SAR and LIDAR data, *IEEE Trans. Geosci. Remote Sensing*, 38(4), 1959–1968.

Gamba, P., Houshmand, B., and Saccini, M., 2000. Detection and extraction of buildings from interferometric SAR data, *IEEE Trans. Geosc. Remote Sensing*, 38(1), 611–618.

Meier, E., Frei, U., and Nuesch, D. 1993. Precise terrain corrected geocoded images. In: Schreier, G. (Ed.), *SAR Geocoding: Data and Systems*, Karlsruhe: Wichmann. pp. 173–185.

Roth, A., 2003. TerraSAR-X: a new perspective for scientific use of high-resolution space-borne SAR data, *2nd GRSS/ISPRS Joint Workshop Remote Sensing Data Fusion Urban Areas*, URBAN 2003, pp. 22–23.

Schreier, G., 1993. Geometrical properties of SAR images. In: Schreier, G. (Ed.), *SAR Geocoding: Data and Systems*, Karlsruhe: Wichmann, pp. 103–134.

Shinozuka, M., Ghanem, R., Houshmand, B., and Mansuri, B., 2000. Damage detection in urban areas by SAR imagery, *J. Eng. Mech.*, 126(7), 769–777.

Slatton, K.C., Crawford, M.M., and Evans, B.L., 2000. Combining interferometric radar and laser altimeter data to improve estimates of topography, *Proc. Int. Geosci. Remote Sensing Soc. (IGARSS) Symp.* 2000 (on CD ROM).

Soergel, U., Schulz, K., and Thoennessen, U., 2001. Phenomenology-based segmentation of InSAR data for building detection. In: Radig, B. and Florczyk, S. (Eds.), *Pattern Recognition*, 23rd DAGM Symp., Berlin: Springer, pp. 345–352.

Soergel U., Thoennessen U., and Stilla U., 2003. Iterative building reconstruction in multi-aspect InSAR data. In: Maas, H.G., Vosselman, G., and Streilein, A. (Eds.), 3-D reconstruction from airborne laserscanner and InSAR data, *Int. Arch. Photogrammetry Remote Sensing*, 34, Part 3/W13, 186–192.

Soergel U., Schulz K., Thoennessen U., and Stilla U., 2005. Integration of 3D data in SAR mission planning and image interpretation in urban areas, *Inf. Fusion*, Elsevier B.V., 6(4), 301–310.

Stilla, U. and Jurkiewicz, K., 1999. Reconstruction of building models from maps and laser altimeter data. In: Agouris, P. and Stefanidis, A. (Eds.), *Integrated Spatial Databases: Digital Images and GIS*, Berlin: Springer, pp. 34–46.

Stilla U., Soergel U., and Thoennessen U., 2003. Potential and limits of InSAR data for building reconstruction in built-up areas, *ISPRS J. Photogrammetry Remote Sensing*, 58(1–2), 113–123.

Stilla, U., Rottensteiner, F., and Hinz, S., (Eds.), 2005. Object extraction for 3D city models, road databases, and traffic monitoring — concepts, algorithms, and evaluation (CMRT05), *Int. Arch. Photogrammetry Remote Sensing*, 36 (3/W24).

Takeuchi, S., Suga, Y., Yonezawa, C., and Chen, C.H., 2000. Detection of urban disaster using InSAR — a case study for the 1999 great Taiwan earthquake, *Proc. IGARSS 2000* (on CD ROM).

Tupin, F., Maitre, H., Mangin, J.-F., Nicolas, J.-M., and Pechersky, E., 1998. Detection of linear features in SAR images: application to road network extraction, *IEEE Trans. Geosci. Remote Sensing*, 36(2), 434–453.

Tupin, F. and Roux, M., 2003. Detection of building outlines based on the fusion of SAR and optical features, *ISPRS J. Photogrammetry Remote Sensing*, 58, 71–82.

Part II

Urban Composition and Structure

4 Subpixel Analysis of Urban Landscapes

Qihao Weng and Dengsheng Lu

CONTENTS

4.1 INTRODUCTION

Urban landscapes are typically composed of features that are smaller than the spatial resolution of sensors — a complex combination of buildings, roads, grass, trees, soil, water, and so on. Strahler et al. (1986) defined H- and L-resolution scene models based on the relationship between size of the scene elements and the resolution cell of the sensor. The scene elements in the L-resolution model are smaller than the resolution cells, and are not detectable. When the objects in the scene become increasingly smaller than the resolution cell size, they may no longer be regarded as objects individually. Hence, the reflectance measured by the sensor may be treated as a sum of interactions among various classes of scene elements as weighted by their relative proportions (Strahler et al., 1986).

This is what happens with medium resolution satellite imagery, such as those of Landsat Thematic Mapper (TM) or Enhanced Thematic Mapper Plus (ETM+), applied for mapping the components of urban environments. As the spatial resolution interacts with the fabric of urban landscapes, a special problem of mixed pixels is created because several land-use and land-cover (LULC) types are contained in one pixel. Such a mixture becomes especially prevalent in residential areas, where buildings, trees, lawns, concrete, and asphalt can all occur within a pixel.

Mixed pixels have been recognized as a major problem affecting the effective use of remotely sensed data in LULC classification and change detection (Fisher, 1997; Cracknell, 1998). When mixed pixels occur, pure spectral responses of specific features are confused with the pure responses of other features, leading to the problem of composite signatures (Campbell, 2002). The low accuracy of LULC classification in urban areas is largely attributed to the mixed pixel problem because convectional per-pixel classifiers, such as maximum-likelihood classifiers (MLCs), cannot effectively handle this problem.

One of the major advances in urban landscape analysis in recent years is Ridd's (1995) vegetation–impervious surface–soil (V-I-S) model (Figure 4.1). It assumes that land cover in urban environments is a linear combination of three components: vegetation, impervious surface, and soil. This model may provide a guideline for decomposing urban landscapes and a link for these components to remote sensing spectral data. Ridd believed that this model could also be applied to spatial–temporal analyses of urban morphology and biophysical and human systems. Due to advances in remote sensing digital analysis technology since the 1990s — especially spectral mixture analysis (SMA) (Smith et al., 1990; Settle and Drake, 1993; Adams et al., 1995), this conceptual model has now been successfully implemented with digital analysis methods (Ward et al., 2000; Madhavan et al., 2001; Rashed et al., 2001; Small, 2001; Phinn et al., 2002; Wu and Murray, 2003, Lu and Weng, 2004).

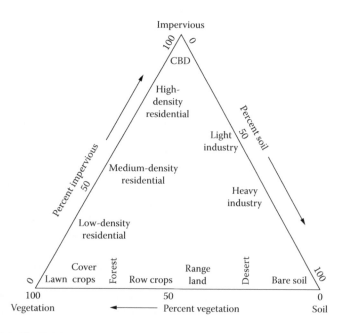

FIGURE 4.1 The V-I-S (vegetation–impervious surface–soil) model illustrating the characteristics of urban landscapes. (From Ridd, M.K., 1995, *Int. J. Remote Sensing*, 16, 2165–2185.)

Ward et al. (2000) applied a hierarchical unsupervised classification approach to a TM image in southeast Queensland, Australia, based on the V-I-S model. An adjusted overall accuracy of 83% was achieved. Madhavan et al. (2001) used an unsupervised classifier to classify TM images in Bangkok, Thailand, and the V-I-S model was found to be useful for improving classification. Phinn et al. (2002) compared traditional image classification and aerial photograph interpretation with a constrained linear SMA; they found that the V-I-S fraction images derived from SMA of a TM image in southeast Queensland, Australia, provided a better classification result than per-pixel classification and aggregated air photograph interpretation. Wu and Murray (2003) used SMA to analyze impervious surface distribution in Columbus, Ohio, and found that impervious surface can be estimated using a linear regression model of low- and high-albedo endmember fractions. Lu and Weng (2004) substantially improved LULC classification accuracy with SMA and developed a conceptual model for characterizing urban LULC patterns based on a case study of Indianapolis, Indiana.

Although the V-I-S model has demonstrated usefulness for characterizing and quantifying urban landscape patterns, its use in practice is still constrained because of the following factors. First, the V-I-S model cannot explain all land cover types, such as water and wetlands. For medium-resolution remotely sensed data, the central business district, light/heavy industry, high/medium density residential, and bare soils are difficult to separate using available digital image processing techniques. Second, impervious surface in the V-I-S model cannot be easily identified as an endmember based on remote sensing images (Wu and Murray, 2003) because impervious surface is a complex mixture of different materials, such as concrete, asphalt, metals, plastic, and so on. Finally, the V-I-S model excludes an important component in the mixed pixels: the shade. Shade caused by tall buildings or trees is an important factor affecting the spectral responses of urban landscapes.

This chapter applies the V-I-S concept for a spatial–temporal analysis of the urban morphology in Indianapolis. By doing this, the potentials and limitations of SMA for characterizing urban landscapes can be further examined.

4.2 BACKGROUND

4.2.1 Spectral Mixture Analysis

SMA is a physically based image analysis procedure that supports repeatable and accurate extraction of quantitative subpixel information (Smith et al., 1990; Roberts et al., 1998; Mustard and Sunshine, 1999). It assumes that the spectrum measured by a sensor is a linear combination of the spectra of all components within the pixel (Adams et al., 1995; Roberts et al., 1998). The mathematical model of linear SMA can be expressed as:

$$R_i = \sum_{k=1}^{n} f_k R_{ik} + ER_i \qquad (4.1)$$

where

$i = 1,\ldots, n$ (number of spectral bands)

$k = 1,\ldots, n$ (number of endmembers)

R_i is the spectral reflectance of band i of a pixel that contains one or more
endmembers

f_k is the proportion of endmember k within the pixel

R_{ik} is the known spectral reflectance of endmember k within the pixel in band i

ER_i is the error for band i

Endmembers are recognizable land cover materials and features that have homogenous spectral properties all over the image. To solve f_k, the following conditions must be satisfied: (1) selected endmembers should be independent of each other; (2) number of endmembers should not be larger than the spectral bands used; and (3) selected spectral bands should not be highly correlated. A constrained least-squares solution assumes that the following two conditions are satisfied simultaneously:

$$\sum_{k=1}^{n} f_k = 1 \quad \text{and} \quad 0 \le f_k \le 1 \tag{4.2}$$

$$\text{RMSE} = \sqrt{\left(\sum_{i=1}^{m} ER_i^2\right) \bigg/ m} \tag{4.3}$$

Estimation of endmember fraction images involves the following steps: image processing, endmember selection, unmixing solution, and evaluation of fraction images. Selecting suitable endmembers is the most critical step in development of high-quality fraction images. Two types of endmembers may be applied: image endmembers and reference endmembers. The former are derived directly from the image and the latter are derived from field measurements or the laboratory spectra of known materials (Roberts et al., 1998). Many remote sensing applications have employed image endmembers because they can be easily obtained and are capable of representing the spectra measured at the same scale as the image data (Roberts et al., 1998). Image endmembers may be derived from the extremes of the image feature space, based on the assumption that they represent the purest pixels in the image (Roberts et al., 1998; Mustard and Sunshine, 1999).

The SMA approach can also be nonlinear. The nonlinear SMA holds the potential to develop into a better approach for vegetation studies because of its ability to account for significant multiple scattering of photons occurring in vegetation (Roberts et al., 1993; Ray and Murray, 1996). However, building and unmixing nonlinear models is a more difficult task (Gong and Zhang, 1999).

4.2.2 Urban Impervious Surface Estimation

SMA has shown the potential for better estimation of impervious surfaces (Rashed et al., 2001; Small, 2001, 2002; Phinn et al., 2002; Wu and Murray, 2003; Lu and Weng, 2004). Because impervious surface is closely related to urban land use patterns, the use of impervious surface may prove to be valuable in urban landscape characterization, LULC classification, and change detection. However, the issue of how to derive impervious surfaces from SMA has not been well addressed and this information has not been used for urban analyses.

In recent years, different methods for impervious surface extraction from remotely sensed imagery based on SMA have been explored. For example, impervious surface may be directly extracted from remotely sensed data as one of the image endmembers (Rashed et al., 2001; Phinn et al., 2002). Impervious surface estimation may be improved by the addition of high- and low-albedo fraction images, both of which were used as endmembers in the spectral unmixing model (Wu and Muarry, 2003). Furthermore, a multiple endmember SMA has also been applied in which several impervious surface endmembers may be extracted (Rashed et al., 2003). However, these SMA-based approaches have a common problem: impervious surfaces may be overestimated in less developed areas and underestimated in well developed areas.

Although nonurban areas such as grassland, forest, and agricultural areas generally do not contain impervious surface, similarity in spectral responses among nonphotosynthetic vegetation (NPV), soil, and various IS materials may confuse each other. This confusion often leads to overestimation of impervious surface. Conversely, shadows caused by tall buildings and large tree crowns in urban areas may result in an underestimation of impervious surface. These problems are aggravated when the method of addition of low- and high-albedo fraction images is applied. Low-albedo fraction images could associate with different kinds of land cover, such as water, canopy shadows, building shadows, moisture in grass or crops, and dark impervious surface materials. High-albedo images tend to be confused with dry soils and bare grounds. Ancillary data or improved techniques are therefore called for to distinguish true impervious surfaces from others.

4.3 DATA AND METHODS

4.3.1 STUDY AREA

Located in Marion County, Indiana, the city of Indianapolis, with a population of over 800,000, was chosen as the study area (Figure 4.2). It is a key center of manufacturing, warehousing, distribution, and transportation. Situated in the middle of the country, Indianapolis possesses several other advantages that make it an appropriate choice. It has a single central city, and other large urban areas in the vicinity have not influenced its growth. The city is located on a flat plain and is relatively symmetrical, with possibilities of expansion in all directions.

Like most American cities, Indianapolis is increasing in population and area. The areal expansion is through encroachment into the adjacent agricultural and nonurban land. Certain decision-making forces, such as density of population, distance to work, property value, and income structure, encourage some sectors of metropolitan Indianapolis to expand faster than others. Examining its urban landscape patterns and dynamics is conducive to understanding and planning its future development.

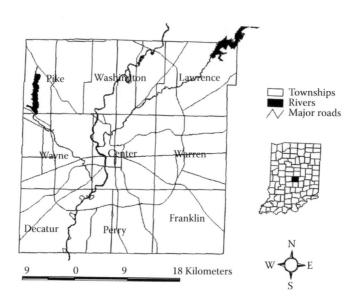

FIGURE 4.2 Study area of Indianapolis, Indiana.

4.3.2 IMAGE PREPROCESSING

Landsat TM images of June 6, 1991 (acquisition time: approximately 10:45 a.m.), and July 3, 1995 (approximately 10:28 a.m.), and a Landsat ETM+ image of June 22, 2000 (approximately 11:14 a.m.) were used in this study. Although the images purchased were geometrically corrected, their geometrical accuracy was not high enough to combine them with other high-resolution data sets. The images were therefore further rectified to a common universal transverse mercator coordinate system based on 1:24,000 scale topographic maps and were resampled to a pixel size of 30 m for all bands, using the nearest-neighbor algorithm. A root mean square error of less than 0.5 pixels was obtained for all the rectifications. These Landsat images were acquired under clear sky conditions, and an improved image-based dark object subtraction model was applied to implement atmospheric corrections (Lu et al., 2002).

4.3.3 FRACTION IMAGE DEVELOPMENT

To identify image endmembers effectively and to achieve high-quality endmembers, different image transform approaches, such as principal component analysis (PCA) and minimum noise fraction (MNF), may be used to transform the multispectral images into a new data set (Green et al., 1988; Boardman and Kruse, 1994). Endmembers can then be selected from the feature spaces of the limited transformed images (Garcia–Haro et al., 1996; Cochrane and Souza, 1998; van der Meer and de Jong, 2000; Small, 2001). The MNF transform contains two steps: (1) decorrelation and rescaling of the noise in the data based on an estimated noise covariance matrix, producing transformed data in which the noise has unit variance and no band-to-band correlations; and (2) implementation of a standard PCA of the noise-whitened data.

The result of MNF is a two-part dataset: one part is associated with large eigenvalues and coherent eigenimages and a complementary part with near-unity eigenvalues and noise-dominated images (ENVI, 2000). When an MNF transform is performed, noise can be separated from the data by saving only the coherent portions, thus improving spectral processing results. In this research, image endmembers were selected from the feature spaces formed by the MNF components.

The first three components accounted for the majority of the information (approximate 99%) and were then used for selection of endmembers. The scatterplots among MNF components 1, 2, and 3 are illustrated in Figure 4.3, showing the potential endmembers. Four endmembers — vegetation, high albedo, low albedo, and soil — were finally selected. A constrained least-squares solution was applied to unmix the multispectral ETM+ image into fraction images. The same procedures were applied to the Landsat TM 1991 and 1995 images to derive fraction images. Figure 4.4 shows four fraction images by year.

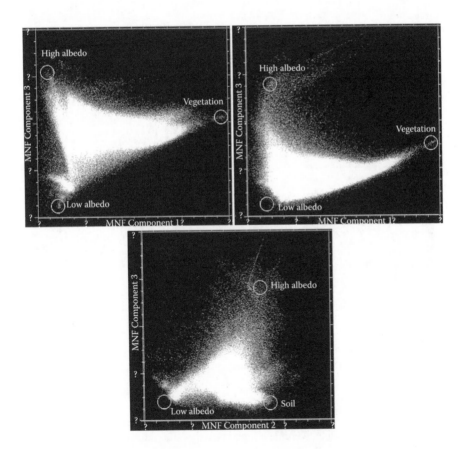

FIGURE 4.3 Feature spaces between the MNF components illustrating the potential endmembers of the Landsat ETM+ image.

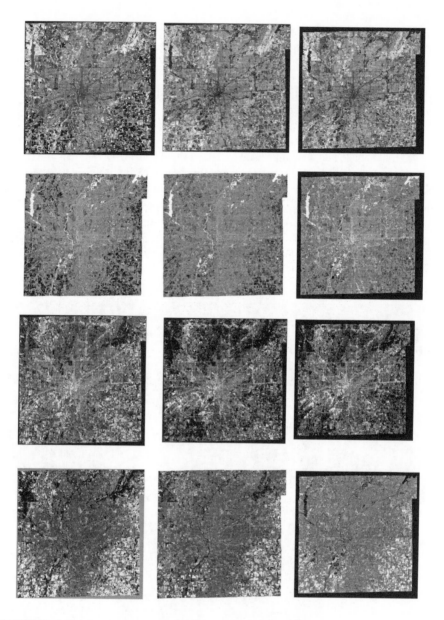

FIGURE 4.4 Fraction images derived from spectral mixture analysis of each year. First row: green vegetation; second row: low albedo; third row: high albedo; and fourth row: soil.

4.3.4 DEVELOPMENT OF IMPERVIOUS SURFACE IMAGES

Impervious surface images were developed based on the addition of high- and low-albedo fraction images. Figure 4.4 shows that high-albedo fraction images related mainly to bright objects such as buildings and roads, but also included some information of dry soils due to similar spectral responses between impervious surfaces and dry soils. Low-albedo fraction images highlighted information on water and shadows, such as different types of water bodies, shadows from forest canopy and tall buildings, and moisture in crops or pastures. However, some impervious surface information was also contained in the low-albedo images, especially for dark impervious surface objects.

To extract impervious surface information from the low-albedo images, land surface temperature images were used to differentiate nonimpervious surface from impervious surface covers. For the high-albedo fraction images, impervious surfaces were predominantly confused with dry soils. Therefore, high-albedo images were thresholded by using soil fraction images. After these processing procedures, an impervious surface image was produced by adding low- and high-albedo images. Figure 4.5 displays resultant impervious surface images for the 2000 ETM+, 1995, and 1991 TM images.

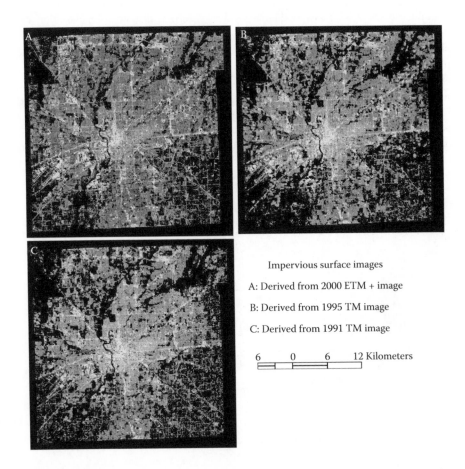

Impervious surface images

A: Derived from 2000 ETM + image

B: Derived from 1995 TM image

C: Derived from 1991 TM image

6   0   6   12 Kilometers

FIGURE 4.5 Impervious surface images developed from the three dates of Landsat TM/ETM+ images.

4.3.5 IMAGE CLASSIFICATION

Fraction images were used for LULC classification via a hybrid procedure that combined maximum likelihood and decision tree classifiers. Sample plots were identified from high spatial resolution aerial photographs, covering initially ten LULC types: commercial and industrial, high-density residential, low-density residential, bare soil, crop, grass, pasture, forest, wetland, and water. On average, 10 to 16 sample plots for each class were selected. A window size of 3×3 pixels was applied to extract the fraction value for each plot. The mean and standard deviation values were calculated for each LULC class and the characteristics of fractional composition for selected LULC types were then examined.

Next, the maximum likelihood classification algorithm was applied to classify the fraction images into ten classes, generating a classified image and a distance image. A distance threshold was selected for each class to screen out pixels that probably did not belong to that class; this was determined by examining interactively the histogram of each class in the distance image. Pixels with a distance value greater than the threshold were assigned a class value of zero. A distance tree classifier was then applied to reclassify these pixels. The parameters required by the distance tree classifier were identified based on the mean and standard deviation of the sample plots for each class.

Finally, the accuracy of the classified image was checked with a stratified random sampling method against the reference data of 150 samples collected from large-scale aerial photographs. Six LULC types were identified, including: (1) commercial and industrial urban land; (2) residential land; (3) agricultural and pasture land; (4) grassland; (5) forest; and (6) water. Figure 4.6 shows the classified LULC maps of the three years.

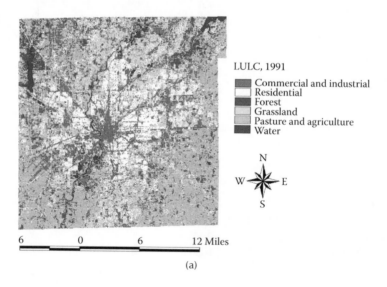

(a)

FIGURE 4.6 (Color Figure 4.6 follows page 240.) LULC maps of 1991, 1995, and 2000.

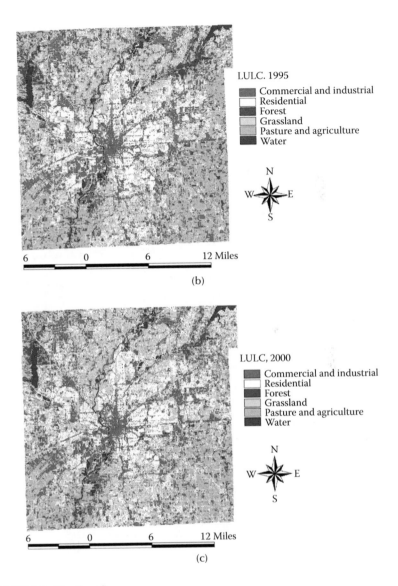

FIGURE 4.6 (Continued).

4.4 RESULTS

Three images in the first row of Figure 4.4 show the geographic patterns of vegetation fractions. These images display a large dark area (low values) at the center of the study area corresponding to the central business district of Indianapolis. Bright areas of high GV values were found in the surrounding areas. Various types of crops were still at the early stage of growth or before emergence, as indicated by medium gray to dark tones of the GV fraction images in the southeastern and southwestern parts of the city. Table 4.1 indicates that forest had the highest GV fraction values, followed by grassland.

TABLE 4.1

V-I-S Compositions of LULC Types in Indianapolis in 1991, 1995, and 2000

Land Cover Type	1991 TM Image			1995 TM Image			2000 ETM+ Image		
	Mean Vegetation (Standard Deviation)	Mean Impervious Surface (Standard Deviation)	Mean Soil (Standard Deviation)	Mean Vegetation (Standard Deviation)	Mean Impervious Surface (Standard Deviation)	Mean Soil (Standard Deviation)	Mean Vegetation (Standard Deviation)	Mean Impervious Surface (Standard Deviation)	Mean Soil (Standard Deviation)
Commercial and industrial	0.167 (0.128)	0.709 (0.190)	0.251 (0.193)	0.127 (0.097)	0.679 (0.178)	0.273 (0.177)	0.125 (0.092)	0.681 (0.205)	0.276 (0.191)
Residential	0.314 (0.132)	0.558 (0.138)	0.198 (0.152)	0.371 (0.115)	0.508 (0.108)	0.149 (0.092)	0.298 (0.095)	0.467 (0.124)	0.247 (0.137)
Grassland	0.433 (0.176)	0.451 (0.135)	0.268 (0.208)	0.553 (0.145)	0.366 (0.096)	0.155 (0.131)	0.442 (0.099)	0.276 (0.083)	0.305 (0.119)
Agriculture and pasture	0.304 (0.213)	0.374 (0.112)	0.602 (0.285)	0.388 (0.191)	0.291 (0.091)	0.378 (0.236)	0.371 (0.168)	0.275 (0.072)	0.407 (0.222)
Forest	0.654 (0.162)	0.436 (0.128)	0.182 (0.166)	0.716 (0.085)	0.388 (0.065)	0.046 (0.052)	0.584 (0.075)	0.327 (0.074)	0.175 (0.055)
Water	0.226 (0.186)	0.730 (0.197)	0.188 (0.178)	0.176 (0.210)	0.805 (0.167)	0.094 (0.068)	0.111 (0.120)	0.891 (0.136)	0.078 (0.071)

In contrast, commercial and industrial land displayed the lowest GV values. Little vegetation was found in water bodies, as indicated by the GV fraction values. Residential land and pasture-agricultural land yielded a mediate GV fraction value, subject to the impact of the date of the image acquired. In all the years of observation, pasture–agricultural land exhibited a large standard deviation value, suggesting that pasture and agricultural land may hold various amounts of vegetation.

The percentage of land covered by impervious surfaces may vary significantly with LULC categories and subcategories (Soil Conservation Service, 1975). This study shows a substantially different estimate for each LULC type because it applied a spectral unmixing model to the remote sensing images, and the modeling introduced some errors. For example, a high impervious fraction value was found in water because water related to low-albedo fractions, and the latter were included in the computation of impervious surfaces. Generally speaking, an LULC type with a higher GV fraction appeared to have a lower impervious fraction. Commercial and industrial land detected very high impervious fraction values of around 0.7 in all years. Residential land followed with fraction values of around 0.5. Grassland, agricultural-pasture land, and forestland detected lower values of impervious surface, owing largely to their exposure to bare soil, confusion with commercial/industrial and residential land, and computational errors.

Soil fraction values were generally low in the majority of the urban area, but high in the surrounding areas. Especially in agricultural fields located in the southeastern and southwestern parts of the city, soil fraction images appeared very bright because various types of crops were still at the early stage of growth. Table 4.1 shows that agricultural–pasture land observed a fraction value close to 0.4 at all times. Grassland possessed medium fraction values averaging 0.25 — substantially higher than fraction values of forestland and residential land. Commercial and industrial land displayed fraction values similar to those of grassland, which has much to do with its confusion with dry soils in the high-albedo images. Water generally had a minimal impervious fraction value. Like GV fraction, soil fraction displayed the highest standard deviation values in agricultural-pasture land because of various amounts of emerged vegetation.

The V-I-S composition may be examined by taking samples along transects. Figure 4.7 shows ternary plots of transects running across the geometric center of the city from southwest to northeast in each year. Errors from spectral unmixing modeling are not included in these diagrams due to their low values clustering to near zero. The majority of the pixels sampled showed GV fraction values of larger than 0.6. Soil fraction values were mostly below 0.6; impervious fractions largely possessed a value above 0.2. A clustering pattern was apparent if impervious fraction values were observed in the range from 0.3 to 0.7. The 1991 and 1995 transects exhibited a more dispersed pattern of pixel distribution, suggesting a variety of V-I-S composition types. When mean signature values of the fractions of each LULC type are plotted, quantitative relationships among the LULC types in terms of the V-I-S composition can be examined. Figure 4.8 shows the V-I-S composition by LULC in 2000, with an area delimiting one standard deviation from the mean fraction value.

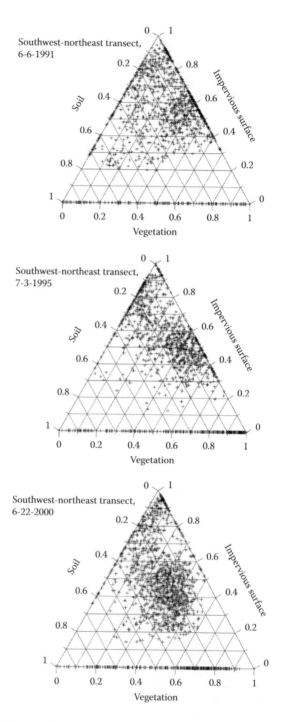

FIGURE 4.7 V-I-S composition along the transect from southwest to northeast in each year.

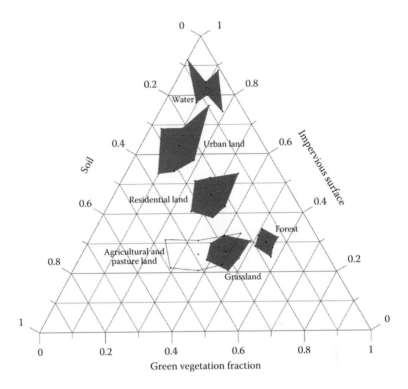

FIGURE 4.8 Quantitative relationships among the LULC types with respect to the V-I-S model, based on the Landsat ETM+ 2000 image.

Three dates of fraction images were classified into three thematic maps. Figure 4.6 shows the classified maps. Table 4.2 shows the composition of LULC by year and changes that occurred between any two years. In 1991, residential use and pasture–agriculture accounted equally for 27% of the total land; grassland shared another 20%. The combination of commercial and industrial land used 13% of the total area, and forestland had a close match, yielding another 10%. Water bodies occupied the remaining 3%, and this percentage remained unchanged from 1991 to 2000.

However, LULC dynamics occurred in all other categories, as seen in the last three columns of Table 4.2. The most notable increment was observed in residential use, which grew from 27% in 1991 to 33% in 1995, leveling out in 2000 to 38%. Associated with this change, grassland increased from 20 to 23%. Highly developed land, mainly for commercial, industrial uses, transportation, and utilities, continued to expand. In 2000, it accounted for over 15,000 ha, or 15%, thus generating a 2% increase over the 9-year period. These results suggest that urban land dispersal in Indianapolis was more related to urban population increase than to economic growth during the study period.

TABLE 4.2
Changes in Land Use and Land Cover in Indianapolis, 1991 to 2000

LULC Type	Area, 1991 (ha)	Area, 1995 (ha)	Area, 2000 (ha)	Change, 1991–1995	Change, 1995–2000	Change, 1991–2000
Commercial and industrial	13,322.10	16,706.50	15,489.00	3384.40 (25.40%)	−1217.50 (−7.29%)	2,166.90 (16.27%)
Residential	28,708.90	34,123.70	40,771.70	5414.80 (18.86%)	6648 (19.48%)	12,062.80 (42.02%)
Grassland	21,132.50	21,356.60	23,976.40	224.10 (1.06%)	2619.80 (12.27%)	2,843.90 (13.46%)
Pasture and agriculture	28,466.00	20,853.80	14,272.50	−7612.20 (−26.74%)	−6581.30 (−31.56%)	−14,193.50 (−49.86%)
Forest	9,965.58	8,547.71	7,100.77	−1417.87 (−14.23%)	−1446.94 (−16.93%)	−2,864.81 (−28.75%)
Water	2,894.21	2,903.96	2,903.06	9.75 (0.34%)	−0.90 (−0.03%)	8.85 (0.31%)

In contrast, a pronounced decrease in pasture and agricultural land was seen from 1991 (27%) to 1995 (20%). This decrease was also evident between 1995 and 2000, when pasture and agricultural land shrank further by 6581.30 ha (31.56%). Forestland in a city like Indianapolis was understandably limited in size. Our remote sensing-GIS analysis indicates, however, that forestland continued to disappear with a stable marked rate. Between 1991 and 2000, forestland was reduced by 2864.81 ha (i.e., 28.75%) and leveled down to approximately 7100 ha. The cross-tabulation of the 1991 and 2000 LULC maps reveals that most of the losses in pasture, agricultural, and forestland were converted to residential and other urban uses because of the process of urbanization and suburbanization. GIS overlay of the two maps further shows the spatial occurrence of urban expansion to be mostly on the edges of the city. It is worth noting that high commercial and industrial use in 1995 appeared to be an anomaly due to its confusion with dry, bare soils in agricultural land and pasture. These changes in LULC have led to changes in composition of image fractions.

Figure 4.9 shows changes of V-I-S compositions (mean fraction values) by each LULC type in Decatur Township, located in the southwestern part of the city. Our current research includes examining the use of multitemporal remote sensing data for urban time–space modeling and comparison of urban morphology in different geographical settings.

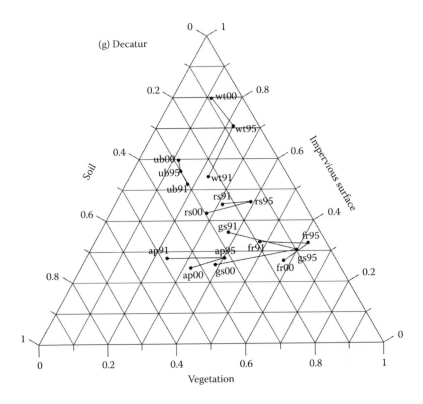

FIGURE 4.9 Changes in V-I-S composition by LULC type in Decatur Township, Indianapolis, Indiana.

4.5 DISCUSSION AND CONCLUSIONS

This study has developed an effective approach for characterizing urban landscape patterns and may provide a physically based solution to quantifying spatial and temporal changes in urban landscape compositions. Fraction images not only can be used for describing urban morphology and biophysical conditions, but also can be applied to LULC classification and change detection. A combined analysis of landscape components and LULC and their changes makes it possible for subpixel analysis of the urbanization process in Indianapolis, Indiana. SMA provides a suitable model to decompose the spectral mixtures of L-resolution data such as Landsat TM/ETM+.

Thus, a more realistic representation of the true nature of a surface is possible, as compared with that provided by assignment of a single dominant class to every pixel by statistical models. This research indicates that an SMA approach is suitable to solving the mixture problem in L-resolution data and provides better classification results than conventional "hard" classifiers. Moreover, stable and reliable fraction estimates derived from the multitemporal image data were found to be more effective for an LULC change detection than pixel-by-pixel comparison methods because fractional characteristics of LULC types at one date were comparable with other dates of fraction images.

Spectral unmixing models need to be further calibrated and validated — especially in the area of endmember selection — for a better characterization of urban morphology and quantification of urban dynamics. For instance, because of the complexity of impervious surfaces, different urban areas may have substantially different impervious surface types. Identifying a single endmember to represent all types of impervious surfaces is therefore unrealistic. This research shows that impervious surface was overestimated in the less developed areas, but underestimated in well-developed areas. In the less developed areas, impervious surface was confused with various land cover types, including vegetation, soil, and water. Impervious surface may have similar spectral responses as nonphotosynthesis vegetation and dry soils.

On the other hand, shadows from tall buildings, large tree crowns, and dark objects may cause impervious surface to be underestimated. Previous research has not examined the possibility of extracting different impervious surface materials from image fractions. This is probably due to the limitation in spectral channels of medium spatial resolution imagery and the high correlation among the image bands. Both factors limit the use of a large number of endmembers with the standard SMA approach. Three possible approaches may be taken to overcome these problems: (1) use of reference endmembers; (2) use of multiple endmembers; and (3) use of hyperspectral imagery.

Shade is an important component captured by optical remote sensors that is not included in the V-I-S model. In this model, vegetation, impervious surface, and soil are regarded as three fundamental components; however, in practice, impervious surface is a complex combination of different materials and is difficult to identify. Therefore, the shade component was included in the Lu–Weng model (PE&RS, 2004, 1053–1062). Impervious surface, shade, and vegetation were considered essential components in urban areas; in nonurban areas, soil, vegetation, and shade were thought to be able to account for the spectral variability. However, our 2004 model was not able to separate soil and impervious surface.

Given the importance of impervious surfaces in urban landscapes and their confusion with soils, further research efforts are necessary to examine the fractional characteristics of soils and impervious surfaces. In this study, the low-albedo fraction was found mainly related to objects with very low reflectance, such as water, canopy shadow, tall building shadows, or dark impervious surface materials. Future study should examine how a low-albedo fraction image relates to shade and water. Because we combined high- and low-albedo images into the impervious fractions, water was found to have a high impervious coverage.

The image endmember method assumes that true endmembers are always contained in the data set used. In practice, this assumption is not always true and is subject to the influence of the extent and characteristics of the study area. Hence, it is necessary to use reference endmembers to relate image endmembers to actual target materials (Roberts et al., 1993, 1998; Adams et al., 1995). Selecting reference endmembers from a spectral library or from field measurements is flexible, but it is difficult to account for all possible features and processes because of the many factors influencing the data spectra in a spectral library. For example, image spectra may not be accurately correspondent with the library spectra due to the effect of atmospheric corrections.

Furthermore, the units of the image spectra (e.g., reflectance, radiance, or DN) are not usually equivalent to reflectance values of reference endmembers. Thus, a conversion of spectral library data into the unit of the image spectra or vice versa is desirable. It is also important to make sure that the library spectra and the image spectra match in terms of wavelengths, bandwidths, and band shapes (Roberts et al., 1998). Another important point to note is that endmembers selected for the spectral mixture analysis should not be significantly correlated to each other.

ACKNOWLEDGMENTS

This research is supported by the National Science Foundation (BCS-0521734) for a project entitled "Role of Urban Canopy Composition and Structure in Determining Heat Islands: A Synthesis of Remote Sensing and Landscape Ecology Approach" and by the USGS IndianaView program and NASA's Indiana Space Grant program for a project entitled "Indiana Impervious Surface Mapping Initiative." We further acknowledge helpful suggestions and comments by the reviewers.

REFERENCES

Adams, J.B., Sabol, D.E., Kapos, V., Filho, R.A., Roberts, D.A., Smith, M.O., and Gillespie, A.R. (1995). Classification of multispectral images based on fractions of endmembers: application to land-cover change in the Brazilian Amazon. *Remote Sensing Environ.*, 52, 137–154.

Boardman, J.W. and Kruse, F.A. (1994). Automated spectral analysis: a geological example using AVIRIS data, north Grapevine Mountains, Nevada. In *Proc., ERIM 10th Thematic Conf. Geologic Remote Sensing*, Ann Arbor, MI, 407–418.

Campbell, J.B. (2002). *Introduction to Remote Sensing*, 3rd ed., The Guilford Press, New York, 621 pp.

Cochrane, M.A. and Souza, C.M., Jr. (1998). Linear mixture model classification of burned forests in the eastern Amazon. *Int. J. Remote Sensing*, 19, 3433–3440.

Cracknell, A.P. (1998). Synergy in remote sensing — what's in a pixel? *Int. J. Remote Sensing*, 19, 2025–2047.

ENVI (2000). ENVI User's Guide. Research Systems Inc., Boulder, CO.

Fisher, P. (1997). The pixel: a snare and a delusion, *Int. J. Remote Sensing*, 18, 679–685.

Garcia–Haro, F.J., Gilabert, M.A., and Melia, J. (1996). Linear spectral mixture modeling to estimate vegetation amount from optical spectral data. *Int. J. Remote Sensing*, 17, 3373–3400.

Gong, P. and Zhang, A. (1999). Noise effect on linear spectral unmixing. *Geogr. Inf. Sci.*, 5, 52–57.

Green, A.A., Berman, M., Switzer, P., and Craig, M.D. (1988). A transformation for ordering multispectral data in terms of image quality with implications for noise removal. *IEEE Trans. Geosci. Remote Sensing*, 26, 65–74.

Lu, D., Batistella, M., and Moran, E. (2002). Linear spectral mixture analysis of TM data for land-use and land-cover classification in Rondonia, Brazilian Amazon, *Proc. ISPRS Commission IV Symp.: Geospatial Theory, Process. Applications*, (C. Armenakis and Y.C. Lee, Eds.), Published by Center for Topographic Information Mapping Services Branch, Geomatics Canada, Department of Natural Resources Canada, Ottawa, Canada, pp. 557–562.

Lu. D. and Weng, Q. (2004). Spectral mixture analysis of the urban landscapes in Indianapolis with Landsat ETM+ imagery. *Photogrammetric Eng. Remote Sensing*, 70, 1053–1062.

Madhavan, B.B., Kubo, S., Kurisaki, N., and Sivakumar, T.V.L.N. (2001). Appraising the anatomy and spatial growth of the Bangkok Metropolitan area using a vegetation-impervious-soil model through remote sensing. *Int. J. Remote Sensing*, 22, 789–806.

Mustard, J.F. and Sunshine, J.M. (1999). Spectral analysis for earth science: investigations using remote sensing data. In *Remote Sensing for the Earth Sciences: Manual of Remote Sensing*, 3rd ed., vol. 3, New York: John Wiley & Sons, Inc., A.N. Rencz (Ed.), 251–307.

Phinn, S., Stanford, M., Scarth, P., Murray, A.T., and Shyy, P.T. (2002). Monitoring the composition of urban environments based on the vegetation-impervious surface-soil (VIS) model by subpixel analysis techniques. *Int. J. Remote Sensing*, 23, 4131–4153.

Rashed, T., Weeks, J.R., Gadalla, M.S., and Hill, A.G. (2001). Revealing the anatomy of cities through spectral mixture analysis of multispectral satellite imagery: a case study of the Greater Cairo region, Egypt. *Geocarto Int.*, 16, 5–15.

Rashed, T., Weeks, J.R., Roberts, D., Rogan, J., and Powell, R. (2003). Measuring the physical composition of urban morphology using multiple endmember spectral mixture models. *Photogrammetric Eng. Remote Sensing*, 69, 1011–1020.

Ray, T.W. and Murray, B.C. (1996). Nonlinear spectral mixing in desert vegetation. *Remote Sensing Environ.*, 55, 59–64.

Ridd, M.K. (1995). Exploring a V-I-S (vegetation-impervious surface-soil) model for urban ecosystem analysis through remote sensing: comparative anatomy for cities. *Int. J. Remote Sensing*, 16, 2165–2185.

Roberts, D.A., Smith, M.O., and Adams, J.B. (1993). Discriminating green vegetation, non-photosynthetic vegetation, and soils in AVIRIS data. *Remote Sensing Environ.*, 44, 255–269.

Roberts, D.A., Batista, G.T., Pereira, J.L.G., Waller, E.K., and Nelson, B.W. (1998). Change identification using multitemporal spectral mixture analysis: applications in eastern Amazonia. In *Remote Sensing Change Detection: Environmental Monitoring Methods and Applications*, Lunetta, R.S. and Elvidge, C.D., Eds., Chelsea, MI: Ann Arbor Press, 137–161.

Settle, J.J. and Drake, N.A. (1993). Linear mixing and the estimation of ground cover proportions. *Int. J. Remote Sensing*, 14, 1159–1177.

Small, C. (2001). Estimation of urban vegetation abundance by spectral mixture analysis. *Int. J. Remote Sensing*, 22, 1305–1334.

Small, C. (2002). Multitemporal analysis of urban reflectance. *Remote Sensing Environ.*, 81, 427–442.

Smith, M.O., Ustin, S.L., Adams, J.B., and Gillespie, A.R. (1990). Vegetation in deserts: I. A regional measure of abundance from multispectral images. *Remote Sensing Environ.*, 31, 1–26.

Soil Conservation Service. (1975). Urban hydrology for small watersheds. USDA Soil Conservation Service Technical Release No. 55. Washington, D.C.

Strahler, A.H., Woodcock, C.E., and Smith, J.A. (1986). On the nature of models in remote sensing, *Remote Sensing Environ.*, 70, 121–139.

Van der Meer, F. and de Jong, S.M. (2000). Improving the results of spectral unmixing of Landsat Thematic Mapper imagery by enhancing the orthogonality of end-members. *Int. J. Remote Sensing*, 21, 2781–2797.

Ward, D., Phinn, S.R., and Murray, A.L. (2000). Monitoring growth in rapidly urbanizing areas using remotely sensed data. *Professional Geogr.*, 53, 371–386.

Wu, C. and Murray, A.T. (2003). Estimating impervious surface distribution by spectral mixture analysis. *Remote Sensing Environ.*, 84, 493–505.

5 Bayesian Spectral Mixture Analysis for Urban Vegetation

Conghe Song

CONTENTS

5.1 INTRODUCTION

Vegetation is a critical structure component in the urban environment because quality of urban life and surface biophysical processes are closely related to the presence of vegetation. Therefore, the amount of vegetation in the urban environment is frequently sought from remotely sensed imagery. Due to the complexity of urban structure, traditional classification of remotely sensed imagery, in which each pixel is assigned to a unique land-cover class, cannot provide an accurate estimate of vegetation content. Spectral mixture analysis (SMA) is often used to derive vegetation fractions at the subpixel scale based on the following assumptions (Sabol et al., 1992): (1) the landscape is composed of a few spectrally distinctive endmembers; (2) the spectral signature for each endmember is constant within the spatial extent of interest; (3) the remote sensing signal for a pixel is a linear combination of end-member signatures.

Identifying the appropriate number of endmembers and their corresponding spectral signatures is the key step for SMA (Tompkins et al., 1997; Elmore et al., 2000; Theseira et al., 2003). In theory, the number of endmembers can be as many as the number of bands. In practice, the number of endmembers is usually lower because multispectral remotely sensed data are usually highly correlated across the spectrum. The number of endmembers cannot exceed the dimensionality of remotely sensed data (Mustard, 1993; Radeloff et al., 1999). Though Landsat Thematic Mapper/Enhanced Thematic Mapper Plus (TM/ETM+) imagery has six bands in the solar spectrum, only three or four endmembers are feasible for SMA with imagery from these sensors.

The vegetation–impervious surface–soil (VIS) is one well known three-endmember model developed by Ridd (1995) to characterize urban structure for Salt Lake City, Utah. The model was later applied to Bangkok, Thailand, and Brisbane, Australia (Madhavan et al., 2001; Phinn et al., 2002). Wu and Murray (2003) modified the VIS model to a four-endmember model, vegetation–low albedo–high albedo–soil (VLHS), to characterize the urban structure for Columbus, Ohio. The impervious component in the VIS model was broken into low- and high-albedo components. Small (2001) developed a three-endmember model similar to the VIS model, vegetation–low albedo–high albedo (VLH), in which the soil component was omitted. Even with hyperspectral remotely sensed images, the number of endmembers that can be used in SMA is limited. Roberts et al. (1993) found that 98% of the spectral variations in an airborne visible infrared imaging spectrometer (AVIRIS) image could be explained by three endmembers: green vegetation, shade, and soil.

Spectral signatures of endmembers in general may be derived from a reference endmember library in which endmember spectral signatures are measured on the ground (Adams et al., 1995; Roberts et al., 1998; Smith et al., 1990) or the image to be unmixed (Elmore et al., 2000; Ridd, 1995; Wessman et al., 1997). The advantage of using reference endmember spectral signatures is that they are generally very accurate, but the accuracy of spectral signature does not necessarily guarantee high accuracy in endmember fractions in spectral mixture analysis due to the difference in measurements on the ground and in space (Gong and Zhang, 1999). Use of reference endmember spectral signature for spectral mixture analysis requires a complicated calibration process (Roberts et al., 2002). Spectral signatures for image endmembers are usually obtained from the pure pixels in the image or from the feature space.

The advantage of using an image endmember is that the endmember signature and the image to be unmixed are already at the same measurement scale. Therefore, no calibration is needed. However, finding pure pixels or the endmember points in the feature space may not be a straightforward process because: (1) the purity of a pixel is difficult to determine from the image to be unmixed; and (2) the spectral signatures in the feature space usually change continuously as seen in Figure 5.1. It is difficult to determine which points in the continuous space are the best choices for endmembers.

Recently, Song (2004) developed a third approach to obtain endmember spectral signatures from high-resolution images. The approach requires cross-sensor calibration, and a high spatial resolution image must be collected relatively closely in time with the coarser resolution imagery to be unmixed. The high spatial resolution remotely sensed imagery can help identify endmembers well.

FIGURE 5.1 (**Color Figure 5.1 follows page 240.**) The red/near-infrared feature space (horizontal = red, vertical = near-infrared) for an Ikonos image collected over Bangkok, Thailand, on November 27, 2002. At 4-m spatial resolution, most of the pixels are pure pixels. Vegetation and other urban components separate well in the feature space, but each varies continuously in the spectral space.

Traditional SMA using constant spectral signatures for endmembers cannot accommodate the natural variation of endmember reflectance properties. It is legitimate to argue in the VLH and VLHS models for an intermediate albedo feature, and this argument can be recursively made. Recently, new approaches that account for endmember spectral signature variation began to emerge in the literature. Roberts et al. (1998) developed a multiple endmember spectral mixture analysis (MESMA) to derive the subpixel proportion of chaparral in the Santa Monica Mountains, California, where endmembers were dynamically selected from a spectral library containing hundreds of reference endmembers.

Bateson et al. (2000) introduced the concept of endmember bundle for SMA in which an endmember is no longer represented by a single vector in the multidimensional spectral space, but with a bundle of vectors. The average of minimum and maximum endmember fractions within the bundle is used as the mean fraction for each endmember; this usually occurs at the outer profile of the vector bundle. The vectors in the middle of the bundle are not used. Asner and Heidebrechet (2002) recently developed an approach that randomly selects vectors from the endmember bundles to estimate the uncertainty caused by endmember signature variation.

In this chapter, I introduce a new approach, the Bayesian spectral mixture analysis (BSMA), in which endmember spectral signatures are no longer assumed as constants, but rather represented by probability density functions. Therefore, BSMA can incorporate the natural variability of endmember spectral signatures. Although the theory is tested with two endmember spectral mixture analyses, it can be used with more numbers of endmembers (Song, 2005).

5.2 BAYESIAN SPECTRAL MIXTURE ANALYSIS MODEL

Traditional spectral mixture analysis with constant endmember spectral signatures can be expressed in the following:

$$S_{mi} = \sum_{j=1}^{n} f_j \times S_{ij} + \varepsilon_i \, , \, i = 1, 2, ..., k \tag{5.1}$$

and

$$\sum_{j=1}^{n} f_j = 1 \tag{5.2}$$

where
S_{mi} is the mixture remote sensing signal in the ith spectral dimension
S_{ij} is the jth endmember spectral signature in the ith spectral dimension
$n \leq k$
f_j is the jth endmenber fraction within the field of view
ε_i is error

If $n = 2$, we only need one spectral measurement over a given area, and the working solution is

$$S_m = f \times S_1 + (1 - f)S_2 \tag{5.3}$$

It is clear from Equation (5.3) that if S_1 is overestimated, the solution for f will be smaller than it actually is and vice versa. The simple mixing model concept is limited because it fails to account for variation of endmembers within the field of view (Roberts et al., 1998). Assuming the variation of endmember signatures can be described by probability density functions p_1 and p_2, the mixture signal will not be unique. The endmember fraction for a given mixture signal will not be unique, but rather has the following probability density function

$$P_m(f|S_m) = \frac{p(S_m|f) \times \pi(f)}{\int_0^1 p(S_m|u)\pi(u)du} \tag{5.4}$$

where $\pi(f)$ is the prior probability density function of endmember fractions, which can be assumed to be uniform between 0 and 1 before we know anything about a pixel. $p(S_m|f)$ is the probability density function for the mixture signature at given endmember fractions, which can be derived from the probability density functions of the endmember signatures as

$$p(S_m|f) = p_1(S) \otimes p_2(S_m - f \times S) \tag{5.5}$$

where $\otimes$ denotes convolution, and $p_1(S)$ is the probability density for endmember 1 taking a spectral signature S. Due to the constraint that the two endmembers produce the mixed signal S_m, the second endmember must take a signature $(S_m - f \times S)$ if the first one occupies f fractions of the pixel with spectral signature S.

Unlike traditional SMA with constant endmember signatures, BSMA produces a probability density function of subpixel fractions for each endmember in a pixel. Therefore, the expected subpixel endmember fractions as well as the standard deviations can be estimated. An additional advantage of BSMA is that it can be recursively used to refine the endmember fraction estimates as

$$P'_m (f|S_m) = \frac{p(S'_m \,|f) \times \lambda(f)}{\int\limits_0^1 p(S'_m \,|u)\lambda(u)du} \tag{5.6}$$

where $\lambda(f)$ is the probability of subpixel endmember fractions obtained from the first spectral measurement from Equation (5.4).

5.3 MODEL TESTING

5.3.1 Testing with Simulated Imagery

The unmixing concept of the BSMA model can be tested with simulated imagery. The advantage here is that the endmember signature probability density functions and the endmember subpixel fractions are known a priori. Therefore, the concept can be well validated. Based on Figure 5.1, an urban environment primarily comprises two kinds of materials: vegetation and nonvegetation. For BSMA with two endmembers, we only need one spectral measurement, as shown in Equation (5.5). Because it is sensitive to the presence of vegetation, simulated images are based on the normalized difference vegetation index (NDVI).

The probability density of NDVI for vegetation increases from 0 at NDVI = 0.1 to its peak at NDVI = 0.5 and linearly decreases to 0 at NDVI = 0.7. The probability density for nonvegetation linearly increases from 0 at NDVI = –0.3 to its peak at NDVI = –0.22, and linearly decreases to 0 at NDVI = –0.1. For both probability density functions, the area under the curve integrates to unity (Figure 5.2). The mixture image is generated in two steps. First, a vegetation fraction image is generated using Monte Carlo simulation, assuming a Gaussian distribution for the subpixel vegetation fractions with a mean of 0.5 and a standard deviation of 0.15 for 100 × 100 pixels. Second, based on the subpixel vegetation fraction image generated, a mixture NDVI image is produced using Equation (5.3) in which endmember signatures are randomly drawn from the probability density functions as shown in Figure 5.2. For the mixture image, the subpixel vegetation fractions are known for each pixel. The mixture image is then unmixed with Equation (5.4).

As a comparison, the mixture image is also unmixed with traditional spectral mixture analysis using the means and the modes from the endmember probability density functions as the endmember signature. The approach of model testing sounds circular.

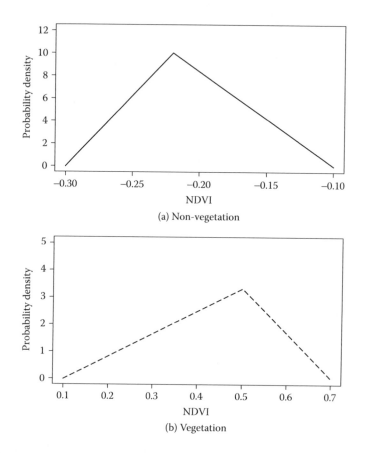

FIGURE 5.2 Probability density function of NDVI for vegetation and nonvegetation. (From Song, C., *Remote Sensing Environ.*, 95, 254, 2005, with permission from Elsevier.)

In fact, it is not reversible because the subpixel vegetation fractions are unmixed from a probability density function, while each of the mixture pixels is generated with a specific endmember signature randomly drawn from its probability density function. During the unmixing process, the exact signatures used to generate the mixture image are unknown. Therefore, simulated imagery is a very effective source of data for testing the concept of BSMA. Figure 5.3 shows that traditional SMA using modes from Figure 5.2 produced a very biased estimate of subpixel vegetation fractions.

Using the means of the probability density function leads to much improved subpixel vegetation fraction estimates. This implies that the use of most commonly seen vegetation as endmembers for spectral mixture analysis may not be the best choice. Other vegetation types need to be taken into consideration. This also implies that using signatures from the "extreme" pure pixels identified from the feature space for SMA may also lead to biased estimates of subpixel vegetation fractions.

BSMA produced the best estimates for subpixel vegetation fractions among the three unmixing approaches. In addition, BSMA also provides uncertainty information

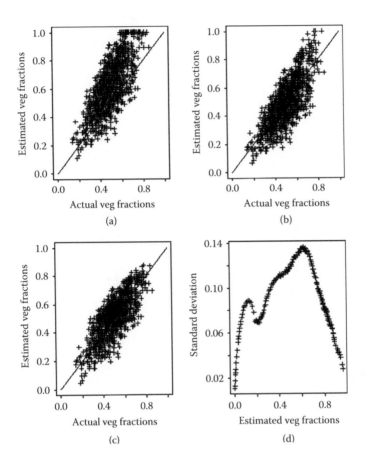

FIGURE 5.3 Subpixel vegetation fractions from spectral mixture analysis: (a) SMA using modes of signature distributions in Figure 5.2 as endmember signatures; (b) SMA using means of signature distributions in Figure 5.2 as endmember signatures; (c) BSMA using the entire probability density functions in Figure 5.2; (d) standard deviations of the subpixel vegetation fractions from BSMA. (From Song, C., *Remote Sensing Environ.*, 95, 257, 2005, with permission from Elsevier.)

for each pixel that is unavailable from traditional SMA. It is interesting to note that the uncertainty is much lower at low and high vegetation fractions. Under such circumstances, the remotely sensed signals are almost entirely from one endmember. The uncertainty is highest with intermediate vegetation fractions. The situation allows many different combinations of vegetation and nonvegetation, each of which can contribute significantly to the mixture signal.

5.3.2 TESTING WITH IKONOS AND LANDSAT ETM+ IMAGERY

After testing BSMA with simulated images, I continue to test the concept with real images. Obtaining adequate endmember spectral signatures is the major challenge in traditional spectral mixture analysis. The challenge is greater to obtain endmember spectral signature probability distributions. In fact, it is not feasible to obtain the endmember spectral signature probability distributions from the image to be unmixed. Testing BSMA with real images is made possible with the recent availability of high-resolution images. Here, I demonstrate BSMA with Landsat ETM+ imagery with endmember spectral signature probability distributions obtained from the Ikonos image.

The Ikonos image was collected on November 27, 2002, over Bangkok, Thailand. The Landsat ETM+ imagery was collected for the same place on November 16, 1999. Though the two images are 3 years apart, they are quite close in time of year. Assuming no significant change on the ground during this time, the reflectance measured by the satellites should be comparable. Due to the difference in sensor and radiometric resolution, the raw remotely sensed data from the two sensors are not comparable. Therefore, the first step in obtaining the endmember spectral signature probability distribution function from Ikonos for BSMA with Landsat ETM+ imagery is to convert the digital numbers into at-satellite reflectance as:

$$\rho_i = \frac{\pi L_i d^2}{E_i \cos(\theta)} \tag{5.7}$$

where
ρ_i is the at-satellite reflectance for band i
E_i is the solar constant for band i
solar zenith angle is θ
Sun–Earth distance in astronomical units is d
L_i is radiance measured by the satellite for band i

For Ikonos imagery, L_i can be obtained from the raw Ikonos image as

$$L_i = \frac{DN_i}{K_i} \tag{5.8}$$

where DN_i is Ikonos digital numbers (DNs) for band i, and K_i is the DN-to-radiance conversion coefficient for Ikonos imagery. For Landsat ETM+ imagery, L_i can be obtained from the raw image as

$$L_i = G_i \times DN_i + B_i \tag{5.9}$$

where G_i and B_i are Landsat ETM+ sensor gain and bias for band i, respectively. Similarly to the generated imagery, I continue to test BSMA with NDVI calculated from both sensors as

$$NDVI = \frac{\rho_{NIR} - \rho_{RED}}{\rho_{NIR} + \rho_{RED}} \tag{5.10}$$

It is clear from Equation (5.10) that the Sun–Earth distance can be ignored when calculating NDVI because it is cancelled.

Though NDVIs calculated as in Equation (5.10) are based on absolute physical measurements from Ikonos and Landsat ETM+ imagery, they may still not be comparable due to sensor differences and atmospheric effects. Recent studies found systematic differences between Ikonos imagery and Landsat ETM+ imagery. Compared to the latter, an Ikonos sensor produces higher reflectance in red and lower reflectance in near-infrared bands (Goward et al., 2003). Therefore, additional calibration is needed to bring NDVI values from Ikonos imagery to that of Landsat ETM+ imagery at the same measurement scale.

Two approaches are available in the literature for cross-sensor calibration: absolute and relative calibration (Song et al., 2001). Absolute calibration leads to surface reflectance. Relative calibration does not necessarily produce surface reflectance, but it results in a set of images that are consistent on a common measurement scale. Because conversion of at-satellite to surface reflectance is a linear process, spectral mixture analysis is also a linear process. Whether the process is based on at-satellite reflectance or surface reflectance does not affect the quality of the product as long as the endmember spectral signatures are at the same measurement scale as the image to be unmixed. Therefore, relative calibration can be as effective as absolute calibration for spectral mixture analysis.

The advantage of a relative calibration is that it is much simpler than the absolute calibration because the atmospheric condition data for a quality absolute calibration are usually not available. The approach taken here is to use information from all pixels in the relative calibration process, i.e., histogram matching. Because of the difference in spatial resolution, the histograms of NDVI for Ikonos and Landsat ETM+ imagery are quite different (Figure 5.4). An NDVI histogram from Ikonos is multimodal, indicating major land cover classes: urban, vegetation, and water. Vegetation and other urban materials do not separate in the NDVI histogram from Landsat ETM+ imagery.

To match the NDVI histograms from the two sensors effectively, Ikonos imagery needs to be rescaled to the same spatial resolution as that of the Landsat imagery. After rescaling, the NDVI histograms from the two sensors have almost identical shapes (Figure 5.5). The systematic differences between the histograms are caused by the combined effects of sensor response and atmosphere. Matching the histograms in Figure 5.5 is now much easier than matching those in Figure 5.4. However, the modes for land and water cannot be matched at the same time due to other factors. In this study, the histograms are matched primarily based on the pixels for land.

To obtain the endmember spectral signature probability distributions, the four-band multispectral Ikonos image at 4- × 4-m spatial resolution is classified into vegetation and nonvegetation, and their probability distributions are derived from the NDVI histogram in the area they occupy. The NDVI probability distribution functions at 4- × 4-m spatial resolution for vegetation and nonvegetation are used as endmember signature distribution functions for BSMA of the relatively calibrated Landsat ETM+ imagery at 30- × 30-m spatial resolution (Figure 5.5). Though there is a mismatch in spatial resolution, the mismatch in scale should not influence SMA results because the endmember spectral signature should be scale invariant. The endmember spectral signature is the signature when a pixel is entirely occupied by

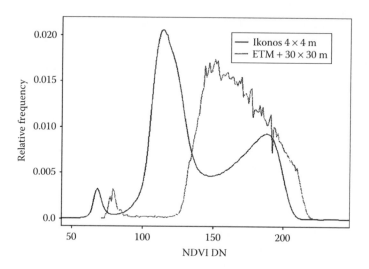

FIGURE 5.4 NDVI histograms from Ikonos and Landsat ETM+ imagery over Bangkok, Thailand. The multimodal histogram from Ikonos imagery indicates most Ikonos multispectral pixels are pure pixels. (From Song, C., *IEEE Geosci. Remote Sensing Lett.*, 1(4): 272–276, 2004, with permission from IEEE.)

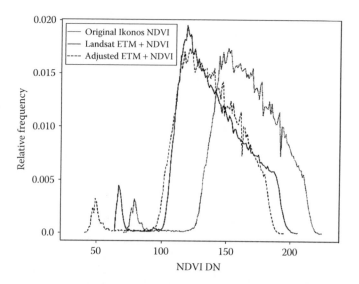

FIGURE 5.5 Matching an NDVI histogram from Landsat ETM+ with that from Ikonos. The two histograms at the same spatial resolution have almost identical shape. (From Song, C., *IEEE Geosci. Remote Sensing Lett.*, 1(4): 272–276, 2004, with permission from IEEE.)

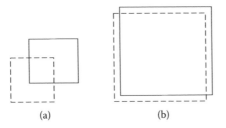

FIGURE 5.6 Impact of registration error with scale on validation: (a) pixels; (b) windows.

the endmember. In a spatial uniform landscape, the spectral signature should not change with spatial resolution of measurement.

Validation of BSMA with real images is further complicated by errors in geometric registration (Townshend et al., 1992; Dai and Khorram, 1998). Though the vegetation and nonvegetation information is available for the same area at 4- × 4-m spatial resolution, a pixel-by-pixel comparison of BSMA subpixel vegetation information with Ikonos imagery may not lead to legitimate validation results. Regardless of how small the root mean squared error is in the image-to-image registration, the vertical and horizontal directions can potentially be half a pixel off because chances are very low for pixels from the two images to overlay exactly on the ground. On a pixel-by-pixel basis, we can only reliably count on 25% overlap for pixels from two images at the same spatial resolution (Figure 5.6a). In a complex urban environment, validation with pixel-by-pixel comparison is not acceptable. The strategy adopted here is to validate with a larger window (Figure 5.6b). The impact of registration error decreases as the size of the window increases. The mismatch areas can be 75% on a per-pixel basis, while the impact reduced to less than 2.5% for a window in 40 × 40 pixels.

Figure 5.7 shows validation of BSMA with the Landsat ETM+ image over the Ikonos image for subpixel vegetation fractions based on 40 × 40 Landsat pixels. A 40- × 40-pixel window in Landsat image is equivalent to a 300- × 300-pixel window in an Ikonos image at 4- × 4-m spatial resolution. Within this 300- × 300-pixel window in the Ikonos image, every pixel has already been classified as vegetation or nonvegetation. An ensemble vegetation fraction for the entire window is calculated based on the classification at 4- × 4-m spatial resolution. Eight 40- × 40-pixel windows in the Landsat image were identified for validation. These windows were selected so that they contained a wide range of vegetation fractions. As a comparison, the subpixel vegetation fractions using the means and modes of the endmember NDVI probability distributions were also used.

Using the modes of the NDVI probability distributions as endmember spectral signatures in traditional spectral mixture analysis continues to have a biased estimate of subpixel vegetation fractions; BSMA and traditional SMA using the means have much improved estimates of subpixel vegetation fractions. This indicates that using the statistical means of the endmember signatures in traditional spectral mixture analysis may obtain subpixel endmember fractions as good as BSMA. However, no uncertainty information is available from traditional SMA. This implies that using

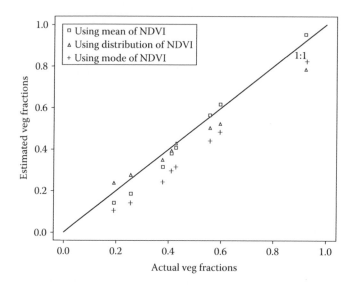

FIGURE 5.7 Validation of BSMA with Ikonos imagery for subpixel vegetation fractions based on eight 40×40 windows at Landsat spatial resolution. (From Song, C., *Remote Sensing Environ.*, 95, 259, 2005, with permission from Elsevier.)

extremely pure pixels identified from feature space as endmembers may also lead to biased estimates of endmember fractions. The results here indicate that representative endmember signatures are important in SMA. The most commonly seen (mode) or the extremely pure pixels may not represent endmember spectral signatures well.

5.4 BSMA WITH THREE ENDMEMBERS

BSMA with two endmembers is very flexible because a large variety of materials in the nonvegetation can be incorporated. Two-endmember SMA is the simplest case. In many other situations, more endmembers are needed. In fact, BSMA can be expanded to include more endmembers. For BSMA with three endmembers, we need at least two independent spectral measurements, S_{m1} and S_{m2}. The linear mixture equations for three endmembers can be written as

$$\begin{cases} S_{m1} = f_1 S_{11} + f_2 S_{12} + (1 - f_1 - f_2) S_{13} \\ S_{m2} = f_1 S_{21} + f_2 S_{22} + (1 - f_1 - f_2) S_{13} \end{cases} \qquad (5.11)$$

where S_{ij} are the endmember signatures for the jth endmember in the ith spectral measurements and f_1 and f_2 are the subpixel endmember fractions for the first two endmembers.

Due to the unity constraint in subpixel endmember fractions, we know the subpixel fractions for the third endmember once we know the subpixel endmember fractions for the first two. For BSMA, each of the S_{ij} takes a probability distribution function. Obtaining subpixel vegetation fractions with BSMA changes to obtaining probability distribution functions of subpixel endmember fractions, given two mixed spectral measurements and the endmember signature probability distribution functions. Based on Bayesian theorem (Carlin and Louis, 1996), we have

$$p(f_i|S_{m1}S_{m2}) = \frac{h(S_{m2}|f_i)g(f_i|S_{m1})}{\int_0^1 h(S_{m2}|u)g(u|S_{m1})du} \qquad (5.12)$$

where

$$g(f_i|S_{m1}) = \frac{h(S_{m1}|f_i)\pi(f_i)}{\int_0^1 h(S_{m1}|u)\pi(u)du} \qquad (5.13)$$

where $h(S_{m1}|f_i)$ and $h(S_{m2}|f_i)$ can be obtained from Equation (5.11) through convolution similar to Equation (5.5).

Theoretically, the expansion can continue with more endmembers. In reality, this expansion is limited by dimensionality of the remotely sensed data and distinctiveness of endmember spectral signature probability distribution functions. If the endmember probability density fractions overlap significantly, BSMA will lead to very uncertain subpixel endmember fractions.

5.5 DISCUSSION

Variation of endmember spectral signatures has been a major challenge in traditional spectral mixture analysis (Roberts et al., 1998). On the one hand, the spectral signature of endmembers usually varies continuously within a certain spectral space; on the other, the number of endmembers that can be used in SMA is very limited. The limited dimensionality in remotely sensed data prevents the possibility of incorporating the continuous variation in endmember spectral signature in traditional SMA.

BSMA is a flexible model that can incorporate the variation of endmember spectral signatures based on Bayesian theorem. The model output is no longer a single subpixel endmember fraction, but rather a probability distribution of subpixel endmember fractions. Figure 5.8 shows a few examples of endmember fraction probability for different NDVI values. When NDVI is low, the probability decreases with increasing subpixel endmember fractions and eventually becomes zero; this means that, given the low NDVI, it is impossible to have subpixel vegetation fractions higher. At intermediate NDVI values, the subpixel vegetation fractions take a wide range of values. This is why the uncertainty is high with intermediate vegetation fractions in Figure 5.3. For high NDVI values, the probability of higher subpixel vegetation fractions increases.

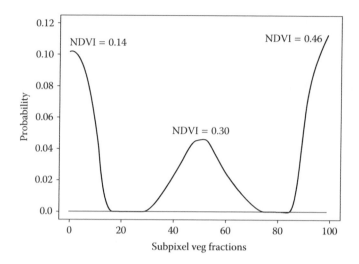

FIGURE 5.8 Probability distribution functions for subpixel vegetation fractions at different NDVI values. High and low NDVI values lead to a narrower range of subpixel vegetation fractions. Intermediate NDVI values correspond to a wide range of subpixel vegetation fractions. (From Song, C., *Remote Sensing Environ.*, 95, 260, 2005, with permission from Elsevier.)

Such information is not available from traditional spectral mixture analysis. Though this chapter focused on extracting subpixel fractions of vegetation in the urban environment, BSMA can be used for extracting subpixel fractions for other objects (e.g., impervious surface) as long as the object signature probability distributions can be uniquely defined.

This study demonstrated that NDVI can be used for extracting subpixel vegetation fractions in the urban environment despite earlier concerns about the nonlinear relationship between NDVI and vegetation fractions (Huete et al., 1985; Ray and Murray, 1996; Borel and Gerstl, 1994). Further analysis indicates that NDVI and vegetation fractions are highly nonlinearly related when the background is dark and vegetation fractions are low. I evaluated the relationship between NDVI and subpixel vegetation fractions using the mean vegetation and nonvegetation reflectance values in the red and near-infrared bands from the Ikonos image over Bangkok, Thailand. The relationship in this case is linear (Figure 5.9). In fact, Jasinski (1990) demonstrated that NDVI and subpixel vegetation cover are linearly related when ($\rho_{red} + \rho_{nir}$) of background is equivalent to that of vegetation.

Figure 5.9 is simulated with the ensemble signature from the top of vegetation, including nonlinear interaction of photons within the entire vegetation layer. The spectral signature at the leaf level may not be used for spectral mixture analysis because multiple scattering of photons between leaves and background can lead to nonlinear mixture signals. Other studies have found that NDVI can be used to extract subpixel vegetation fractions with coarse spatial resolution imagery as the signal is dominated by a horizontal mixture of different endmembers (Defries et al., 2000; Qi et al., 2000).

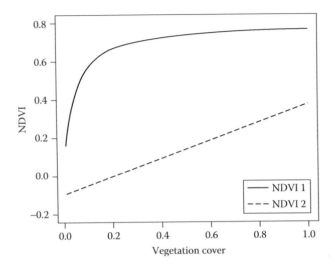

FIGURE 5.9 Relationship between vegetation fraction and NDVI. NDVI1 is simulated with bright vegetation on a dark background; NDVI2 is simulated with the reflectance in the red and near-infrared bands derived from the Ikonos image over Thailand for vegetation and the background.

5.6 CONCLUSIONS

A new SMA model based on Bayesian theorem — Bayesian spectral mixture analysis — is introduced in this chapter. Unlike traditional spectral mixture analysis, which assumes constant spectral signatures for endmembers, BSMA can effectively incorporate variability in endmember spectral signatures that is typical for vegetation in the urban environment. In BSMA, endmember spectral signatures are no longer constants. They are described by probability distribution functions. The output of BSMA is no longer a single subpixel endmember fraction, but a probability distribution of subpixel endmember fractions. Thus, not only the expected endmember fractions, but also the uncertainty of the estimates, can be produced by BSMA.

The model was demonstrated with simulated images and combined use of Ikonos and Landsat ETM+ images based on NDVI. Despite concerns of a nonlinear relationship between NDVI and subpixel vegetation fractions, NDVI was successfully used to extract subpixel vegetation fractions in the urban environment. Remotely sensed data from Ikonos and Landsat ETM+ sensors can be well calibrated so that the probability distribution functions for endmembers can be derived from the Ikonos imagery for BSMA with Landsat imagery. BSMA produces significantly improved estimates of subpixel vegetation fractions compared to traditional spectral mixture analysis using the modes of the spectral signature probability distributions as endmember signatures. Traditional SMA using means of endmember spectral signature distributions can lead to almost identical subpixel vegetation fractions as those in BSMA except without uncertainty information.

ACKNOWLEDGMENTS

This research work was partly supported by Andrew Mellon Foundation grant 02081908 and partly supported by National Science Foundation grant 0351430. The manuscript was completed while the author was a Charles Bullard Fellow in Forest Research at Harvard Forest, Harvard University.

REFERENCES

Adams, J.B., Sabal, D.E., Kapos, V., Filho, R.A., Roberts, D.A., Smith, M.O., and Gillespie, A.R. 1995. Classification of multispectral images based on fractions of endmembers: application to land-cover change in the Brazilian Amazon. *Remote Sensing Environ.*, 52: 137–154.

Asner, G.P. and Heidebrecht, K.B. 2002. Spectral unmixing of vegetation, soil and dry carbon cover in arid regions: comparing multispectral and hyperspectral observations. *Int. J. Remote Sensing*, 23(19): 3939–3958.

Bateson, C.A., Asner, G.P., and Wessman, C.A. 2000. Endmember bundles: a new approach to incorporating endmember variability into spectral mixture analysis. *IEEE Trans. Geosci. Remote Sensing*, 38(2): 1083–1094.

Borel, C.C. and Gerstl, S.A.W. 1994. Nonlinear spectral mixing models for vegetation and soil surfaces. *Remote Sensing Environ.*, 47(3): 403–416.

Carlin, B.P. and Louis, T.A. 1996. *Bayes and Empirical Bayes Methods for Data Analysis*, Chapman & Hall, London.

Dai, X. and Khorram, S. 1998. The effects of image misregistration on the accuracy of remotely sensed change detection. *IEEE Trans. Geosci. Remote Sensing*, 36(5): 1566–1577.

Defries, R.S., Hansen, M.C., and Townshend, J.R.G. 2000. Global continuous fields of vegetation characteristics: a linear mixture model applied to multi-year 8 km AVHRR data. *Int. J. Remote Sensing*, 21(6–7): 1389–1414.

Elmore, A.J., Mustard, J.F., Manning, S.J., and Lobell, D.B. 2000. Quantifying vegetation change in semiarid environments: precision and accuracy of spectral mixture analysis and the normalized difference vegetation index. *Remote Sensing Environ.*, 73: 87–102.

Gong, P. and Zhang, A. 1999. Noise effect on linear spectral unmixing. *Geogr. Inf. Sci.*, 5(1): 52–57.

Goward, S.N., Davis, P.E., Fleming, D., Miller, L., and Townshend, J.R. 2003. Empirical comparison of Landsat 7 and IKONOS multispectral measurements for selected Earth observation system (EOS) validation sites. *Remote Sensing of Environ.*, 88: 80–99.

Huete, A.R., Jackson, R.D., and Post, D.F., 1985. Spectral response of a plant canopy with different soil backgrounds. *Remote Sensing Environ.*, 17: 37–53.

Jasinski, M.F. 1990. Sensitivity of the normalized difference vegetation index to subpixel canoy cover, soil albedo, and pixel scale. *Remote Sensing Environ.*, 32(2–3): 169–187.

Madhavan, B.B., Kubo, S., Kurisaki, N., and Sivakumar, T.V.L.N. 2001. Appraising the anatomy and spatial growth of the Bangkok metropolitan area using a vegetation–impervious soil model through remote sensing. *Int. J. Remote Sensing*, 22(5): 789–806.

Mustard, J.F. 1993. Relationships of soil, grass, and bedrock over the Kaweah Serpentinite Melange through spectral mixture analysis of AVIRIS data. *Remote Sensing Environ.*, 44: 293–308.

Phinn, S., Stanford, M., Scarth, P., Murray, A.T., and Shyy, P.T. 2002. Monitoring the composition of urban environments based on the vegetation–impervious surface–soil (VIS) model by subpixel analysis techniques. *Int. J. Remote Sensing*, 23(20): 4131–4153.

Qi, J., Marsett, R.C., Moran, M.S., Goodrich, D.C., Heilman, P., Kerr, Y.H., Dedieu, G., Chehbouni, A., and Zhang, X.X. 2000. Spatial and temporal dynamics of vegetation in the San Pedro River basin area. *Agric. Forest Meteorol.*, 105(1–): 55–68.

Radeloff, V.C., Mladenoff, D.J., and Boyce, M.S. 1999. Detecting Jack Pine budworm defoliation using spectral mixture analysis: separating effects from determinants. *Remote Sensing Environ.*, 69: 156–169.

Ray, T.W. and Murray, B.C. 1996. Nonlinear spectral mixing in desert vegetation. *Remote Sensing Environ.*, 55: 59–64.

Ridd, M.K. 1995. Exploring a V-I-S (vegetation–impervious surface–soil) model for urban ecosystem analysis through remote sensing: comparative anatomy for cities. *Int. J. Remote Sensing*, 16(12): 2165–2185.

Roberts, D.A., Smith, M.O., and Adams, J.B. 1993. Green vegetation, nonphotosynthetic vegetation, and soils in AVIRIS data. *Remote Sensing Environ.*, 44: 255–269.

Roberts, D.A., Gardner, M., Church, R., Ustin, S., Scheer, G., and Green, R.O. 1998. Mapping chaparral in the Santa Monica Mountains using multiple endmember spectral mixture models. *Remote Sensing Environ.*, 65: 267–279.

Roberts, D.A., Numata, I., Holmes, K., Batista, G., Krug, T., Monteiro, A., Powell, B., and Chadwick, O.A. 2002. Large area mapping of land-cover change in Rondonia using multitemporal spectral mixture analysis and decision tree classifiers. *J. Geophys. Res.*, 107(D20): 8073.

Sabol, D.E., Adams, J.B., Jr., and Smith, M.O. 1992. Quantitative subpixel spectral detection of targets in multispectral images. *J. Geophys. Res.*, 97(E2): 2659–2672.

Small, C. 2001. Estimation of urban vegetation abundance by spectral mixture analysis. *Int. J. Remote Sensing*, 22(7): 1305–1334.

Smith, M.O., Ustin, S.L., Adams, J.B., and Gillespie, A.R. 1990. Vegetation in deserts: I. A regional measure of abundance from multispectral images. *Remote Sensing Environ.*, 31: 1–26.

Song, C., Woodcock, C.E., Seto, K.C., Pax–Lenney, M., and Macomber, S.A. 2001. Classification and change detection using Landsat TM data: when and how to correct atmospheric effects? *Remote Sensing Environ.*, 75: 230–244.

Song, C. 2004. Cross-sensor calibration between Ikonos and Landsat ETM+ for spectral mixture analysis. *IEEE Geosci. Remote Sensing Lett.*, 1(4): 272–276.

Song, C. 2005. Spectral mixture analysis for subpixel vegetation fractions in the urban environment: how to incorporate endmember variability? *Remote Sensing Environ.*, 95(2): 248–263.

Theseira, M.A., Thomas, G., Taylor, J.C., Gemmell, F., and Varjo, J. 2003. Sensitivity of mixture modeling to end-member selection. *Int. J. Remote Sensing*, 24(7): 1559–1575.

Tompkins, S., Mustard, J.F., Pieters, C.M., and Forsyth, D.W. 1997. Optimization of endmembers for spectral mixture analysis. *Remote Sensing Environ.*, 59: 472–489.

Townshend, J.R.G., Justice, C.O., Gurney, C., and McManus, J. 1992. The impact of misregistration on change detection. *IEEE Trans. Geosci. Remote Sensing*, 30(5): 1054–1060.

Wessman, C.A., Bateson, A., and Benning, T.L. 1997. Detecting fire and grazing patterns in tallgrass prairie using spectral mixture analysis. *Ecol. Appl.*, 7(2): 493–511.

Wu, C. and Murray, A.T. 2003. Estimating impervious surface distribution by spectral mixture analysis. *Remote Sensing Environ.*, 84: 493–505.

6 Urban Mapping with Geospatial Algorithms

Soe W. Myint

CONTENTS

6.1 INTRODUCTION

Visual or manual interpretation of aerial photography has traditionally been recognized as an effective tool in urban mapping. In fact, this type of remote sensing technology is still used today in urban mapping. Because of their widespread availability, frequency of updates, and reasonable cost, the use of remote sensing studies for community growth has shifted more towards the use of digital, multispectral images, particularly those acquired by Earth-orbiting satellites (Donnay et al., 2001). This may be partly because the new generation of very high spatial, spectral, and radiometric resolution sensor data (e.g., IKONOS from 1999 and QuickBird from 2001) are extensively available in the market.

When urban information is extracted from remotely sensed data, spatial resolution is generally considered more important than spectral resolution. It is more practical to consider higher spatial resolution (i.e., smaller pixel size) than higher spectral resolution (i.e., greater number of spectral bands). For example, if we need to estimate water use or energy consumption based on the number of dwelling units of each housing type in an area (single family, two family, multifamily), we usually require a minimum resolution of from ≤0.25 to 5 m (Jensen and Cowen, 1999) to identify the type of individual buildings. In general, any visible bands at this range of spatial resolution should provide sufficient spectral contrast between the object to be identified (e.g., single-family house) and its background (e.g., roads, trees, grass). It is also important to note that higher radiometric resolution (pixel depth) may not increase information about smaller objects and features.

However, it should be noted that remote sensors, regardless of the platform (airborne or space borne), can only see the physical layout of features — mostly roof tops, tree tops, roads, and parking lots. It may or may not be possible to observe lower level features, such as grass, road surface, shrubs, scrubs, and water bodies. The information acquired through analysis of remotely sensed data over cities is the physical structure of urban landscapes, but in general not the direct explanations of human activities.

The objective of launching high spatial resolution sensor data was for increased detection of terrestrial features, especially urban objects, by reducing the mixed pixel problem and thereby improving land-cover identification. Clarity is certainly more evident in these finer resolution data than in those from preceding sensors. However, this greater level detail is also translated into many more complex per-pixel spectral combinations (Myint et al., 2006). This is because urban features are composed of spectrally different diverse materials (e.g., plastic, metal, rubber, glass, cement, wood, etc.) concentrated in a small area (Jensen and Cowen, 1999). The complex spatial appearance of urban land-cover features may be a limitation in digitally classifying urban land-use and land-cover classes in high-resolution image data (Myint et al., 2002).

The classification accuracy of remotely sensed data is the result of a trade-off between two main factors: class boundary pixels and within-class variances (Fortin, 1992; Metzger and Muller, 1996). Most common image processing algorithms do not take local structure or spatial arrangement of neighborhood pixels into consideration. To extract the heterogeneous nature of urban features in high-resolution images, texture information contained in a group of neighborhood pixels needs to be considered. Traditional spectral classification algorithms use individual pixel values and ignore spatial information. This spatial information is crucial in urban mapping when high-resolution images are used because most of the urban classes contain a number of spectrally different features or objects. For example, roads, houses, grass, trees, bare soil, shrubs, swimming pools, and sidewalks, each of which may have a completely different spectral response, may need to be considered together as a residential class.

One limitation for supervised classification is that it is difficult to define suitable training sets for many categories within urban environments for land-use information because of variation in the spectral response of their component surface covers in each land-use class. For example, do we want to select rooftops alone as a training sample to represent a residential class or do we need to consider cement roads, tar roads, sidewalks, driveways, swimming pools, grass, shrubs, or trees as part of a residential class? If we select shingle rooftops as a residential training sample, all shingle rooftops will be identified as residential class, and all other land-cover types within residential areas will not be identified as residential class.

Hence, it is important that all land-cover types within a residential area be considered together as a residential class in selection of a residential training sample. Consequently, training statistics may exhibit very high standard deviation. This type of training sample may also violate one of the basic assumptions of statistical approaches, including the widely used maximum-likelihood decision rule that the data are normally distributed (Barnsley et al., 1991; Sadler et al., 1991).

It is generally accepted that accurately classifying digital remote sensing images into urban land-use and land-cover categories from high-resolution image data remains a challenge despite significant advances in geographic information science and technology. As discussed earlier, complex spatial arrangements of urban features, lack of spatial consideration in traditional per-pixel classifiers, difficulty in defining suitable training sets, and the nature of high standard deviation and lack of multivariate normal distribution in training samples mean that traditional image classification approaches such as the maximum-likelihood decision rule are ineffective in classifying the urban land-use and land-cover features in high-resolution images. The key question here is how to construct different signature responses from land-cover features in multispectral bands into organized urban land use categories. The answer to this question is that we need to employ geospatial approaches that extract textures of land-use classes to improve the classification accuracy of urban mapping.

6.2 IMAGE TEXTURE AND PATTERN

Texture plays an important role in the human visual system for pattern recognition and interpretation. In image interpretation, pattern is defined as the overall spatial form of related features; the repetition of certain forms is a characteristic pattern found in many cultural objects and some natural features. Texture is often referred to as the visual impression of coarseness or smoothness caused by the variability or uniformity of image tone or color in digital image processing (Avery and Berlin, 1992). It should be noted that this general description does not represent the complete characterization of a texture, but it is by far the most commonly used explanation of texture in image analysis.

Texture can be generally characterized as fine, coarse, smooth, rippled, mottled, irregular, or lineated. In fact, texture is an inherent property of virtually all surfaces (e.g., the grain of wood, the weave of a fabric) including all types of land-use and land-cover classes (e.g., the pattern of crops in a field, the crown features of trees in a dense forest). It contains important information about the structural arrangement of surfaces and their relationship to the surrounding environment. Although it is quite easy for human observers to recognize and describe in empirical terms, texture has been extremely intractable to precise definition and to analysis by digital computing techniques (Haralick et al., 1973).

Texture is an elusive notion that mathematicians and scientists tend to avoid because they cannot grasp it. Engineers and artists cannot avoid it, but mostly fail to handle it to their satisfaction (Mandelbrot, 1983). There is no universally accepted

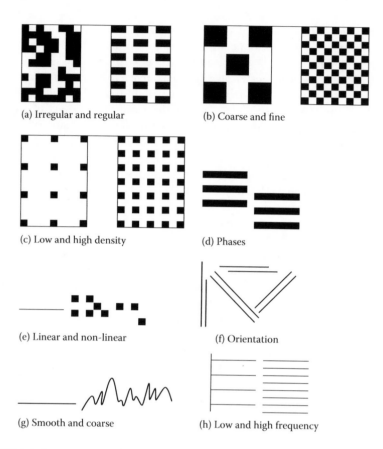

FIGURE 6.1 Some characteristics of texture patterns.

mathematical definition of texture. Textures that are homogeneous at small scale can be heterogeneous at large scale and vice versa. In other words, macrotexture and microtexture can be smooth or coarse. Hence, the question of scale or resolution is fundamental to the study of texture.

Some of the characteristics that may be used to explain texture in different ways may include (1) irregular and regular; (2) coarse and fine; (3) smooth and rough; (4) orientation; (5) high density and low density; (6) phases; (7) linear and nonlinear; and (h) high and low frequency (Figure 6.1). Because spatial arrangements of features or textural properties of images appear to carry useful information for discrimination purposes, it is important to develop algorithms and measures to capture the characteristics of texture. Hence, size of the local window plays an important role in extracting texture information from a given set of data. This is because the minimum distances required to cover different textures may be completely different at a given scale, depending on their complexity, repetition, spatial arrangements, orientation, and size of objects and features. This is also true for same-texture features at different scales.

Lark (1996) described a working definition of texture. Two segments of an image may be regarded as having the same texture if they do not differ significantly with respect to:

Variance of their DN values
Spatial dependence of this variability at a characteristic scale (or scales)
Directional dependence of this variability
Any spatial periodicity of this variation

This working definition does not claim to be comprehensive but it is proposed as a useful basis for image analysis in the context of remote sensing. It is clear from this definition that a simple texture transform such as the sample standard deviation within a local window will depend on local image texture but will not characterize it very precisely because standard deviation does not represent directionality or spatial periodicity. The same or similar value might be given for image segments with different texture appearances.

6.2.1 LOCAL WINDOW SIZE

To analyze spatial information of textures, we need to observe a group of objects and features in a neighborhood. Local moving windows are commonly used in digital image classification approaches to define the local information content around a center pixel. In general, accuracy should increase with a larger local window size because it contains more information and provides more complete coverage of spatial variation, directionality, and spatial periodicity of a particular texture than a smaller window size does. However, this is not necessarily true in a real-world situation because, in most cases, we are dealing with more than two homogeneous texture features while the local window moves across the image (Myint, 2003). Hence, it should be noted that a larger window can produce better accuracy only when dealing with homogeneous texture features (one window does not cover more than one texture).

As mentioned earlier, texture of a residential area may consist of single-family houses, lawns, shrubs, trees, tar roads, concrete roads, concrete sidewalks, swimming pools, etc. If we were to use 4-m resolution multispectral IKONOS data, it would be inappropriate to analyze these residential texture features using a local window size of less than 5 × 5 (i.e., 20 m × 20 m) because the window needs to cover at least one house with all of the previously stated objects or features. It is understood that we need to consider a minimum distance between two pixels (characteristic scale), which covers a particular texture or pattern of a land-use or land-cover type for a minimum window size.

Gong and Howarth (1990) generated edge-density images with the use of sizes varying from 7 × 7 to 31 × 31. The window size of 25 × 25 was selected and the resultant edge-density image was used as the texture-transformed band in their study. Identification of a method for determining optimal window size a priori classification is elusive (Gong and Howarth, 1992). From a computational perspective, the ideal window size is the smallest size that also produces the highest accuracy.

The most common approach to determining appropriate window size is based on empirical results using automated classifications (Hodgson, 1998). The seminal work on second-order texture statistics by Haralick et al. (1973) was based on windows of 64 rows × 64 columns in size or 20 rows × 50 columns. In this effort, the gray-tone spatial dependency matrix was developed along with 14 fundamental measures of texture from this spatial dependency matrix. Pesaresi (2000) experimented with 47 different square window sizes, ranging from 5 × 5 to 99 × 99 and showed the increase of histogram separation index with the increase of window size.

However, minimization of local window size is important in image texture and pattern recognition techniques because larger window size tends to cover more classes and consequently creates mixed-boundary class problems. Another problem of using larger window size is that smaller land-cover features will be lost in classification. In other words, the larger the window size, the smaller the number of segmented regions or land-use and land-cover features identified. It will also maximize missing pixels on the edges (e.g., 12 missing pixels on the left, right, top, and bottom of the image for 25 × 25 window size).

6.3 GEOSPATIAL TECHNIQUES

Local variability in remotely sensed data can be characterized by computing statistics of a group of pixels, e.g., standard deviation or coefficient of variance or autocovariance, or by analysis of fractal similarities or autocorrelation of spatial relationships. Some attempts have been made to improve spectral analysis of remotely sensed data by using texture transforms in which some measure of variability in DN values is estimated within local windows — for example, contrast between neighboring pixels (Edwards et al., 1988), the standard deviation (Arai, 1993), or local variance (Woodcock and Harward, 1992).

It is understood that many different texture sets may share the same standard deviation or variance values. For example, "2, 2, 4, 4, 6, 6," "2, 4, 6, 2, 4, 6," and "6, 6, 8, 8, 10, 10" will share the same standard deviation or variance value. However, the standard deviation and local variance measures are still accepted as simple and common approaches to describe spatial arrangements of objects and features. An example illustrating how a 3 × 3 local window with standard deviation function works on a small image is given in Figure 6.2.

On the other hand, the coefficient of variation seems to be more meaningful and gives a measure of the total relative variation of pixel values in an area because the mean of a spatial set is considered in the calculation. De Jong and Burrough (1995) argue that it does not give much information about spatial patterns. Snow and Mayer (1992), Klinkenberg (1992), and Burrough (1993) criticized many other neighborhood operations such as diversity or variation filters. Their absolute outcome was easy to compare, but they did not reveal any information on spatial irregularities.

One commonly used statistical procedure for interpreting texture uses an image spatial co-occurrence matrix, which is also known as a gray-level co-occurrence matrix (GLCM) (Franklin et al., 2000). A number of texture measures could be applied to spatial co-occurrence matrices for texture analysis (Peddle and Franklin, 1991). Herold et al. (2003) proposed a method of using landscape metrics to classify IKONOS images

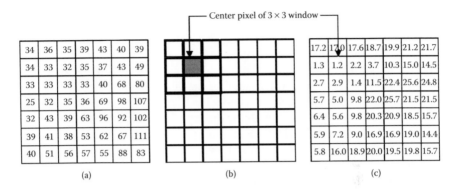

FIGURE 6.2 Example of a 3 × 3 local window with standard deviation function works on an image: (a) original image; (b) 3 × 3 window starts from top left-hand corner and assigns a standard deviation value in the center pixel; and (c) the output image filled with standard deviation values.

and compared it with GLCM. Liu and Herold (2005) further compared spatial metrics, GLCM, and semivariogram in terms of land-use classification.

Lam et al. (1998) demonstrated that the fractal dimension of remote sensing data could yield quantitative insight on the spatial complexity and information content contained within these data. Quattrochi et al. (1997) used a software package known as image characterization and modeling systems (ICAMS) to explore how fractal dimension is related to surface texture. De Jong and Burrough (1995) analyzed variograms of remotely sensed measurements to describe spatial patterns quantitatively. Variogram interpretation of satellite data was also carried out by Woodcock et al. (1988) and Webster et al. (1989). Emerson et al. (1999) analyzed the fractal dimension using isarithm method and the spatial autocorrelation of satellite imagery using Moran's I and Geary's C to observe differing spatial structures of smooth and rough surfaces in remotely sensed images. Myint and Lam (2005a, b) developed a number of lacunarity approaches to characterize spatial features with completely different texture appearances that may share the same fractal dimension values.

Some of the advanced geospatial approaches used to improve urban mapping, including spatial co-occurrence matrix, spatial autocorrelation, fractal analysis, lacunarity, and wavelet approaches, are discussed in this chapter. The texture-transformed images generated by these geospatial approaches alone can be used to perform a supervised or unsupervised classification. We can also layer stack the texture-transformed images and the original images to identify land-use and land-cover classes using a classifier. Before layer stacking texture-transformed images and original images, we need to rescale the former to match the radiometric resolution or pixel depth (e.g., 8 bits) of the original images. It is easier to select training samples from the color display of the original images because texture-transformed images may present complex color combinations and appearances. The procedure for selection of training samples is the same as the way we select them in traditional supervised classification approaches.

6.3.1 Spatial Co-Occurrence Matrix

One commonly used method for interpreting image texture employs an image spatial co-occurrence matrix (SCM), which is also known as a gray-level co-occurrence matrix (GLCM). The method is also referred to as the spatial gray-level dependence method. This approach is by far the most widely used texture- and pattern-recognition technique in the analysis of remotely sensed data, and it has been successful to a certain extent.

Haralick et al. (1973) proposed the spatial co-occurrence procedure in the derivation of textural features, which characterize spatial variability in digital imagery. They assumed that the texture information on an image is contained in the overall, or "average," spatial relationship, with which the gray tones in the image must co-occur. In that algorithm, relationships of adjacent gray tones are captured in spatial co-occurrence matrices for a specified orientation and window size from which a series of texture measures can be computed. In 1987, Franklin and Peddle implemented this algorithm in analyzing texture features. They presented a procedure to characterize spatial variability in elevation.

Several researchers have used gray-level co-occurrence matrices in developing texture images. Construction of the four directional spatial co-occurrence matrices for a 3 × 3 window from an example image normalized to four gray levels (0 to 3) is illustrated in Figure 6.3. The final matrix for a given point location in the image

	1	2	3
	1	1	2
	0	1	2

(a)

	0	1	2	3
0	# (0, 0)	# (0, 1)	# (0, 2)	# (0, 3)
1	# (1, 0)	# (1, 1)	# (1, 2)	# (1, 3)
2	# (2, 0)	# (2, 1)	# (2, 2)	# (2, 3)
3	# (3, 0)	# (3, 1)	# (3, 2)	# (3, 3)

(b)

0	1	0	0
1	2	3	0
0	3	0	1
0	0	1	0

(c)

0	1	0	0
1	4	1	0
0	1	2	1
0	0	1	0

(d)

0	0	0	0
0	4	1	0
0	1	2	0
0	0	0	0

(e)

0	1	0	0
1	0	2	1
0	2	0	0
0	1	0	0

(f)

FIGURE 6.3 A 3 × 3 window with gray tone range 0 to 3 (a); general form of any spatial co-occurrence matrix for window with gray tone range 0 to 3 (b). # (i,j) represents number of times gray tones i and j were neighbors; spatial co-occurrence matrices derived for four angular orientations: horizontal 0° (c); vertical 90° (d); left diagonal 135° (e); and right diagonal 45° (f). (Adapted from Myint, S.W. et al., *Photogrammetric Eng. Remote Sensing*, 70(7): 803–812, 2004.)

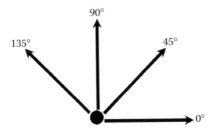

FIGURE 6.4 The selected orientations (e.g., horizontal 0°, vertical 90°, left diagonal 135°, and right diagonal 45°).

contains the number of times each possible pair of pixel values occurred in the selected orientation within the specified neighborhood surrounding that point (Figure 6.4).

A number of texture measures could be applied to spatial co-occurrence matrices for texture analysis. The formulae used to compute some texture measures, such as angular second moment (ASM) or energy; entropy (ENT); homogeneity or inverse difference moment (IDM); contrast (CON) or inertia; and correlation (COR), from the spatial co-occurrence matrix and their performance in general (Franklin and Peddle, 1987) are as follows.

ASM or energy can be computed as

$$ASM = \sum_i \sum_j \left\{ \frac{P(i,j)}{R} \right\}^2 . \tag{6.1}$$

For the angular second moment and four other spatial co-occurrence formulae following, P is the spatial co-occurrence matrix and R is the frequency normalization constant for the selected orientation. R is used to increment or decrement the matrix in order to avoid the numerous divide operations that otherwise would be necessary. The ASM (also termed energy) gives higher results to textures that have order and regularity in the relation between neighboring pixels because that is when the transition probabilities become concentrated in a few places in the matrix and the energy grows. It is a measure of homogeneity. For example, in relatively homogeneous land cover such as grassland, there are only a few tonal changes and hence a few entries in the matrix off the diagonal. The ASM value correspondingly will be close to 1 (its maximum).

ENT can be computed as

$$ENT = -\sum_i \sum_j \left(\frac{P(i,j)}{R} \right) * \log\left(\frac{P(i,j)}{R} \right) . \tag{6.2}$$

Entropy detects the same properties in textures as the energy feature because it also measures the amount of order in the transition between neighboring pixels. It is a measure of variability or randomness. For example, a concentration of brightness value changes in relatively few locations (e.g., water) would be essentially nonrandom and the entropy measure would be low. Thus, we could expect a large negative correlation between ENT and ASM.

Inverse difference moment (IDM) or homogeneity can be computed as

$$IDM = \sum_i \sum_j \frac{1}{1+(i-j)^2} \left(\frac{P(i,j)}{R} \right). \tag{6.3}$$

The inverse difference moment feature characterizes urban features in terms of a lack of variability in digital numbers. It gives less weight to transition probabilities from i to j as the absolute value of $(i-j)$ gets bigger (as the probability is far from the main diagonal of the co-occurrence matrix). By doing so, it gives bigger scores to images containing homogeneous features and having slow changes in gray level values. When high IDM values occur, we should have high ASM values, but low entropy values.

CON or inertia can be computed as

$$CON = \sum_i \sum_j \frac{(i-j)^2 P(i,j)}{R}. \tag{6.4}$$

For all the previously mentioned formulae, P is the spatial co-occurrence matrix and R is the frequency normalization constant for the selected orientation. The contrast gives more weight to transition probabilities that are more distant from the main diagonal in the matrix because contrast increases when the difference between two neighboring pixels gets bigger. In spite of the fact that the contrast is correlated to the inverse difference moment feature, it adds information and improves the segmentation. It gives higher values to images containing homogeneous features and vice versa.

COR can be computed as

$$COR = \sum \sum \frac{(i-\mu_x)(j-\mu_y)P(i,j)}{\sigma_x \sigma_y} \tag{6.5}$$

where $-1 \leq P \leq 1$. COR takes the value 1 if only values on the main diagonal of P are nonzero and the value 0 if the gray values are uncorrelated. This means the spatial arrangement of objects and features in an image under investigation is random. A value approaching -1 represents that there is a dissimilar, or deterministic structure (checkerboard pattern) of spatial arrangements observed in the spatial set. A value of 1 indicates that no similar, regionalized, smooth, or clustered patterns of spatial features exist.

6.3.2 Spatial Autocorrrelation

Two spatial autocorrelation techniques, Moran's I and Geary's C (Cliff and Ord, 1973), measure the degree to which similar or different objects in space are closely associated with each other. In other words, the spatial autocorrelation measures or tests how dispersed, uniformly distributed, or clustered points are in space with regard to their characteristic values. Hence, spatial autocorrelation can be considered as a representative index that characterizes the spatial arrangements of objects and features as rough and smooth surfaces in images. Some analysts favor Moran's I mainly because they believe its distribution characteristics are more desirable. Myint (2003) reports that the effectiveness of both approaches in measuring the spatial arrangements of objects and features in satellite images is about the same. Both statistics are based on a comparison of the attribute values of neighboring units. The statistics should indicate strong positive spatial autocorrelation if the attribute values of areal units over a study area are similar and vice versa. The two statistics use different approaches to compute representative indices of neighboring spatial arrangements.

Moran's I can be computed as

$$I(d) = \frac{n \sum\limits_{i}^{n} \sum\limits_{j}^{n} w_{ij} z_i z_j}{W \sum\limits_{i}^{n} z_i^2} \tag{6.6}$$

where

w_{ij} = weight at distance d so that $w_{ij} = 1$ if point j is within distance d from point i; otherwise, $w_{ij} = 0$

z = deviations (i.e., $z_i = y_i - \mu_y$ for variable y)

n = number of points in the point distribution (number of pixels)

W = sum of all the weights where $i \neq j$

Moran's I varies from +1.0 for perfect positive correlation (clumped pattern or clustered) to −1.0 for perfect negative correlation (checkerboard pattern or dispersed). Geary's C is calculated from the following formula:

$$C(d) = \frac{(n-1) \sum\limits_{i}^{n} \sum\limits_{j}^{n} w_{ij} (y_i - y_j)^2}{2W \sum\limits_{i}^{n} z_i^2}, \tag{6.7}$$

with the same terms listed for Equation 6.6.

For Geary's C index, a value of less than 1 indicates positive correlation, 1.0 indicates no correlation, and values greater than 1.0 indicate negative correlation. Because both spatial autocorrelation approaches measure how dispersed or clustered points are in space with regard to their attribute values, they can be applied to any satellite images.

6.3.3 G INDEX

Moran's I and Geary's C ratios have well-established statistical properties to describe spatial autocorrelation globally. However, they are not effective in identifying different types of clustering spatial patterns. They cannot distinguish "hot spots" and "cold spots." The general G statistic (Getis and Ord, 1992) has the advantage over Moran's I and Geary's C of detecting the presence of hot spots and cold spots over the entire area. The G statistic is defined as:

$$G(d) = \frac{\sum_{i}^{n}\sum_{j}^{n} w_{ij}(d)x_i x_j}{\sum_{i}^{n}\sum_{j}^{n} x_i x_j} \text{ , for } i \neq j. \tag{6.8}$$

The G statistic is defined by a distance, d, within which areal units can be regarded as neighbors of i. The weight, $w_{ij}(d)$, is 1 if areal unit j is within d and is 0 otherwise. This index has not been widely used in image texture analysis. However, this approach could potentially measure spatial arrangements of surface features more effectively because it is designed to capture cold and hot spots. Myint (2006) developed and explored the G index approach in comparison to spatial autocorrelation approaches and found that it gave better accuracy than the autocorrelation approaches.

6.3.4 FRACTAL

Since the beginning of Mandelbrot's effort (Mandelbrot, 1983), fractal analysis has received tremendous attention from researchers in identifying and understanding the processes of natural phenomena. Fractal dimensions may be viewed as a measure of irregularity or heterogeneity of spatial arrangements in many areas of study or physical processes. There has been growing interest in application of fractal geometry to observe the spatial complexity of natural features at different scales. Study of the relationship between physical processes and the effects of scale has become increasingly important in geographic information sciences. A number of studies have been carried out to evaluate the performance of fractals to characterize texture features in remotely sensed images.

Fractals characterize the concept of self-similarity, in which the spatial behavior or appearance of a system is mainly independent of scale (Burrough, 1993). Self-similarity is defined as a property of curves or surfaces in which each part is indistinguishable from the whole or the form of the curve or surface is invariant with respect to scale. In other words, we obtain similar objects and features whether we change scale (large or small) or spatial resolution (fine or coarse) of images. Although the fractal dimension may not be an integer value, an ideal fractal (or monofractal) curve or surface has a constant value over all scales. An example of a scaling relation between COUNTs and STEPs of an ideal fractal set is explained next.

FIGURE 6.5 Koch curve at different levels (ideal fractal).

If we observe the curve as shown in Figure 6.5, we can understand that it is constructed at different resolutions for each level. STEP is related to scale or spatial resolution of images. In this particular example, STEP can be thought of as the inverse of the number of steps it takes to move from one point to another horizontally (a horizontal segment of the preceding feature) on the Koch curve (Figure 6.5). COUNT refers to the total number of segments counted in each STEP in this case.

At level = 0, STEP = $(1/3)^0$ = 1. At level = 1, STEP = $(1/3)^1$ = 1/3, but COUNT = 4. At level = 2, STEP = $(1/3)^3$ = 1/9, but COUNT = 16, and so forth as demonstrated in Table 6.1. It can be observed that COUNT increases very rapidly as STEP decreases. When the logged STEP and COUNT values are plotted, as shown in Figure 6.6, we get a straight line with a slope value of 1.2619. The regression equation yields log(COUNT) = 0 − 1.2619 log(STEP). Stated another way, COUNT = STEP$^{-1.2619}$. We can say that the slope value (b = 1.2619) is a scaling relationship between number of COUNTs and STEPs at each level. Hence, a useful definition of fractal dimension is $D = -b$ and 1.2619 is the fractal dimension value of the Koch curve.

Theoretically, if the digital numbers of a remotely sensed image resemble an ideal fractal surface, then, due to the self-similarity property, the fractal dimension

TABLE 6.1
Relationship between COUNT and STEP
for a Koch Curve

Level	STEP	COUNT	Log(STEP)	Log(COUNT)
0	1	1	0.000	0.000
1	1/3	4	−0.477	0.602
2	1/9	16	−0.954	1.204
3	1/27	64	−1.431	1.806
4	1/81	256	−1.908	2.408
5	1/243	1024	−2.386	3.010

of the image will not vary with scale and resolution. However, most geographical phenomena may not be strictly self-similar at all scales (Goodchild and Mark, 1987). A number of studies have been carried out to evaluate the performance of fractals to characterize texture features in remotely sensed images:

Lovejoy and Schertzer (1990) studied the multi-fractal analyses of satellite and radar images of cloud and rain fields.

Lam and Quattrochi (1992) demonstrated that the fractal dimension of remote sensing (RS) data could yield quantitative insight on the spatial complexity and information content contained within these data.

Some researchers (Mark and Aronson, 1984; Burrough, 1989, 1993; Klinkenberg and Goodchild, 1992; Xia, 1993; De Jong and Burrough, 1995) have argued that the remotely sensed images of the land-cover units may not be true fractals and demonstrate that land surfaces are rarely self-similar, and if so, only exist within limited scales. They have also reported that different objects and land-cover features may share the same fractal dimension value and have completely different textures.

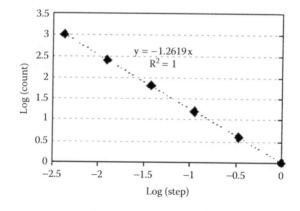

FIGURE 6.6 Scatter plot of log(COUNT) vs. log(STEP).

There have been a few algorithms developed to calculate the fractal dimension of remotely sensed images: the isarithm method (Lam and De Cola, 1993), the variogram (Mark and Aronson, 1984), and the triangular prism methods (Clarke, 1986). Myint (2003) reports that the latter method is the most accurate of all fractal approaches. The procedure to estimate the fractal dimension using a triangular prism approach is described next.

6.3.5 Fractal Triangular Prism

This method calculates the surface areas defined by triangular prisms; to use it, a remote sensing image is interpreted as being located on a grid of x and y coordinates. At each coordinate pair, the value of the pixel is interpreted as the z value. Figure 6.7 illustrates this arrangement in detail. The points E, F, G, and H in the figure are the coordinates of the four pixels on a square grid. The height of the line MO is the average of the digital values from the corner pixels so that MO = (EA + FB + GC + HD)/4. The vertex of this line is joined to the vertex of each of the four vertical lines, resulting in a triangular prism structure. The prism has four sides formed by joining the vertex of the center vertical line to the vertices of the vertical lines of the four corners. Hence, for a specific step size, a series of triangular prism grids are created from the entire image, and their surface areas can be calculated. The computation is repeated for all step sizes. The logarithm of the total surface area is plotted against the logarithm of the squares of varying step sizes. The fractal dimension is computed as $D = 2.0 -$ slope of the preceding regression (Jaggi et al., 1993).

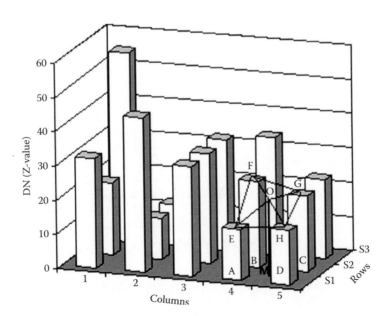

FIGURE 6.7 Example of the triangular prism area method to estimate fractal dimension value.

6.3.6 LACUNARITY

Mandelbrot (1995) reports that fractal dimensions may not be very accurate in providing a complete characterization of a set's texture. In other words, different fractal sets may share the same fractal dimension values but have different appearances or textures (Mandelbrot, 1983; Voss, 1986; Dong, 2000), just as different texture appearances of classes may share the same variance or mean value. As an initial step toward quantifying texture or spatial arrangements of features, Mandelbrot (1983) introduced the term "lacunarity" (*lacuna* is Latin for "gap") to characterize different texture appearances that may have the same fractal dimension value. Different fractal sets that have the same dimension value may be constructed, but they look completely different because they may have different lacunarity.

Lacunarity represents the distribution of gap sizes: low-lacunarity geometric objects are homogeneous because all gap sizes are the same, whereas high-lacunarity objects are heterogeneous (Dong, 2000). Lacunarity is a scale-dependent measure of spatial complexity or texture of a landscape (Plotnick et al., 1993). Unlike most other landscape indices and measures (Haines–Young and Chopping, 1996; Gustafson, 1998), the computed values of lacunarity are not sensitive to map boundaries but are sensitive to scale. It measures the deviation of a geometric structure from translational invariance or "gappiness" of geometric structure (Gefen et al., 1983). Some of the preceding researchers report that because remote sensing images of urban landscape are full of textures and hardly self-similar, traditional spectral classification techniques, as well as fractals and other spatial indices applied at a single-scale level, may not be able to capture the gaps and complexity of the landscape.

Methods for calculating lacunarity were first introduced in general form by Mandelbrot (1983) and several other algorithms of computing lacunarity have been developed (Gefen et al., 1983; Lin and Yang, 1986; Voss, 1986; Allain and Cloitre, 1991; Dong, 2000; Myint and Lam, 2005a, b). Allain and Cloitre (1991) initiated a conceptually straightforward and computationally simple "gliding box" algorithm for calculating lacunarity and reported that lacunarity appears to be a new tool for identifying the geometry of deterministic and random sets. Because lacunarity measures heterogeneity or degree of contagion, a higher index value of lacunarity indicates a higher heterogeneous feature or a more complex spatial arrangement and vice versa. Plotnick et al. (1996) emphasized the concept and utilization of lacunarity for characterization of spatial features, which may not be fractals. The gliding box algorithm has been used for calculating lacunarity value of binary images as well as gray-scale images. Description of the binary algorithm to compute lacunarity follows.

6.3.6.1 Binary Approach (Gliding Box Method)

The gliding box of a specific size (r, length of a square box) is first placed at the top left-hand corner of an image in which each and every pixel is filled with either 1 or 0 (Allain and Cloitre, 1991; Plotnick et al., 1993). Binary images can be created by converting each gray-scale image (each band) into four quartile images with value 1s and 0s. For example, the binary image for the first quartile will turn pixels that have values above 75% of all pixels as 0s and the rest as 1s, and so on. If the first quartile breakpoint has DN value of 41, all digital values above 41 will be converted to 0 and the rest will be assigned 1.

Then, the box mass "S," the number of occupied pixels (1s), is computed. The gliding box is systematically moved through the binary image one pixel at a time and the box mass value is determined for each of the overlapping boxes. For a given box size r, the probability of box mass S is:

$$P(S,r) = \frac{n(S,r)}{N(r)} \qquad (6.9)$$

where $n(S,r)$ is the number of gliding box size r with mass S, and $N(r)$ is the total number of boxes of size r. The first and second moment of this distribution, $E(S)$ and $E(S^2)$ are:

$$E(S) = \sum SP(S,r) \quad \text{and} \qquad (6.10)$$

$$E(S^2) = \sum S^2 P(S,r). \qquad (6.11)$$

Lacunarity for gliding box size r, $\Lambda(r)$, is defined as:

$$\Lambda(r) = \frac{E(S^2)}{E^2(S)}. \qquad (6.12)$$

Based on a random binary image, which has only two values — 0 for empty and 1 for filled — it can be described as

$$E(S^2) = \text{var}(S) + E^2(S). \qquad (6.13)$$

Plotnick et al. (1993) extended Equation 6.5 into:

$$\Lambda(r) = \frac{\text{var}(S)}{E^2(S)} + 1 \qquad (6.14)$$

where $E(S)$ is the mean and var(S) the variance of the number of occupied pixels per box.

Figure 6.8 shows six hypothetical 15×15 binary image patterns; white pixels represent 1s and black pixels represent 0s. The lacunarity indices at different scales — $r = 3, 5, 7, 9,$ and 11 — are computed for each pattern (Figure 6.9). It can be observed that lacunarity measures the geometry of spatial features. For instance, pattern B (Figure 6.8) has a bigger gap than the other patterns, and lacunarity value

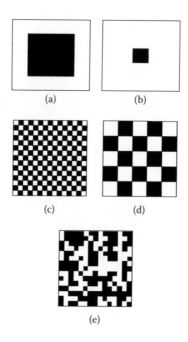

FIGURE 6.8 Binary images of six hypothetical spatial features.

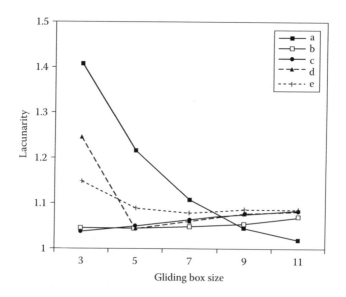

FIGURE 6.9 Lacunarity curves of the six spatial features shown in Figure 6.8 using the binary method.

is larger than that of others for almost all sizes because it contains many empty boxes. On the other hand, pattern C — small checkerboard — is a regular pattern (i.e., translationally invariant), and the lacunarity value is close to 1 because the number of occupied pixels and empty pixels is constant at any location within a neighborhood.

Texture-based image classification can be performed by using the preceding approaches to convert original bands into texture-transformed images followed by traditional image classification techniques.

6.3.7 WAVELET TRANSFORMS

The previous advanced spatial methods alone may not provide satisfactory accuracy when they are applied to fine-resolution remote sensing images for urban classification. That is, most of them focus primarily on coupling between features and objects at single scale and cannot determine the effective representative value of particular texture features according to their directionality, spatial arrangements, variations, edges, contrasts, and the repetitive nature of object and features. New developments in the mathematical theory of wavelet transform approaches based on multichannel or multiresolution analysis have received enormous attention. A number of reports in spatial frequency analysis of mathematical transforms have provided multiresolution analyses. Recent developments in spatial/frequency transform, Wigner distribution, discrete cosine transform, and wavelet transform have provided good multiresolution analytical tools. Of all transformation approaches, wavelets play the most critical part in texture analysis.

Mallat (1989a, b) developed a multiresolution analysis theory using the orthonormal wavelet basis. A wavelet is orthogonal when all of the pairs formed from the basis functions, $\psi_{j,k}$, are orthogonal to each other. An orthogonal wavelet normalized to 1 is called an orthonormal wavelet (Fukuda and Hirosawa, 1998). The multiresolution wavelet transform decomposes a signal into low-frequency approximation and high-frequency detail information at a coarser resolution. In satellite image analysis using two-dimensional wavelet transform techniques, rows and columns of image pixels are considered signals. The approximation and details of an image $f(x,y)$ at resolution 2^j can be defined by the coefficients computed by the following convolutions:

$$A_{2^j}^d f = ((f(x,y) * \phi_{2^j}(-x)\phi_{2^j}(-y))(2^{-j}n, 2^{-j}m))_{(n,m)\in Z^2} \, , \qquad (6.15)$$

$$D_{2^j}^1 f = ((f(x,y) * \phi_{2^j}(-x)\psi_{2^j}(-y))(2^{-j}n, 2^{-j}m))_{(n,m)\in Z^2} \, , \qquad (6.16)$$

$$D_{2^j}^2 f = ((f(x,y) * \psi_{2^j}(-x)\phi_{2^j}(-y))(2^{-j}n, 2^{-j}m))_{(n,m)\in Z^2} \, , \text{ and} \qquad (6.17)$$

$$D_{2^j}^3 f = ((f(x,y) * \psi_{2^j}(-x)\psi_{2^j}(-y))(2^{-j}n, 2^{-j}m))_{(n,m)\in Z^2} \, . \qquad (6.18)$$

where

integer j is a decomposition level

m, n are integers

$\phi(x)$ is a one-dimensional scaling function

$\psi(x)$ is a one-dimensional wavelet function

In general, $\phi(x)$ is a smoothing function that provides low-frequency information (low-pass filter), and $\psi(x)$ is a differencing function that provides high frequency information (high-pass filter). $A_{2^{j+1}}^d f$ can be perfectly reconstructed from $A_{2^j}^d f$, $D_{2^j}^1 f$, $D_{2^j}^2 f$, $D_{2^j}^3 f$. Expressions (6.1) through (6.4) show that in two dimensions, $A_{2^j}^d f$ and $D_{2^j}^k f$ are computed with separable filtering of the signal along the abscissa and ordinate. The wavelet decomposition can thus be interpreted as a signal decomposition in a set of independent, spatially oriented frequency channels (Mallat, 1989a). $\phi(x)$ and $\psi(x)$ can be defined as:

The dilation equation related to the low-pass filter is:

$$\phi(t) = \sqrt{2} \sum_k c(k)\phi(2t - k). \tag{6.19}$$

The wavelet equation related to the high-pass filter is:

$$\omega(t) = \sqrt{2} \sum_k d(k)\phi(2t - k) \tag{6.20}$$

where $c(k)$ and $d(k)$ are the coefficients.

The Haar wavelet transform has coefficients:

$$c(0) = \frac{1}{\sqrt{2}} \;, \; c(1) = \frac{1}{\sqrt{2}} \;, \; d(0) = \frac{1}{\sqrt{2}} \;, \; \text{and } d(1) = -\frac{1}{\sqrt{2}}.$$

Its dilation equation can be expressed as:

$$\phi(t) = \phi(2t) + \phi(2t - 1)$$

and its wavelet equation as:

$$\omega(t) = \phi(2t) - \phi(2t - 1).$$

The Haar wavelet transform is the simplest orthonormal basis. Several other wavelet transform approaches are available to extract spatial features at different scales.

For example, a series of Daubechies and Coiflets approaches can be used to perform the task. The following are the coefficients for Daubechies4 (Db4) and Coiflets6 (Coif6) wavelets, respectively:

$$c(o) = \frac{1+\sqrt{3}}{4\sqrt{2}} \, , \, c(1) = \frac{3+\sqrt{3}}{4\sqrt{2}} \, , \, c(2) = \frac{3-\sqrt{3}}{4\sqrt{2}} \, , \, c(3) = \frac{1-\sqrt{3}}{4\sqrt{2}} \, , \, d(o) = \frac{1-\sqrt{3}}{4\sqrt{2}} \, ,$$

$$d(1) = \frac{\sqrt{3}-3}{4\sqrt{2}} \, , \, d(2) = \frac{3+\sqrt{3}}{4\sqrt{2}} \, , \, d(3) = \frac{-1-\sqrt{3}}{4\sqrt{2}} \quad \text{(Db4 coefficients).}$$

$$c(o) = \frac{1-\sqrt{7}}{16\sqrt{2}} \, , \, c(1) = \frac{5+\sqrt{7}}{16\sqrt{2}} \, , \, c(2) = \frac{14+2\sqrt{7}}{16\sqrt{2}} \, , \, c(3) = \frac{14-2\sqrt{7}}{16\sqrt{2}} \, , \, c(4) = \frac{1-\sqrt{7}}{16\sqrt{2}} \, ,$$

$$c(5) = \frac{-3+\sqrt{7}}{16\sqrt{2}} \, , \, d(o) = \frac{-3+\sqrt{7}}{16\sqrt{2}} \, , \, d(1) = \frac{\sqrt{7}-1}{16\sqrt{2}} \, , \, d(2) = \frac{14-2\sqrt{7}}{16\sqrt{2}} \, ,$$

$$d(3) = \frac{2\sqrt{7}-14}{16\sqrt{2}} \, , \, d(4) = \frac{5+\sqrt{7}}{16\sqrt{2}} \, , \, d(5) = \frac{\sqrt{7}-1}{16\sqrt{2}} \quad \text{(Coif6 coefficients)}$$

More details can be observed in Strand and Nguyen (1997).

Approximation of a signal, $A_{2^j}^d f$, also known as trend, can be obtained by convolving the input signal, $A_{2^{j+1}}^d f$, with the low-pass filter (L). First, the rows of an image are convolved with a one-dimensional L. Next, the filtered signals are down sampled. In the first step, down sampling is performed by keeping one column out of two. Then the resulting signals are convolved with another one-dimensional low-pass filter, retaining every other row. To obtain a horizontal detail image, first the rows of the input image are convolved with a low-pass filter L, and the filtered signals are down sampled by keeping one column out of two, as we do in processing approximation images. However, for the next stage, the columns of the signals are convolved with a high-pass filter H and again every other row is retained.

For vertical details, the original signals are convolved first with a high-pass filter H and then with a low-pass filter L, following the preceding procedure. For the diagonal detail image, the same down sampling procedure is carried out, using two high-pass filters consecutively. The algorithm for the application of the filters and down sampling for computing the approximation and detail coefficients is illustrated in Figure 6.10. Figure 6.11 represents a standard orthonormal wavelet decomposition with two levels of an example image. The pyramid decomposition can be continuously applied to the approximation image until the desired coarseness resolution 2^{-j} ($-1 \geq j \geq -J$) is reached.

From the standard wavelet decomposition, we know that further decomposition is carried out in the low-frequency subimage. However, further wavelet decomposition can also be performed with the detail subimages. Hence, four different decomposition

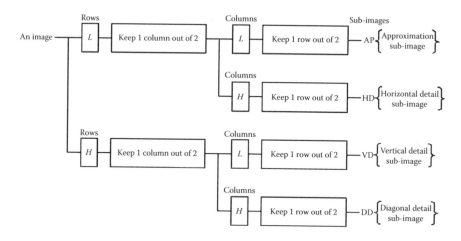

FIGURE 6.10 Decomposition procedure of an image by the multiresolution analysis. $L =$ lowpass filter; $H =$ highpass filter; AP = approximation subimage; HD = horizontal detail subimage; VD = vertical detail subimage; DD = diagonal detail subimage.

approaches will be examined in this study: (1) standard decomposition; (2) decomposition with horizontal details; (3) decomposition with vertical details; and (4) decomposition with diagonal details with the use of minimum distance classifier. Figure 6.12(a) and (b) illustrates an image and the first level decomposition of the original image; Figure 6.12(c) through (f) represents four different decomposition procedures with three levels of an original image.

Gong and Howarth (1992) demonstrated the efficiency of occurrence frequency methods with the use of several measures: mean, standard deviation, skewness, kurtosis, range, and entropy. Zhu and Yang (1998) used information entropy as a measure to identify texture features in 25 types of aerial relief samples selected from remote sensing images. Sheikholeslami et al. (1999) calculated the mean and variance of wavelet coefficients to represent the contrast of the image. Any of the following four feature measures — log energy (LOG), Shannon's index (SHAN),

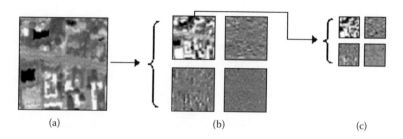

(a)   (b)   (c)

FIGURE 6.11 Multiresolution wavelet decomposition of a subset: (a) original image; (b) wavelet representation at level 1; (c) wavelet representation at level 2 using approximation subimage.

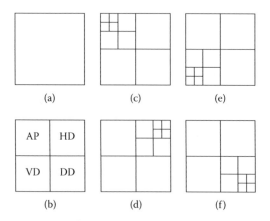

FIGURE 6.12 Four decomposition procedures with three levels of an original image: (a) an image; (b) decomposition at the first level; (c) standard; (d) horizontal; (e) vertical; and (f) diagonal decompositions at the second level.

ENT, and *ASM* — can be used to characterize the texture of urban land-use and land-cover classes:

$$(1) \ LOG = \sum_{i=1}^{K} \sum_{j=1}^{K} \log(P(i,j)^2) \,, \tag{6.21}$$

$$(2) \ SHAN = -\sum_{i=1}^{K} \sum_{j=1}^{K} P(i,j) * \log(P(i,j)) \,, \tag{6.22}$$

$$(3) \ ASM = \sum_{i=1}^{K} \sum_{j=1}^{K} P(i,j)^2 \,, \text{ and} \tag{6.23}$$

$$(4) \ ENT = -\sum_{i=1}^{K} \sum_{j=1}^{K} Q(i,j) * \log |Q(i,j)| \ ; \ Q(i,j) = |P(i,j)|^2 \bigg/ \left(\sqrt{\sum_{i,j} |P(i,j)|^2} \right) \,, \tag{6.24}$$

where $P(i,j)$ is the (i,j)th pixel wavelet coefficient value of a decomposed image at a particular level. Any classification rule including the Euclidean distance classifier and Mahalanobis distance rule can be employed for texture classification using the computed texture feature value of the subimages (decomposed images at different levels).

6.4 CONCLUSION

In general, it may be difficult to say which approach performs best among all geospatial techniques. Each approach has its advantages and limitations. No single approach may be considered best for all applications. Accuracy depends on several factors, such as spatial resolution (scale), nature of study area, number of classes, local window size, training sample selection, spatial technique used, knowledge of analyst, and spectral resolution of data. There is no optimal window size that can be used for all applications. As mentioned earlier, identification of a method for determining optimal window size a priori classification is vague. Again, window size depends on the nature of the classes and the spatial resolution (scale) of the data.

An ideal approach is to use several geospatial techniques with different classification options (i.e., different training samples, different window sizes, different band combinations) to compare results for a particular type of satellite data to achieve specific objectives. It can be concluded that, under real-world conditions, success of the preceding geospatial approaches for urban mapping may be highly variable and may not be easily controlled. Of all geospatial approaches, wavelet is the only approach that considers texture features at multiple scales to identify different urban land-use and land-cover classes. It can be argued that a fractal analysis approach is also a multiscale analysis approach. In this approach, information on the self-similarity of features at different scales is used to estimate fractal dimension values. The fractal approach does not split or transform any signal, image, or local window into different spatial features at multiple scales.

Any one of the preceding geospatial approaches can be very powerful, but they need to be handled carefully with awareness of their limitations and uncertainties. More importantly, it should be noted that the success of these geospatial approaches largely depends on a user's knowledge of spatial analysis and the selected approach. In other words, any of these approaches may not provide satisfactory accuracy unless the user is well aware of the theoretical background of the spatial concept in general and a selected geospatial approach in particular.

REFERENCES

Allain, C. and Cloitre, M., Characterizing the lacunarity of random and deterministic fractal sets. *Phys. Rev., A*, 44: 3552–3558, 1991.

Arai, K., A classification method with a spatial–spectral variability. *Int. J. Remote Sensing*, 14: 699–709, 1993.

Avery, T.E. and Berlin, G.L., *Fundamentals of Remote Sensing and Airphoto Interpretation.* Macmillan Publishing Co., New York, 472, 1992.

Barnsley, M.J., Barr, S.L., and Sadler, G.J., Spatial reclassification of remotely sensed images for urban land-use monitoring, in *Proc. Spatial Data 2000*, Oxford, 17–20 September, Remote Sensing Society, Nottingham: 106–117, 1991.

Burrough, P.A., Fractals and geochemistry, in *The Fractal Approaches to Heterogeneous Chemistry*, D. Avnir, Ed., Wiley & Sons Ltd., New York, 383–405, 1989.

Burrough P.A., Soil variability: a late 20th century view. *Soils Fert.*, 529–562, 1993.

Clarke, K.C., Computation of the fractal dimension of topographic surfaces using the triangular prism surface area method. *Computers Geosci.*, 12(5): 713–722, 1986.

Cliff, A.D. and Ord, J.K., *Spatial Autocorrelation*. Pion Press, London, 1973.

De Jong S.M. and Burrough, P.A., A fractal approach to the classification of Mediterranean vegetation types in remotely sensed images. *Photogrammetric Eng. Remote Sensing*, 61: 1041–1053, 1995.

Dong, P., Lacunarity for spatial heterogeneity measurement in GIS. *Geogr. Inf. Sci.*, 6(1): 20–26, 2000.

Donnay, J.-P., Barnsley, M.J., and Longley, P.A., Remote sensing and urban analysis, in *Remote Sensing and Urban Analysis*. P.A. Longley, J.-P. Donnay, and M.J. Barnsley, Eds., Taylor & Francis, London, 3–18, 2001.

Edwards, G., Landary, R., and Thomson, K.P.B., Texture analysis of forest regeneration sites in high-resolution SAR imagery, in *Proc. Int. Geosci. Remote Sensing Symp.* (IGARSS 88), ESA SP-284 (Paris: European Space Agency): 1355–1360, 1988.

Emerson, C.W., Lam, N.S.N., and Quattrochi, D.A., Multiscale fractal analysis of image texture and pattern. *Photogrammetric Eng. Remote Sensing*, 65(1): 51–61, 1999.

Fortin, M.J., Detection of ecotones: definition and scaling factors, Ph.D. dissertation, State University of New York, 129, 1992.

Franklin, S.E. and Peddle, D., Texture analysis of digital image data using spatial co-occurrence. *Computer Geosci.*, 13(3): 293–311, 1987.

Franklin, S.E. et al. Incorporating texture into classification of forest species composition from airborne multispectral images. *Int. J. Remote Sensing*, 21(1): 61–79, 2000.

Fukuda, S. and Hirosawa, H., Suppression of speckle in synthetic aperture radar images using wavelet. *Int. J. Remote Sensing*, 19(3): 507–519, 1998.

Gefen, Y., Meir, Y., and Aharony, A., Geometric implementation of hypercubic lattices with noninteger dimensionality by use of low-lacunarity fractal lattices. *Phys. Rev. Lett.*, 50: 145–148, 1983.

Getis, A. and Ord, J.K., The analysis of spatial association by use of distance statistics. *Geogr. Ana.*, 24(3): 1269–1277, 1992.

Gong, P. and Howarth, P.J., The use of structural information for improving land-cover classification accuracies at the rural urban fringe. *Photogrammetric Eng. Remote Sensing*, 56(1): 67–73, 1990.

Gong, P. and Howarth, P.J., Frequency based contextual classification and gray level vector reduction for land use identification. *Photogrammetric Eng. Remote Sensing*, 58(4): 423–437, 1992.

Goodchild, M.F., and Mark, D.M., The fractal nature of geographic phenomena. *Ann. Assoc. Am. Geogr.*, 77(2): 265–278, 1987.

Haines–Young, R. and Chopping, M., Quantifying landscape structure: a review of landscape indices and their application to forested landscapes. *Prog. Phys. Geogr.*, 20: 418–445, 1996.

Haralick, R.M., Shanmugan, K., and Dinstein, J., Textural features for image classification. *IEEE Trans. Syst., Man, Cybernetics*, SMC-3(6): 610–621, 1973.

Herold, M., Liu, X., and Clarke, K., Spatial metrics and local texture for mapping urban land use. *Photogrammetric Eng. Remote Sensing*, 991–1002, 2003.

Hodgson, M.E., What size window for image classification? A cognitive perspective. *Photogrammetric Eng. Remote Sensing*, 64(8): 797–807, 1998.

Jaggi, S., Quattrochi, D.A., and Lam, N.S.N., Implementation and operation of three fractal measurement algorithms for analysis of remote-sensing data. *Computer Geosci.*, 19(6): 745–767, 1993.

Jensen, J.R. and Cowen, D.C., Remote sensing of urban/suburban infrastructure and socio-economic attributes. *Photogrammetric Eng. Remote Sensing*, 65(5): 611–622, 1999.

Klinkenberg B., Fractals and morphometric measures: is there a relationship? *Geomorphology*, 5: 5–20, 1992.

Klingkenberg, B. and Goodchild, M.F., The fractal properties of topography: a comparison of methods. *Earth Surface Processes Landforms*, 17: 217-234, 1992.

Lam, N.S.N. and De Cola, L., Fractal simulation and interpolation, in *Fractals in Geography*, N.S.N. Lam and L. De Cola, Eds., Prentice Hall, Englewood Cliffs, NJ, 56–74, 1993.

Lam, N.S.N. et al., Environmental assessment and monitoring with image characterization and modeling system using multiscale remote sensing data. *Appl. Geogr. Stud.*, 2(2): 77-93, 1998.

Lark, R.M., Geostatistical description of texture on an aerial photograph for discriminating classes of land cover. *Int. J. Remote Sensing*, 17: 2115–2133, 1996.

Liu, X. and Herold, M., Of patterns and processes: spatial Metrics and geostatistics in urban analysis, in *Integrating Geographic Information Systems and Remote Sensing*, V. Mesev (Ed.), John Wiley & Sons, New York, 2005.

Lovejoy, S. and Schertzer, D., Multifractals, universality classes and satellite and radar measurements of cloud and rain fields. *J. Geophys. Res.*, 95: 2021–2034, 1990.

Mallat, S.G., A theory for multiresolution signal decomposition: the wavelet representation. *IEEE Trans. Pattern Anal. Mach. Intelligence*, 11: 674–693, 1989a.

Mallat, S.G., Multifrequency channel decompositions of images and wavelet representation. *IEEE Trans. Acoustics, Speech, Signal Process.*, 37(12): 2091–2110, 1989b.

Mandelbrot, B.B., *The Fractal Geometry of Nature*. New York: Freeman and Co., 1983.

Mandelbrot, B.B., Fractals. *Encyclopedia Phys. Sci. Technol.*, 5: 579–593, 1987.

Mandelbrot, B.B., Measures of fractal lacunarity: Minkowski content and alternatives. *Prog. Probability*, 37: 15–42, 1995.

Mark, D.M., and Aronson, P.B., Scale-dependent fractal dimensions of topographic surfaces: an empirical investigation with applications in geomorphology and computer mapping. *Math. Geol.*, 16: 671–683, 1984.

Metzger, J.P. and Muller, E., Characterizing the complexity of landscape boundaries by remote sensing. *Landscape Ecol.*, 11: 65–77, 1996.

Myint, S.W., Lam, N.S.N., and Tyler, J., An evaluation of four different wavelet decomposition procedures for spatial feature discrimination within and around urban areas. *Trans. GIS*, 6(4): 403–429, 2002.

Myint, S.W., Fractal approaches in texture analysis and classification of remotely sensed data: comparisons with spatial autocorrelation techniques and simple descriptive statistics. *Int. J. Remote Sensing*, 24(9): 1925–1947, 2003a.

Myint, S.W., The use of wavelets for feature extraction of cities in satellite images, in *Remotely Sensed Cities*, V. Mesev, Ed., pp. 109–134, Taylor & Francis, 2003b.

Myint, S.W., Lam, N.S.N., and Tyler, J., Wavelet for urban spatial feature discrimination: comparisons with fractal, spatial autocorrelation, and spatial co-occurrence approaches. *Photogrammetric Eng. Remote Sensing*, 70(7): 803–812, 2004.

Myint, S.W. and Lam, N.S.N., A study of lacunarity-based texture analysis approaches to improve urban image classification. *Computers, Environ. Urban Syst.*, 29(2005): 501–523, 2005a.

Myint, S.W. and Lam, N.S.N., Examining lacunarity approaches in comparison with fractal and spatial autocorrelation techniques for urban mapping. *Photogrammetric Eng. Remote Sensing*, 71(8): 927–937, 2005b.

Myint, S.W., Lam, N.S.N., and Mesev, V., Texture analysis and segmentation using a lacunarity approach based on the differential box counting method. 2006 (under review).

Myint, S.W., Image texture analysis and classification using G index and spatial autocorrelation approaches. 2006 (under review).

Peddle, D.R. and Franklin, S.E., Image texture processing and data integration for surface pattern discrimination. *Photogrammetric Eng. Remote Sensing*, 57: 413–420, 1991.

Pesaresi, M., Texture analysis for urban pattern recognition using fine-resolution panchromatic satellite imagery. *Geogr. Environ. Modeling*, 4(1): 43–63, 2000.

Plotnick, R.E., Gardner, R.H., and O'Neill, R.V., Lacunarity indices as measures of landscape texture. *Landscape Ecol.*, 8: 201–211, 1993.

Plotnick, R.E. et al., Lacunarity analysis: a general technique for the analysis of spatial patterns. *Phys. Rev. E*, 53: 5461–5468, 1996.

Quattrochi, D.A., Lam, N.S.N., Qiu, H., and Zhao, W., Image characterization and modeling system (ICAMS): a geographic information system for the characterization and modeling of multiscale remote sensing data, in *Scale in Remote Sensing and GIS*, D.A. Quattrochi and M.F. Goodchild, Eds., CRC Press, Boca Raton, FL, 295–308, 1997.

Sadler, G.J., Barnsley, M.J., and Barr, S.L., Information extraction from remotely sensed images for urban land analysis, in *Proceedings of the Second European Conference on Geographical Information Systems (EGIS'91)*, Brussels, Belgium, EGIS Foundation, Utrecht: 955–964, April 1991.

Sheikholeslami, G., Zhang, A., and Bian, L., A multiresolution content-based retrieval approach for geographic images. *Geoinformatica*, 3(2): 109–139, 1999.

Snow, R.S. and Mayer, L., Fractals in geomorphology. *Geomorphology*, 5(1/2): 194, 1992.

Voss, R., Random fractals: characterization and measurement, in *Scaling Phenomena in Disordered Systems*, R. Pynn and A. Skjeltorp, Eds., Plenum Press, New York, 1986.

Webster, R., Curran, P.J., and Munden, J.W., Spatial correlation in reflected radiation from the ground and its implication for sampling and mapping by ground-based radiometry. *Remote Sensing Environ.*, 29: 67–78, 1989.

Woodcock, C.E., Strahler, A.H., and Jupp, D.L.B., The use of variograms in remote sensing. II. Real digital images. *Remote Sensing Environ.*, 25: 323–348, 1988.

Woodcock, C. and Harward, V.J., Nested-hierarchical scene models and image segmentation, *Int. J. Remote Sensing*, 13: 3167–3187, 1992.

Xia, Z., The uses and limitations of fractal geometry in digital terrain modeling, Ph.D. thesis, City University of New York, 252 pp., 1993.

Zhu, C. and Yang, X., Study of remote sensing image texture analysis and classification using wavelet. *Int. J. Remote Sensing*, 13: 3167–3187, 1998.

7 Applying Imaging Spectrometry in Urban Areas

Martin Herold, Sebastian Schiefer, Patrick Hostert, and Dar A. Roberts

CONTENTS

7.1 INTRODUCTION

The use of imaging spectrometry for urban applications has made considerable progress over the past few years, in tandem with advances in high-spatial resolution urban remote sensing using sensors such as IKONOS. Imaging spectrometers acquire a large number of spectral bands with narrow bandwidths, and numerous studies have taken advantage of the vast amount of spectral detail for precise identification of chemical and physical material properties (Goetz et al., 1985). Traditionally, the majority of imaging spectrometry, also referred to as hyperspectral remote sensing, has focused on natural targets such as vegetation (e.g., Roberts et al., 1993) and minerals (e.g., Clark, 1999).

Urban environments are characterized by types of materials different from those in natural landscapes (Jensen et al., 1983), only recently has urban spectral complexity been studied and better understood using hyperspectral observations (Ben-Dor et al., 2001; Roessner et al., 2001; Herold et al., 2003a). Common imaging spectrometers include airborne systems such as the airborne visible infrared imaging spectrometer (AVIRIS, Green et al., 1998) and the hyperspectral mapper (HyMap; Cocks et al., 1998), which provide high-spatial and high-spectral resolution data operationally. The HYPERION sensor on the EO-1 satellite is a prominent example for the evolving field of space-borne imaging spectrometry.

As such, imaging spectrometry represents a unique data source that helps to satisfy needs for detailed information about urban environments. Urban areas represent the hub of human activities. Driven by tremendous worldwide expansion of urban population and urbanized area, dynamic urban change processes affect natural and human systems at all geographic scales. Worsening conditions of insufficient or obsolete infrastructure and rising costs for their maintenance, higher demands for resources, increasing urban climatologic and ecological problems, and issues related to natural disasters and urban security, underlie an increasing need for effective management and planning based on high-quality urban data (O'Meara, 1999).

In this chapter, we provide a summary of the current state of knowledge of imaging spectrometry and show how this technology can support urban applications. A thorough understanding of the spectral characteristics of urban environments and their representation in hyperspectral observations forms the basis for successful applications. Mapping efforts must consider the unique characteristics of urban environments that require specific sensor characteristics and raise particular challenges in understanding and analyzing urban spectra (Ben-Dor et al., 2001; Herold et al., 2003a). An overview of urban spectroscopy will be given in the first part of the chapter and several case studies will be presented in the second. Although urban areas are characterized by a large diversity of materials such as man-made features, vegetation, soils, etc., the focus in the introduction and practical examples will be on artificial and man-made surface types.

7.2 UNDERSTANDING URBAN SPECTRA

7.2.1 SPATIAL SCALE

All spectral studies must consider the spatial scale of analysis. Laboratory and field spectra commonly consist of a sample material illuminated under specific conditions. However, even such fine-scale measurements represent intimate spectral mixtures. For example, a small sample of a road surface may consist of different material components (e.g., rocky components and asphalt hydrocarbons) resulting in a spectrally mixed signal. Thus, spectral mixing appears at all observation scales. Airborne and space-borne observations contain even higher within-class variability. This variability stems from the diversity of materials in the observed area, object geometry, illumination effects, atmospheric interactions, and spectral resolution of the sensors as well as spectral mixing effects as a function of the spatial sensor resolution (Price, 1997).

Spatial resolution determines whether the spectral information measured within the instantaneous field of view (IFOV) originates from a single land object of interest (e.g., vegetation patch) or encompasses multiple objects within the IFOV and is a spectral mixture (e.g., of concrete and vegetation or soil and vegetation). Coarse spatial resolution has been commonly cited as a limitation in use of remote sensing in urban areas (e.g., Hepner et al., 1998; Roessner et al., 2001). In general, a spatial resolution of 5 m or finer is considered necessary for accurate spatial representation of common urban land cover objects such as buildings, roads, or urban vegetation patches (Welch, 1982; Woodcock and Strahler, 1987). With recent advances in space-borne systems (e.g., IKONOS and QUICKBIRD), this spatial resolution requirement can be provided operationally (Jensen and Cowen, 1999).

7.2.2 URBAN MATERIALS VERSUS LAND COVER AND USE

In addition to spatial complexity, urban environments also possess a high spectral heterogeneity (Ben-Dor et al., 2001; Roberts and Herold, 2004). From a material perspective, urban areas are composed of different major components. Rocks and soils and green and nonphotosynthetic vegetation are also commonly found in natural environments, as well as a variety of artificial surfaces, including different minerals (e.g., concrete); processed hydrocarbons (i.e., asphalts and plastics); other man-made materials, such as refined oil products like paints; and metal surfaces. Imaging spectrometry samples specific material properties represented by distinct spectral absorption features and thus enables mapping based on physical and chemical surface characteristics. Prominent examples that use small-scale spectral features to derive detailed material properties (e.g., mineral compositions, grain sizes, etc.) are described in Clark (1999), Swayze et al. (2003), and van der Meer (2004).

Depending on the application, remote sensing is often used to map urban land cover or land use rather than materials. Although they are related — i.e., red tile roofs (land cover) may have unique surface properties (material) and be dominant for residential land use — mapping land-cover types requires a different perspective in urban area remote sensing. Land cover considers characteristics in addition to those that come from the material. The surface structure (roughness) affects the spectral signal as much as normal variations within the land-cover type (i.e., cracks in roads, buildings with different roof angles, age, or cover). Two different land-cover types (e.g., asphalt roads and composite shingle or tar roofs) can be composed of very similar materials (hydrocarbons). From a material perspective, these surfaces would map accurately. The spectral discrimination of the land types of roads and roofs, however, may be limited by this similarity. Thus, analysis of urban land cover necessarily raises questions on the spectral separability and the within-class variability.

Land use is characterized by the arrangement, activities, and inputs people have undertaken on a certain land-cover type to produce, change, or maintain it. Common urban land-cover types are residential, industrial, educational, or transportation. There are relationships among material, land cover, and land use. However, experiences in mapping urban land-use types have shown that analyzing spectral signals is insufficient to discriminate them successfully and additional image information (e.g., spatial, textural, contextual) is usually required (Herold et al., 2003b).

Thus, a clear understanding of material vs. land cover and land use is required if imaging spectrometry is to be used in urban areas. Analysis of "spectral classes" often includes a mixture of the three. Sometimes, spectral analysis is practical and there is a clear link between material characteristics and land cover types. However, inappropriate definitions of map targets, based upon spectral analysis, can also hinder the use of imaging spectrometry by producing products that lack the thematic detail and flexibility that many users desire.

7.2.3 TERMINOLOGY AND SPECTRA ACQUISITION

Spectra are typically acquired in three ways: in the laboratory from destructive samples (i.e., leaves harvested in the field and brought into the laboratory), in the field (also called *in situ*), or from an imaging platform. In field spectrometry, the sample measurement is taken along with a calibration signal of a 100% reflectant material. The ratio of both measurements is usually referred to as the reflectance of the sample surface expressed in percent. In this manner, reflectance spectra are standardized and thus the spectrum of an unknown material can be compared to a known material while removing other factors that modify reflected radiation, such as the light source, illumination geometry, or effects of external energy sources such as scattered radiation.

Reflectance can be expressed in a variety of ways depending upon the nature and geometry of the incoming and reflected radiation and the type of spectrometer. The two most common types of reflectance used in remote sensing include directional-hemispherical reflectance and bidirectional reflectance. The former is typically measured in the laboratory using a collimated beam as the directional source and an integrating sphere to capture reflected radiation at all possible view geometries (because infinitesimal measurement angles are not possible, the term "directional" refers to very small conical measurements). Bidirectional reflectance is typically measured in the field or by most imaging spectrometers. In this case, incident energy and reflected energy are measured from a specific set of incident and view angles.

Imaging spectrometers differ from laboratory or field-based spectrometers in that they create an image of spectra. As mentioned before, an important spatial resolution component of an imaging spectrometer is the IFOV, which describes the angle subtended by a single detector element and is typically reported in milliradians. When the IFOV is multiplied by sensor height, this is called the ground instantaneous field of view (GIFOV). The IFOV of the AVIRIS sensor (Green et al., 1998) is 1 mrad, producing a GIFOV of 20 m from a height of 20 km. The majority of imaging spectrometers deployed for terrestrial applications sample reflected and emitted radiation in a spectral range from 350 to 2500 nm. This optical spectral range is separated into the visible (VIS: 350 to 700 nm), photographic near-infrared (NIR: 700 to 1000 nm), and short-wave infrared (SWIR: 1000 to 2500 nm).

Spectrometers on remote sensing platform systems do not directly measure reflectance. Rather, they provide some measure of the amount of radiation (usually reported in radiance) reflected by an object, which varies depending upon sensitivity of the instrument, wavelengths sampled, lighting geometry, and atmospheric conditions. The radiance measurements must be converted to reflectance values by correcting for atmospheric influences and wavelength-dependent changes in solar irradiance. This atmospheric correction process can be quite challenging and is one of the main reasons why image spectra are, in general, noisier than ground measurements. In particular, atmospheric water vapor is a major source of contamination in imaging spectrometry — a problem often neglected in multispectral data sets that do not sample water vapor bands.

Other important components during the acquisition of spectral measurement with imaging spectrometers and field instruments include the quality of the instrumentation and calibration, the number of spectral bands, and the spectral resolution. Spectral sampling includes number of spectral bands, wavelengths covered by these bands (reported as band center wavelength), and full-width-half-maximum. The (FWHM) of the bands. The FWHM typically assumes that each spectral band has a Gaussian-shaped response function and is equal to the spectral width of the response function at half the peak response. The AVIRIS sensor, for example, measures 224 bands with a nominal FWHM of about 10 nm over the spectral range of 350 to 2500 nm (Green et al., 1998). Field spectrometers may provide a FWHM of 1 nm or less and more spectral bands, respectively. Such data allow for more detailed representation of spectral features.

However, spectral resolution is most properly defined as the minimum spectral separation required to discriminate two spectral features of interest; thus, it can be less or greater than the actual FWHM. Additional considerations include the precision and stability of spectral calibration, system noise, and potential of artifacts in the measured spectra (i.e., second-order contamination or out-of-band transmission, both of which entail measuring radiation from outside the wavelength region of interest, as well as atmospheric contamination). An ideal instrument is radiometrically stable and has a sufficiently high signal-to-noise ratio (ratio between meaningful information and background noise) and quantization to discriminate two materials based on subtle differences in reflectance (Swayze et al., 2003).

7.2.4 SPECTRAL CHARACTERISTICS OF URBAN MATERIALS

After discussing details on the nature and acquisition of spectral signatures, this section will show example spectra typically found in urban areas. All spectra are bidirectional with units of reflectance and were taken from the Santa Barbara urban spectral library (Herold et al., 2004).

The spectral signatures shown in Figure 7.1 are best explained and understood by electronic and vibrational absorption processes that represent the specific material properties. Electronic absorption processes typically are high energy, involving the transition of an electron from one valence state to another or to a nonvalence state such as a conduction band. The high energy of these interactions typically results in strong absorptions in the UV and shorter wavelength portions of the VIS and NIR. Vibrational bands represent lower energy absorptions occurring when the energy of photons matches the frequency of vibration of molecular bonds. In many instances, the fundamental vibrational frequency of absorption is beyond 2500 nm, but overtones and combinations can produce higher frequency vibrations that produce absorptions within the NIR and SWIR. Vibrational bands occur at longer wavelengths than those due to electronic processes and typically are less broad. For a more detailed discussion of spectroscopy and absorption processes, see Clark (1999).

The diagrams in Figure 7.1 are sorted by major material absorption types — i.e., minerals, hydrocarbons, and vegetation. However, because of the way in which urban

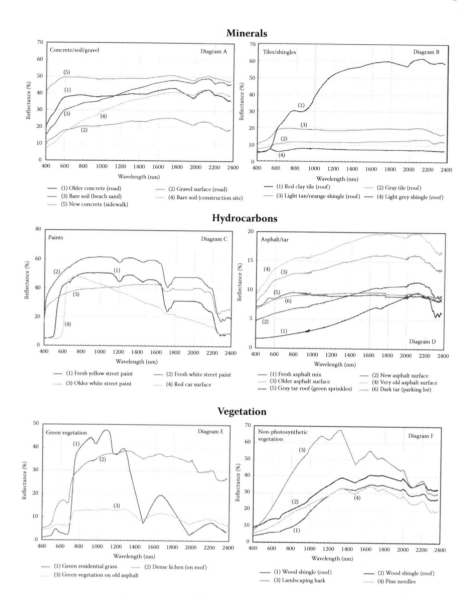

FIGURE 7.1 Spectra of typical materials found in urban areas for three dominant types of surface constituents (minerals, hydrocarbons, vegetation). All spectra were acquired *in situ* with an ASD full-range spectrometer in the Santa Barbara urban area within 2 hours of solar noon. The presented spectral sampling interval is 2.0 nm, resulting in an overall number of 1075 bands. The major water absorption bands (1340 to 1480 nm and 1770 to 1970 nm) are excluded. Note the difference in scale of the y-axis.

materials are altered by manufacturing processes and are combined, it is not always easy to identify a specific absorption feature and relate it to a diagnostic absorption process. Furthermore, urban spectra are modified by processes that alter the original material. For example, road surfaces and roofs change with age, leading to changes in their spectra. Urban materials can also be altered by coatings such as algae, lichen, dirt, dust, rubber tire marks, etc. Thus, urban spectra often represent a mixed spectral signature.

Diagram A in Figure 7.1 shows typical spectra of minerals. Such urban surfaces include gravel, concrete, and soils. Mineral absorptions are quite well understood (Clark, 1999). Prominent electronic absorptions in the VIS and NIR are caused by the presence of iron oxide with broad absorption bands centered at 520, 670, and 870 nm (visible in all spectra shown in diagram A). The red tile roof spectrum in diagram B best emphasizes these iron-oxide absorptions that cause the typical red color.

Minerals show significant vibrational features in the SWIR beyond 2000 nm. They are generally much narrower than electronic absorptions but allow for identification of specific minerals (Clark, 1999). The presence of calcium carbonates is suggested by features near 2330 nm, specifically for calcite (2300 nm) and dolomite (2370 nm). For example, many of the spectra published by Ben-Dor et al. (2001) for Tel Aviv show prominent calcium carbonate absorptions that are missing in the Santa Barbara area. The potential impacts of local construction customs and geologic source materials, as well as other factors such as dust, can cause prominent differences in urban spectra due to varying mineral composition.

Interestingly, red tile roofs indicate a strong increase in reflectance towards longer wavelengths. Liquid water and hydroxyl absorptions, typically found in clays, are lacking in fired brick and the reflectance is higher towards longer wavelengths due to loss of water in the production firing process. Diagram B further shows three low reflectance spectra (5 to 20%) that, in many cases, are nearly featureless. Exceptions are in the visible region emphasizing color differences and some minor absorption features caused by the presence of minerals (calcium carbonates) for the tile roof.

Spectra 3 and 4 (diagram B) are from shingle roofs that represent mixtures of both minerals and tar (hydrocarbons). In general, urban hydrocarbon materials are good examples of absorptions that are not readily found in natural materials — for example, paint, which has no analog in the natural world (diagram C). Spectra 1 and 2 (diagram C) show prominent vibrational hydrocarbon absorption features beyond 1700 nm. Several vibrational overtones and combination bands cause the feature near 1730 nm. Due to the combination of different absorption processes, the absorption feature is asymmetric and reflects a doublet with the strongest absorption at 1720 nm and a second less deep one at 1750 nm. The region between 2200 and 2500 nm is also affected by numerous overlapping combination and overtone bands. The absorption is strong in the 2300-nm region with a well-developed doublet at 2310 and 2350 nm; the 2310-nm feature usually is the stronger one (Cloutis, 1989; Hoerig et al., 2001).

Depending on abundance and type of hydrocarbon, these features may be less developed (spectra 3 and 4, diagram C). The visible region for diagram C spectra reflects the color of the paints. Spectrum 2 is a white color and strongly reflects in all parts of the visible spectrum. The material of spectrum 1 absorbs the blue wavelengths, resulting in a yellow color, and the material of spectrum 4 only reflects red wavelengths (absorbing blue and green), causing the surface to appear red.

Spectrum 1 in diagram D shows a typical asphalt hydrocarbon spectrum that is also found for bitumen, oil, coals, and shales (Cloutis, 1989). The minimum reflectance is near 350 nm with a linear rise towards longer wavelengths. Hydrocarbon compounds exhibit electronic transitions arising from excitation of bonding electrons that cause very strong absorptions in the UV and the visible that lack resolvable absorption bands and decrease in strength beyond the visible. Typical hydrocarbon features are prominent between 1720 and 1750 nm and 2300 and 2350 nm. Spectra 2, 3, and 4 (diagram D) also show asphalt road surfaces. The spectra show a transition from pure hydrocarbon (spectrum 1) to mineral absorption features (spectrum 4). This change is caused by road aging and deterioration and will be further discussed in one of the application examples.

Spectra 5 and 6 (diagram D) provide further examples of low-reflectance targets. Similar to spectra 2 through 4 (diagram B), their low signal is caused by the presence of hydrocarbons with some minor absorption features in the visible part and the SWIR caused by minerals. Such spectral signals are quite common for specific types of roofs (composite shingle, asphalt, and tar roofs) and transportation surfaces (asphalt roads and parking lots). These classes are spectrally very similar on the material scale, resulting in difficulties for mapping them with remote sensing methods, even using the full spectral range (Herold et al., 2003a).

Prominent spectral features of green vegetation surface are evident in diagram E. All spectra have pronounced chlorophyll electronic absorptions at 450 and 680 nm, a green peak centered at 550 nm, and vibrational water absorptions at 980, 1200, 1400, and 1900 nm. The grassland vegetation (spectrum 1) emphasizes a pure vegetation signal. Vegetation is often mixed with other surfaces. In the case of vegetation next to an asphalt road surface, the vegetation reflection is dampened by hydrocarbon absorption. Lichen on top of roofs (spectrum 3) show the signal of green vegetation, the roof material (composite shingle), and nonphotosynthetic vegetation (NPV). The chlorophyll features are less intense (smaller absorptions near 450 and 680 nm and smaller red edge) and significant ligno-cellulose vibrational absorption bands near 2100 and 2300 nm that clearly identify them as vegetation (Roberts et al., 1993).

Further examples of NPV are presented in diagram F. Roofs with wood shingles show the presence of lignin and cellulose (2100- and 2300-nm features). Reflectance differences between the two wood-shingle spectra are in part caused by different ages and degrees of weathering. One wood-shingle target (spectrum 2, diagram F) shows subtle absorptions and a slight red edge due to the presence of chlorophyll in moss, algae, and lichen on top of the roof. Similar findings are reported by Ben-Dor et al. (2001), who point out the important role of coatings on modifying urban spectra. In addition to the NPV character, pine needles (spectrum 4, diagram 5) further exhibit typical hydrocarbon features (1720 to 1750 nm and 2300 to 2350 nm) described earlier.

7.2.5 SPECTRAL SENSOR REQUIREMENTS

Imaging spectrometers have considerable potential for development of optimal techniques and sensors for urban remote sensing by sampling a large number of wavelengths at a fine spectral resolution (Herold et al., 2003a). As an example, hyperspectral observations can be used to identify the most important spectral bands for discriminating a wide diversity of urban materials. Knowledge of optimal wavelengths is also important for assessing spectral capabilities and limitations of existing operational multispectral satellite remote sensing systems, such as IKONOS or Landsat types of satellites. Finally, imaging spectrometry can aid in new sensor development by defining spectral and spatial data requirements for urban mapping. Although hyperspectral sensors may be criticized as collecting an excessive number of bands at high cost, their role in developing more cost-effective sensors and new analysis techniques for urban remote sensing makes them excellent research tools.

Herold et al. (2004) provide a systematic and quantitative assessment of the spectral separability of urban surface types. They use a measure of spectral separability, the Bhattacharya distance, to evaluate the extent to which various urban materials are spectrally distinct and identify a set of optimal wavelengths for urban mapping (Figure 7.2). Examples of materials shown in Figure 7.1 that are difficult to separate include: (1) bare soil surfaces vs. concrete roads; (2) asphalt roads vs. composite shingle, tar, and gray-tile roofs; (3) gray-tile roofs vs. composite shingle and tar roofs; and (4) asphalt roads vs. parking lots (Herold et al., 2003a, 2004).

The most suitable bands for differentiating major urban land-cover types are presented in Figure 7.2. They appear in nearly all parts of the spectrum with a fair number in the visible region. Narrow spectral bands are important in resolving small-scale spectral contrast in the VIS (e.g., from color, iron absorption features, etc.). Additional important bands appear in the near- and short-wave infrared. They represent the larger dynamic range of reflectance values related to the increase in object brightness towards longer wavelengths for several land-cover types (e.g., tile roofs, wood-shingle roofs, vegetation, soils, gravel surfaces). Also, specific absorption features correspond to some of the most suitable bands (Herold et al., 2003a).

Distribution of most suitable bands indicates that some of them are located outside or near the edges of the Landsat ETM+ spectral configuration (Figure 7.2). This results in lower map accuracies by IKONOS and Landsat ETM+ when compared to a sensor that samples the most suitable bands (e.g., AVIRIS). Although IKONOS and Landsat TM are broadband multispectral systems, the narrow hyperspectral bands can resolve small-scale absorption features and the increasing number of bands required to discriminate more cover types. The IKONOS information, for example, does provide sufficient information for mapping green vegetation and urban surfaces with unique color characteristics (e.g., red tile roofs or swimming pools). Landsat-type spectral information has a strong advantage over IKONOS through the availability of SWIR bands. The interpretation of urban spectra has emphasized that important absorption patterns for minerals, hydrocarbons, and vegetation are located in this region (Herold et al., 2003a).

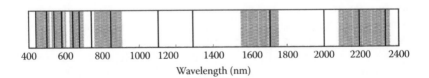

FIGURE 7.2 Most suitable spectral bands (black vertical lines) for urban mapping derived from the ground spectral library and hyperspectral AVIRIS data using spectral separability measures (Herold, M. et al., 2003, *IEEE Trans. Geosci. Remote Sensing*, 41(9), 1907–1919) compared to the spectral coverage of LANDSAT ETM satellite sensor (gray in the background).

In general, the spectral sensor characteristics of IKONOS and Landsat TM were designed for mapping a variety of land surfaces, especially for acquisition of natural and quasi-natural environments. However, different spectral configurations may be needed to resolve the unique spectral properties and complexity of urban environments and a sensor targeted at urban applications should consider this. For example, when comparing map accuracies generated using AVIRIS to maps produced by IKONOS or ETM+, Herold and Roberts (2006) found that AVIRIS was less sensitive to degradation of spatial resolution than the broadband sensors. If only coarser spatial resolution data are available, imaging spectrometry provides better information on the diversity of materials within each pixel and should be preferred for urban land-cover mapping.

7.2.6 Directional Reflectance

The spectra and assumptions on spectral sensor requirements presented in Section 7.2.4 and Section 7.2.5 emphasize the complexity of urban environments. However, additional factors are important to understand spectral signals from imaging spectrometers fully. If specific materials or land-cover types are to be mapped, the heterogeneity of the spectral signal within each category must be considered. If, for example, the application requires mapping of asphalt roads, the full spectral variability of asphalt road surfaces with different ages, stages of deterioration, origins of generic materials, etc. is present in the image data and the analysis algorithm must be tuned to account for within-class variability (Herold et al., 2004).

Another source of within-class variability in imaging spectrometry data is the phenomenon of directional reflectance. Ideal Lambertian reflectance (i.e., independent of view angle) does not apply to many surfaces viewed by imaging spectrometers. The optical properties of surface materials and their geometric assembly in the sensor's field of view (FOV) lead to anisotropic directional reflectance and thus brightness differences. In general, this phenomenon is completely described for individual surfaces by the bidirectional reflectance-distribution function (BRDF) according to Nicodemus et al. (1977) and depends on illumination and viewing geometry — i.e., the sun's and the sensor's positions in relation to the target surface.

Unlike laboratory conditions, illumination and viewing directions often vary between images from different acquisition times, affecting their intercomparability. In the case of airborne images from sensors like AVIRIS or HyMap, the large FOV and adherent view angle changes result in brightness gradients even within individual data sets. These are most dominant when the flight direction is perpendicular to the sun–target–observer plane, also referred to as principal plane (Beisl, 2001).

To illustrate directional reflectance factors, multiple spectroscopic field measurements at different view angles are shown in Figure 7.3. Identical urban material samples have been measured *in situ* at nadir view and at view zenith angles between $-30°$ and $+30°$ in the principal plane (negative angles indicate the sensor looking towards the sun). This way, extreme brightness differences in the 61.4° FOV of the HyMap sensor were simulated.

Most rough surfaces, such as vegetation canopies or densely built-on areas, back scatter radiation. Due to the complex structure, the BRDFs of vegetation canopies are driven by optical properties of individual leaves and plant material, by the three

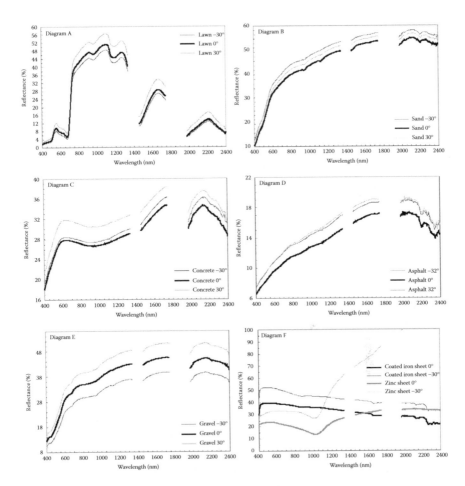

FIGURE 7.3 Spectra from urban materials at varying view angles in the principal plane. *In situ* measurements were taken in summer 2004 and 2003 in Berlin, Germany, with an ASD field spectrometer. Sun zenith angle was between 40 and 46° (in the case of the metal roof materials, at 33°). Water bands (1340 to 1450 nm; 1750 to 1970 nm) were removed and spectra were smoothed using a Savitzky–Golay filter with a third-degree polynomial based on nine wavelengths (From Hostert, P. and A. Damm, 2003, *3rd EARSeL Workshop on Imaging Spectroscopy*, Herrsching, 13th–16th May 2003, 215–219.).

dimensional assembly of the plant components (i.e., leaf angles and leaf area index), and by the geometric distribution of the plants on the surface (for example, see Kimes, 1983; Pinty et al., 2002).

The bidirectional behavior of an irrigated lawn surface in a park area is displayed in diagram A (Figure 7.3). The surface is back scatter dominant; the spectrum at 30° is brightest and the one at –30° darkest. Actually, a measurement at –15° (not shown) appears even darker, indicating an overall minimum near nadir, and a further increase of the signal towards more negative angles can be expected (Deering, 1989). This behavior is in accordance with findings by Kimes (1983), who describes two effects:

at first, a linear increase in brightness towards large, positive view angles caused by the decreasing fraction of shade in the IFOV, and next, a curve with a minimum around nadir induced by the amount of background signal measured. The combination of both effects results in a curved gradient slightly shifted to negative view angles. In the case of the measured, very dense lawn surface, the second effect (and hence the increase towards negative view angles) is less obvious.

Little is known about the directional reflectance of man-made surfaces (Meister, 2000). The need to differentiate between material and land cover or land use also applies to the description of directional reflectance on urban areas: the bidirectionality of materials is driven by their composition, constituent homogeneity, and, to a great extent, surface roughness. Bidirectional effects can thus vary between measurements from materials of the same type. Depending on the land cover, these effects can be significantly influenced — e.g., by irregularly inclined roofs or vertical facades in densely settled areas. In fact, it has been shown that the directional reflectance properties of urban materials affect spectral separability at off-nadir viewing geometries with positive and negative impacts on the discrimination between surface types (Mackay and Barr, 2002).

Sand surfaces are known to be nearly Lambertian, which is underlined by minor differences between the overall very bright spectra from a beach volleyball field (diagram B). Reflectance of a rather fine asphalt (new road) and deteriorated concrete (used sidewalk) is more anisotropic (diagrams C and D). Both materials appear brighter at 30° (back scatter) than at nadir view. This is caused by microshading within the sensor's IFOV, similar to the first effect mentioned for vegetation. This effect can be expected to be less dominant for fresh concrete surfaces.

On the other hand, negative view angles also result in spectra brighter than at nadir view (forward scatter). This effect seems to be very significant for asphalt, which, despite its rougher surface, appears brighter than concrete at negative view angles compared to the spectrum at nadir view and almost as bright as at positive view angles. This is explained by the greater amount and size of forward scattering constituents at the surface of the asphalt material mix. Therefore, directional reflectance properties of asphalt surfaces can be expected to depend on the material composition and its state of deterioration (compare Section 7.2.4 and Section 7.3.3).

The gravel spectra in diagram E underline the influence of surface roughness. The spectra at negative view angles appear darker than at nadir view and the effect of microshading dominates over other possible effects.

For all previous spectra, brightness differences appear to be multiplicative — i.e., the higher the reflectance is, the higher is the offset for brighter spectra. This is not the case for two metal-roof materials, a polished iron sheet with a coating and a dull zinc sheet, as displayed in diagram F. The sun zenith angle is at around 33°. Measurements at 30° were not possible because of instrument shade. The −30° measurements are close to perfect specular geometry — i.e., illumination angle equals measured reflectance angle. The coated iron sheet experiences a systematic offset in brightness due to forward scattering, but no specular behavior is shown, presumably because of the narrow focus of the specular angle. The zinc sheet's brightness, on the other hand, increases extremely in the NIR and it is higher than the reference in the SWIR region. It appears that extreme forward scattering of the dull zinc exists even at angles that differ slightly from the specular condition.

The findings of the spectral analysis correspond to an analysis of HyMap image data acquired over Berlin, Germany, on July 30, 2003. The flight direction was at 70° relative to the principal plane and the solar zenith was at 34°. Prior to analysis, the data were corrected for atmospheric effects, including a pixel-wise estimation of water vapor, and georectified. Severe brightness gradients are evident (Figure 7.4).

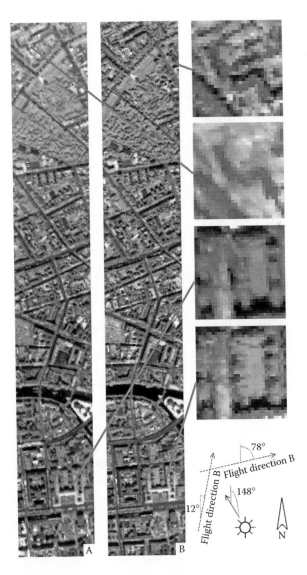

FIGURE 7.4 (Color Figure 7.4 follows page 240.) Subsets from two HyMap images of the same area in Berlin, Germany, flown at 70° (A) and 44° (B) to the principal plane. A severe brightness gradient can be observed in A, whereas the corresponding nadir region of B is free of such phenomena. The zoom areas show that the effect is greatest for vegetated surfaces.

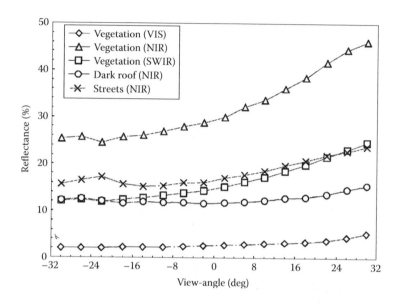

FIGURE 7.5 Brightness gradients by surface type HyMap data from Berlin, Germany. Spectrally pure pixels of three land-cover classes were selected by a restrictive spectral angle mapping. Values were averaged over 4° intervals.

Numerous saturated pixels exist on rooftops at negative view angles and result from extreme forward-scattering materials like sheet metal. After masking out saturated pixels and shadows, pure pixels of vegetation, dry vegetation with a soil background, dark roofs, and streets were derived by spectral angle mapping, and across-track brightness gradients were calculated for the individual land-cover types (Figure 7.5).

As expected, the brightness gradients of vegetation are most complex and differ in shape and height among VIS, NIR, and SWIR. The overall shape is in accordance with the effects described by Kimes (1983) and corresponds to results from vegetation BRDF models (Jacquemoud et al., 2000). The gradients of streets and dark roofs are different: brightness differences by view angle are less distinct and the gradients' shape is similar in all spectral regions (only NIR shown). At this point, the influence of the material's geometric assembly in the IFOV comes into play, underlining the need to differentiate material and land cover; the inclination of dark roofs superposes possible directional reflectance of roof materials. Well-lit and shaded surfaces are viewed in all regions of the image and only a slight symmetric gradient can be observed. The shape of the street gradient shows the dominance of back scatter, but is also influenced by urban geometry, where illuminated surfaces exist in the northern part of streets due to bright reflecting facades and less shade.

In summary, directional reflectance properties of urban surfaces range from nearly perfect forward scattering materials to back scattering surfaces. Lambertian reflectance properties are unusual for most materials (Meister, 2000) and reflectance anisotropy affects image analysis. A correction of these effects based on the documented gradients may be required (for specific applications) and needs to be performed independently for different surface types (Schiefer et al., 2006).

7.3 APPLICATION CASE STUDIES

7.3.1 FIRE DANGER ASSESSMENT

In several regions worldwide, the need for improved maps for fire danger assessment is growing (Woycheese et al., 1997; Cohen, 2000; Medina, 2000). Detailed information about different materials and, in particular, where possible fuels are located is essential to preparing for an emergency response to wildland fires and for fire prevention. One important fuel type is NPV, which is usually dry, burns readily, and can act to propagate and spread fires (Roberts et al., 1999). Within the Santa Barbara urban area, an important roof type for fire risk assessment is wood shingle, which is easy to ignite and can spread fire through spotting (Cohen, 2000). Although wood shingles are now prohibited for new roofs, they remain common in many parts of Santa Barbara. A thorough fire risk assessment for the wildland–urban interface profits from a detailed map of existing wood shingle roofs.

Based on the spectra of wood shingle roofs (Figure 7.1), it is possible to use AVIRIS data with 4-m spatial resolution to map these roof types using a matched filter (ENVI image processing software). Matched filter analysis compares the spectrum in each image pixel to a known reference spectrum from a spectral library — in this case, from a representative wood shingle roof. If the spectral shape of a specific pixel and the reference spectrum match perfectly, the matched filter score would be 100; if they are absolutely dissimilar, the score would be 0.

Figure 7.6 shows results of matched filter analysis. It indicates that only a few areas have material characteristics similar to the reference wood shingle roof. The areas with high matched filter scores actually are wood shingle roofs. However, a few areas in the left part of the image show intermediate matched filter scores. The most significant false positives are open fields consisting of senesced grass in more

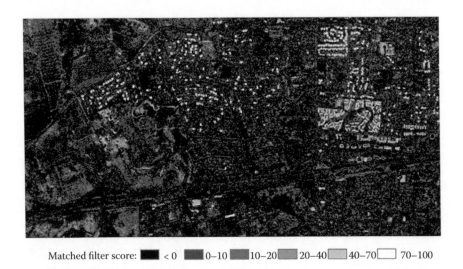

Matched filter score: ■ < 0 ■ 0–10 ■ 10–20 ▨ 20–40 ▢ 40–70 ▢ 70–100

FIGURE 7.6 Results of spectral matched filter analysis for wood shingle roofs from high-resolution AVIRIS data with 4-m spatial resolution.

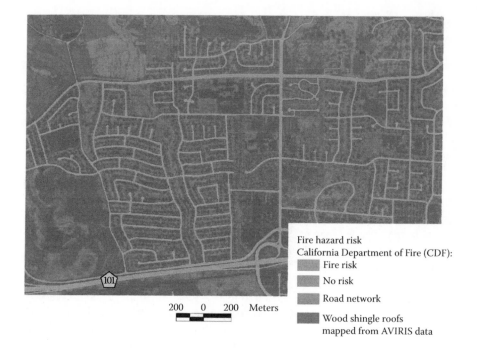

Fire hazard risk
California Department of Fire (CDF):
■ Fire risk
■ No risk
■ Road network
■ Wood shingle roofs
mapped from AVIRIS data

200   0   200   Meters

FIGURE 7.7 (Color Figure 7.7 follows page 240.) Distribution of wood shingle roofs in the Fairview region of the Santa Barbara urban area derived from AVIRIS high-resolution data with 4-m spatial resolution compared to the California Department of Fire risk map.

rural areas. Wood shingle roofs and senesced grasslands contain NPV and therefore their spectral signals are somewhat similar. However, using an image classification approach, it is possible to derive a detailed and accurate map of wood shingle roofs for the area (Herold et al., 2003a).

Figure 7.7 compares the location of these roofs, mapped from AVIRIS at high accuracy, to a California Department of Fire risk map. The map shows no risk for some urban areas with high density of wood shingle roofs that can serve as "bridges" for a fire from the surrounding rural land into the urban environment. Given the result of urban material mapping, the fire risk map should consider these newly found high-risk areas to increase the level of detail. Urban areas with higher fire hazard vulnerability should especially get this specific attention for fire hazard prevention and protection.

7.3.2 INDICATORS IN URBAN ECOLOGY

During the last decade, several indicators have been developed to describe hetero-geneous environmental conditions and processes that are, for example, driving forces in urban climatology and hydrology. One universal indicator, or "meta-indicator," that has proven its usefulness is surface imperviousness. Mapping impervious surfaces, con-sisting of roofs, roads, parking lots, and other materials, is a key input for understanding the urban hydrological system and for flood control (Schueler, 1994; Ridd, 1995).

Thus, imperviousness reflects direct and indirect impacts on manifold urban environmental conditions and serves as a target variable, among others, for urban and environmental planning and assessment.

Little progress has been made in impervious surface mapping using very high spatial resolution multispectral data, due to spectral limitations and the complicated geometry of urban surfaces. Imaging spectrometry data offer an alternative in mapping impervious surfaces on the basis of spectrally driven analysis concepts. Sensor limitations usually do not allow for spatial resolutions similar to multispectral cameras, but high-spectral and moderate-spatial resolution facilitate alternative analysis concepts (Segl et al., 2000; Heiden et al., 2001; Herold et al., 2003a).

Two subsets from a HyMap image acquired over Berlin, Germany, on July 30, 2003, at a GIFOV of 3.9 m were analyzed for mapping imperviousness. Prior to the analysis, the HyMap images were atmospherically corrected and georectified. Data include densely built-up, open residential, and industrial areas. A three-step analysis strategy has been adopted to respect the spectral complexity of urban areas.

In a first step, dark surfaces are separated and individually analyzed — a processing step that can hardly be performed on data with low spectral resolution. All pixels with average reflectance below 10% are masked and used as input for a minimum noise fraction (MNF) transformation. This procedure transforms the original reflectance feature space into optimally separable dimensions of MNF space. MNF components 1 to 12 are chosen as the basis for an unsupervised iterative clustering into 15 classes, which can easily be divided into water, shaded vegetation, and shaded sealed surfaces (Figure 7.8B). The remaining image is free of shaded areas and can be analyzed subsequently using a different approach, such as linear spectral mixture analysis (SMA). An endmember combination similar to the vegetation–impervious surface–soil model (Ridd, 1995) serves as the basis for distinguishing among photosynthetically active vegetation, soil, concrete, and asphalt (Figure 7.8C).

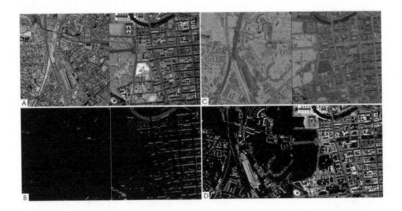

FIGURE 7.8 (Color Figure 7.8 follows page 240.) Processing steps of impervious surface mapping based on HyMap data from two areas in Berlin, Germany. A: original image data (red: 828 nm; green: 1647 nm; blue: 661 nm); B: separation of shaded areas (red) and water (blue); C: results from SMA (red: asphalt; green: vegetation; blue: soil); D:- mask of impervious surfaces with same band combination as in A.

More complex models are possible, but would also involve more ambiguities in the analysis scheme. Reference endmembers were derived from a spectral library of urban materials. Areas dominated by vegetation or pure soil are characterized well and can hence be identified with high accuracy. Given the spectral detail, an accurate differentiation between bright soils and concrete surfaces is easily achieved.

The third step comprised an MNF transformation of the remaining image proportions — soils with residual or dry vegetation cover and most impervious urban surfaces — and an iterative unsupervised clustering in 30 classes. They were labeled afterwards, and the three analysis masks comprising dark surfaces, unmixed vegetation, and soil, and the classified remaining parts of the image were categorized as pervious and impervious surfaces (Figure 7.8D).

A comparison of imperviousness values, derived from the image analysis, to data from the Urban Environmental Information System Berlin (UEIS) reveals a good correlation between both data sets (Figure 7.9). The UEIS offers block-wise information and it is hence only possible to compare results in an integrated manner. Average imperviousness values from the UEIS and from image analysis reveal a good correlation. The analysis is characterized by low residuals with a near 1:1 relationship.

7.3.3 Management of Transportation Infrastructure

Detailed and accurate information about road infrastructure is the foundation for management and planning of transportation assets. Because the cost of frequent, comprehensive inspection is high, many jurisdictions limit their surveys to major roads,

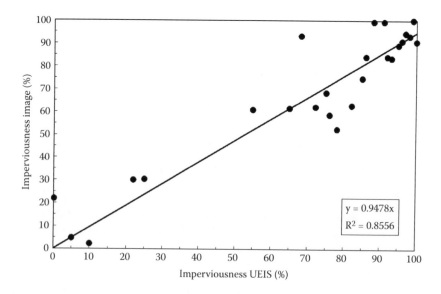

FIGURE 7.9 Comparison of UEIS imperviousness reference information and results from hyperspectral image analysis (values based on blocks of homogeneously structured urban regions).

and minor roads are surveyed in 3- or 4-year cycles. For this purpose, a number of survey technologies have been applied to road condition mapping. Recent advances in imaging spectrometry offer the potential of improved road quality mapping over large areas. In fact, there is spectral evidence for the aging and degradation processes of *in situ* asphalt pavements. As emphasized by diagram D in Figure 7.1, new asphalt pavements are dominated by hydrocarbon absorptions. Pavement aging and erosion of the asphalt mix results in a gradual transition from hydrocarbon to mineral absorption characteristics with a general increase in brightness and changes in distinct small-scale absorption features (Herold and Roberts, 2005).

The use of imaging spectrometry for detailed road surveys requires fine spatial resolution data on the order of 0.5- to 1-m spatial resolution. Herold and Roberts (2005) found a significant relationship with a hyperspectral band ratio. Band ratios are able to emphasize specific absorption features — in this case, the 830- and 490-nm iron oxide absorptions reflecting the presence of minerals. The image information was successfully used to estimate pavement health indicators (e.g., pavement condition index, PCI) from imaging spectrometry data (Figure 7.10).

The PCI patterns highlight recently paved roads with high values (blue colors). Road surfaces in poorer condition show lower PCI values (and higher variability yellow-green colors). Specific cracking patterns are revealed in the highlighted area of Fairview that point at a specific problem of the approach. Structural road damage (e.g., cracks) indicates a somewhat contrary spectral variation to asphalt deterioration.

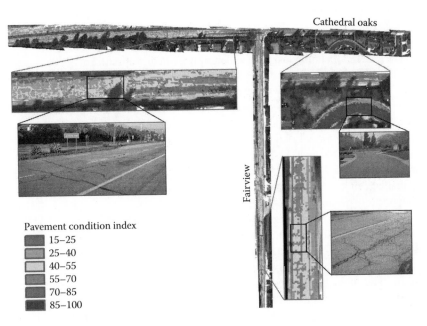

FIGURE 7.10 (Color Figure 7.10 follows page 240.) Spatial distribution of pavement condition index for selected asphalt roads in the Goleta, California, area derived from hyperspectral HyperSpectir data with 0.5-m spatial resolution using band ratios.

This and other problems remain to be resolved, but the case study shows the potential of imaging spectrometry for new and innovative avenues in transportation infrastructure surveys.

7.3.4 DISASTER RESPONSE

Imaging spectrometry has considerable potential to help in emergency disaster response. Potential applications of an imaging spectrometer include mapping thermal sources and environmental contaminants. Clark et al. (2001) demonstrated the potential of imaging spectrometry for mapping environmental contaminants in the aftermath of the World Trade Center attack. A particular concern there was the potential of widespread dissemination of potentially cancer-causing asbestiform dusts. However, other dusts originating from crushed urban materials are also of concern due to their potential to produce respiratory problems. In the case in which diagnostic absorptions are required to map such specific minerals, an imaging spectrometer may be the only practical means. The example study shown here is derived from an open file report published by the U.S. Geological Survey (Clark et al., 2001). The reader is encouraged to read the full report.

AVIRIS imaged the World Trade Center and environs four times between September 15 and 23, 2001. Because of time constraints for emergency relief efforts, the data were radiometrically calibrated, georectified, and corrected to surface reflectance as quickly as possible after image data were acquired, then analyzed at the USGS. For example, data acquired on September 16 were shipped to the USGS by September 17 and fully analyzed between September 17 and 19 using real-time feedback from field crews. Spectra from 33 dust samples were acquired in the vicinity of the two towers between September 17 and 19, 2001.

The spectral mapping task was to determine whether abestiform dust was present at harmful levels based on the presence of chrysotile-specific absorption features in image data. Chrysotile (an asbestiform mineral) contains prominent iron absorptions in the VNIR and prominent vibrational bands in the SWIR. The latter include a strong liquid water band centered at 1940 nm, a weaker water/OH absorption at 1450 nm, and a triplet of diagnostic absorptions for gypsum between 1420 and 1540 nm (Clark et al., 2001). The spectral subset of several dusts was determined to have variable amounts of chrysotile asbestos and thus allowed derivation of related material maps from the AVIRIS data (using the Tetracorder program; Clark et al., 2003) to support the disaster relief efforts.

A different study by Bhaskaran et al. (2004) used HyMap data in connection with geographic information systems (GISs) to develop a vulnerability map for decision making in hailstorm disaster emergency operations. There is a high correlation of damaged roofs to the material composition of the roofing material, which in turn determines their resistance to hailstones. The study used a spectral library of surface materials from Sydney, Australia, urban areas. The spectral angle-mapper method allowed for automated comparison between image spectra and laboratory spectra and was used to map distribution of roofing materials with different resistances to hailstones. Using a GIS environment, several types of cartographic data, such as special hazard locations, population density, data about less mobile people, and the street network, were integrated with the material map to derive spatially explicit risk and vulnerability data.

7.3.5 Urban Climate, Air Quality, and Human Health

Similarly to the examples presented previously, other applications require precise measurements of the physical properties of urban environments. For example, the materials present in urban areas affect solar radiation absorption, surface temperature, evapotranspiration, water vapor, and pollutant concentration, which are directly linked to urban climate, air quality, and human well-being (Clarke, 1972; Roth et al., 1989). In that context, hyperspectral data can be used to link between short-wave reflection and long-wave emission properties of surfaces to better understand surface energy fluxes and urban heat island effects (Medina, 2000; Ozawa et al., 2004). Studies used day- and night-time imaging spectrometry data acquired over urban areas (Barducci et al., 2003; Ozawa et al., 2004).

There are examples of how imaging spectrometry can be used to map aerosol concentrations to support air quality studies over urban areas. In this case, absorption features sensitive to atmospheric aerosol content (e.g., at 550 nm) are suitable to deriving such concentrations. One of the challenges for this technique to work over land is the terrestrial background signal. A solution is to use the SWIR spectral information (low aerosol radiative influence) to estimate terrestrial reflectance for spectral regions where the aerosol signal is the strongest and allow for quantitative aerosol retrieval (Bojinski et al., 2004).

These case studies provide a flavor of imaging spectrometry applications in urban areas. Many other applications of urban imaging spectrometry have been excluded here, including urban health (Wilson, 2002; Jacquez et al., 2002), urban climatology, ecology, pollution, etc. (Curran, 2001; Brando and Dekker, 2003; Xiao et al., 2004). Thus, the field of imaging spectrometry applications is wide and, certainly, many of them have yet to evolve.

7.4 SUMMARY

Remote sensing is currently undergoing a renaissance for improved urban area mapping, in large part due to the availability of sensors that provide unprecedented spatial and spectral detail. Urban areas are particularly challenging for remote sensing because of the requirement of fine spatial resolution, high diversity of materials, complex lighting geometries (and bidirectional reflectance properties of many surfaces), and potentially variable atmospheric properties due to airborne pollutants. By sampling a large number of wavelengths, imaging spectrometers are particularly qualified to map urban materials at high accuracies.

In this chapter, we have provided a brief overview of some important aspects of urban imaging spectrometry. We presented basic spatial and spectral concepts with a focus on how they apply to urban areas. We illustrated important aspects of urban spectra using a field spectral library. Issues of spectral sensor requirements and effects of bidirectional reflectance properties were outlined and discussed in the context of urban mapping. We concluded the chapter with several examples of applications, including fire risk mapping, urban hydrology and climatology, transportation management, disaster response, and others.

Overall, applying imaging spectrometry in urban environments has great potential if specific challenges are known and considered in data acquisition and analysis. A variety of applications can directly benefit from the unique detail of information provided by such data. In fact, imaging spectrometry is a rapidly evolving field in which new sensors and new analysis methods are continually developed. Urban remote sensing represents a rather new and exciting application for this technology.

Using multiple sensors for analyzing urban environments also has considerable potential. One very promising technology is LIDAR, which provides a three-dimensional description of the urban landscape. Similarly, analyzing geometric and spectral very high resolution data offers new opportunities in urban mapping applications. The strength of hyperspectral data analysis certainly lies in the possibilities of optimizing feature space in a thematically driven way, while high-detail geometric data may contribute with their strength concerning image structures analysis. It can be anticipated that future analysis concepts will focus on integrating different sensor types.

REFERENCES

Barducci, P., Marcoionni, I., Pippi, and Poggesi, M., 2003. Effects of light pollution revealed during a nocturnal aerial survey by two hyperspectral imagers. *Appl. Opt.*, 42(21), 159.

Beisl, U., 2001. New method for correction of bidirectional effects in hyperspectral images. *Proc. 8th Int. Symp. Remote Sensing* (SPIE), Toulouse, France.

Ben-Dor, E., Levin, N., and Saaroni, H., 2001. A spectral based recognition of the urban environment using the visible and near-infrared spectral region (0.4 to 1.1 m). A case study over Tel-Aviv. *Int. J. Remote Sensing*, 22(11), 2193–2218.

Bhaskaran S., Datt, B., Forster, B., Neal, T., and Brown, M., 2004. Integrating imaging spectroscopy (445 to 2543 nm) and geographic information systems for postdisaster management: a case of hailstorm damage in Sydney. *Int. J. Remote Sensing*, 25(13), 2625–2639.

Bojinski, S., Schläpfer, D., Schaepman, M., Keller, J., and Itten, K.I., 2004. Aerosol mapping over land with imaging spectroscopy using spectral autocorrelation. *Int. J. Remote Sensing*, 25(22), 5025–5047.

Brando, V.E. and A.G. Dekker, 2003. Satellite hyperspectral remote sensing for estimating estuarine and coastal water quality. *IEEE Trans. Geosci. Remote Sensing*, 41, 1–10.

Clark, R.N., 1999. Spectroscopy of rocks and minerals and principles of spectroscopy, in Rencz, A.N. (Ed.), *Manual of Remote Sensing*, John Wiley & Sons, New York, chap. 1, 3–58.

Clark, R.N., Green, R.O., Swayze, G.A., Meeker, G., Sutley, S., Hoefen, T.M., Livo, K.E., Plumlee, G., Pavri, B., Sarture, C., Wilson, S., Hageman, P., Lamothe, P., Vance, J.S., Boardman, J., Brownfield, I., Gent, C., Morath, L.C., Taggart, J., Theodorakos, P.M., and Adams, M., 2001. Environmental studies of the World Trade Center area after the September 11, 2001 attack. U.S. Geological Survey, open file report OFR-01-0429, URL: http://speclab.cr.usgs.gov/wtc/ (access: February 2004).

Clark, R.N., Swayze, G.A., Livo, K.E., Kokaly, R.F., Sutley, S.J., Dalton, J.B., McDougal, R.R., and Gent, C.A., 2003. Imaging spectroscopy: Earth and planetary remote sensing with the USGS Tetracorder and expert systems. *J. Geophys. Res.*, 108(E12), 5131, doi:10.1029/2002JE001847.

Clarke, J.F., 1972. Some effects of the urban structure on heat mortality. *Environ. Res.*, 5, 93–104.

Cloutis, E., 1989. Spectral reflectance properties of hydrocarbons: remote-sensing implications. *Science*, 4914, 165–168.

Cocks, R., Jensen, A., Stewart, I., Wilson, and Sheilds, T., 1998. The HyMap airborne hyperspectral sensor: the system, calibration and performance. Paper presented at 1st EARSEL Workshop on Imaging Spectroscopy, Zurich, October, 1998.

Cohen, J.D., 2000. Preventing disaster, home ignitability in the wildland–urban interface. *J. Forestry*, 98, 15–21.

Curran, P.J., 2001. Imaging spectrometry for ecological applications. *Int. J. Appl. Earth Obs. Geoinf.*, 3(4), 305–312.

Deering, D.W., 1989. Field measurements of bidirectional reflectance, in Asrar, G. (Ed.), *Theory and Applications of Optical Remote Sensing*, John Wiley & Sons, New York, 734 pp.

Goetz, A.F.H., Vane, G., Solomon, J.E., and Rock, B.N., 1985. Imaging spectrometry for earth remote sensing. *Science*, 228, 4704, 1147–1153.

Green, R.O., Eastwood, M.L., Sarture, C.M., Chrien, T.G., et al., 1998. Imaging spectroscopy and the airborne visible infrared imaging spectrometer (AVIRIS). *Remote Sensing Environ.*, 65(3), 227–248.

Heiden, U., Rössner, S., and Segl, K., 2001. Potential of hyperspectral HyMap data for material-oriented identification of urban surfaces. *Remote Sensing Urban Areas*, Regensburg, Germany, 69–77.

Hepner, G.F., Houshmand, B., Kulikov, I., and Bryant, N., 1998. Investigation of the integration of AVIRIS and IFSAR for urban analysis. *Photogrammetric Eng. Remote Sensing*, 64(8), 813–820.

Herold, M., Gardner, M., and D.A. Roberts, 2003a. Spectral resolution requirements for mapping urban areas. *IEEE Trans. Geosci. Remote Sensing*, 41(9), 1907–1919.

Herold, M., Liu X., and Clarke, K.C., 2003b. Spatial metrics and image texture for mapping urban land use. *Photogrammetric Eng. Remote Sensing*, 69(9), 991–1001.

Herold, M., Roberts, D., Gardner, M., and Dennison, P., 2004. Spectrometry for urban area remote sensing — development and analysis of a spectral library from 350 to 2400 nm. *Remote Sensing Environ.*, 91(3–4), 304–319.

Herold, M. and D.A. Roberts, 2005. Spectral characteristics of asphalt road aging and deterioration: implications for remote sensing applications. *Appl. Opt.*, 44, 20, 4327–4334.

Herold, M. and Roberts, D.A., 2006. Multispectral satellites — imaging spectrometry — LIDAR: spatial–spectral trade-offs in urban mapping, *Int. J. Geoinf.*, 2, 1, 1–14.

Hoerig, B., Kuehn, F., Oschutz, F., and Lehmann, F., 2001. HyMap hyperspectral remote sensing to detect hydrocarbons. *Int. J. Remote Sensing*, 22(8), 1213–1422.

Hostert, P. and A. Damm, 2003. Sensitivity analysis of multi-source spectra from an urban environment. 3rd EARSeL Workshop on Imaging Spectroscopy, Herrsching, 13th–16th May 2003, 215–219.

Jacquemoud, S., Bacour, C., Poilvé, H., and Frangi, J.-P., 2000. Comparison of four radiative transfer models to simulate plant canopies reflectance: direct and inverse mode. *Remote Sensing Environ.*, 74, 471–481.

Jacquez, G.M., Marcus, W.A., Aspinall, R.J., and Greiling, D.A., 2002. Exposure assessment using high-spatial resolution hyperspectral (HSRH) imagery. *J. Geogr. Syst.*, 4, 1–14.

Jensen, J.R., Bryan, M.L., Friedman, S.Z., Henderson, F.M., Holz, R.K., Lindgren, D., Toll, D.L., Welch, R., and Wray, J.R., 1983. Urban/suburban land use analysis, in Colwell, R.N. (Ed.), *Manual of Remote Sensing*, 2nd ed, vol. II, 30, 1571–1661.

Jensen, J.R. and D.C. Cowen, 1999. Remote sensing of urban/suburban infrastructure and socioeconomic attributes. *Photogrammetric Eng. Remote Sensing*, 65(5), 611–622.

Kimes, D.S., 1983. Dynamics of directional reflectance factor distributions for vegetation canopies. *Appl. Opt.*, 22(9), 1364–1372.

Mackay L. and S. Barr, 2002. Assessing the potential of hyperspectral angular imaging of urban built form surfaces for the inference of fine-scale urban land cover information. *Proc. 22nd EARSeL Symp. Gen. Assembly*, Prague, Czech Republic.

Medina, M.A., 2000. Effects of shingle absorptivity, radiant barrier emissivity, attic ventilation flowrate, and roof slope on the performance of radiant barriers. *Int. J. Energy Res.*, 24(8), 665–678.

Meister, G., 2000. Bidirectional reflectance of urban surfaces, Ph.D. thesis, University of Hamburg, II. Institut für Experimentalphysik, 186 pp., URL: http://kogs-www.informatik.uni-Hamburg.de/PROJECTS/censis/publications.html (accessed September 2002).

Nicodemus, F.E., Richmond, J.C., Hsia, J.J., Ginsberg, I.W., and Limperis, T., 1977. Geometrical considerations and nomenclature for reflectance. U.S. Department of Commerce, National Bureau of Standards, Washington, D.C.

O'Meara, M., 1999. Reinventing cities for people and the planet. Washington, D.C., Worldwatch Institute, 68 pp., http://www.worldwatch.org/pubs/paper/147/, accessed September 2003.

Ozawa A., Madhavan, B.B., Okada, H., Mishra, K.K., Tachibana, K., and Sasagawa, T., 2004. Airborne hyperspectral and thermal information for assessing the heat island in urban areas of Japan. *Proc. 20th ISPRS Congr.*, Istanbul, Turkey. http://www.isprs.org/ist anbul2004/comm7/papers/9.pdf.

Pinty, B., Widlowski, J.-L., Gobron, N., Verstraete, M.M., and Diner, D.J., 2002. Uniqueness of multiangular measurements — part I: an indicator of subpixel surface heterogeneity from MISR. *IEEE Trans. Geosci. Remote Sensing*, 40(7), 1560–1573.

Price, J.C., 1997. Spectral band selection for visible-near infrared remote sensing: spectral–spatial resolution trade-offs. *IEEE Trans. Geosci. Remote Sensing*, 35(5), 1277–1285.

Ridd, M.K., 1995. Exploring a V-I-S (vegetation–impervious surface–soil) model for urban ecosystem analysis through remote sensing: comparative anatomy for cities. *Int. J. Remote Sensing*, 16, 2165–2185.

Roberts, D.A., Smith, M.O., and Adams, J.B., 1993. Green vegetation, nonphotosynthetic vegetation, and soils in AVIRIS data. *Remote Sensing Environ.*, 44 (2–3), 255–269.

Roberts, D.A., Dennison, P.E., Morais, Gardner, M.E., Regelbrugge, J., and Ustin, S.L., 1999. Mapping wildfire fuels using imaging spectrometry along the wildland urban inter-face. *Proc. 1999 Joint Fire Scie. Conf. Workshop*, June 17–19, Boise, Idaho, 212–223. http://jfsp.nifc.gov/conferenceproc/Ma-08Robertsetal.pdf.

Roberts, D.A. and M. Herold, 2004. Imaging spectrometry of urban materials, in King, P., Ramsey, M.S., and G. Swayze (Eds.), *Infrared Spectroscopy in Geochemistry, Exploration and Remote Sensing*, Mineral Association of Canada, short course series vol. 33, London, Ontario, 155–181. URL: http://www.ncgia.ucsb.edu/ncrst/research/ pavementhealth/urban/imaging_spectrometry_of_urban_materials.pdf.

Roessner, S., Segl, K., Heiden, U., and Kaufmann, H., 2001. Automated differentiation of urban surfaces based on airborne hyperspectral imagery. *IEEE Trans. Geosci. Remote Sensing*, 39, 7, 1525–1532.

Roth, M., Oke, T.R., and Emery, W.J., 1989. Satellite-derived urban heat islands from three coastal cities and the utilization of such data in urban climatology. *Int. J. Remote Sensing*, 10, 1699–1720.

Schiefer, S., Hostert, P., and Damm, A., 2006. Correcting brightness gradients in hyperspectral data from urban areas. *Remote Sensing Environ. 101*, 25–37.

Schueler, T.R., 1994. The importance of imperviousness. *Watershed Protection Techniques*, *1*, 3, 100–111.

Segl, K., Rößner S., and Heiden, U., 2000. Differentiation of urban surfaces based on hyperspectral image data and a multitechnique approach. *Proc. IEEE IGARSS 2000* (IGARSS, Honolulu), 1600–1602.

Swayze, G.A., Clark, R.N., Goetz, A.F.H., Chrien, T.G., and Gorelick, N.S., 2003. Effects of spectrometer band pass, sampling, and signal-to-noise ratio on spectral identification using the Tetracorder algorithm. *J. Geophys. Res.*, 108(E9), 5105, doi: 1029/ 2002JE001975, 30 pp.

van der Meer, F.D., 2004. Analysis of spectral absorption features in hyperspectral imagery. *Int. J. Appl. Earth Obs. Geoinf. JAG*, 5(2004), 55–68.

Welch, R., 1982. Spatial resolution requirements for urban studies. *Int. J. Remote Sensing*, 3(2), 139–146.

Wilson, M.L., 2002. Emerging and vector-borne diseases: role of high spatial resolution and hyperspectral images in analyses and forecasts. *J. Geogr. Syst.*, 4, 31–42.

Woodcock, C.E. and A.H. Strahler, 1987. The factor scale in remote sensing. *Remote Sensing Environ.*, 21, 311–332.

Woycheese, J.P., Pagni, P.J., and Liepmann, D., 1997. Brand lofting above large-scale fires. *Proc. 2nd Int. Conf. Fire Res. Eng. (ICFRE2)*, August 3–8, 1997, Gaithersburg, MD, 137–150.

Xiao, Q., Ustin, S.L., and McPherson, E.G., 2004. Using AVIRIS data and multiple-masking techniques to map urban forest trees species. *Int. J. Remote Sensing*, 25(24), 5637–5654.

Part III

Urban Land Dynamics

8 Urban Land Use Prediction Model with Spatiotemporal Data Mining and GIS

Weiguo Liu, Karen C. Seto, Zhanli Sun, and Yong Tian

CONTENTS

8.1 INTRODUCTION

Huge amounts of data have posed great challenges to traditional data analysis methods for information and knowledge extraction. Data mining (application of low-level algorithms for revealing hidden information in a database) (Klosgen and Zytkow, 1996) has emerged as a new research field and a new technology in the last decade. Data mining represents the interdisciplinary research of several fields, including machine learning, neural network, statistics, database, visualization, and information theory (Koperski et al., 1996).

Similar to the fields using relational and transactional databases, geography has changed from a data-poor and computation-poor to a data-rich and computation-rich environment. Traditional spatial analytical methods cannot be used to discover hidden information from huge amounts of spatially related data sets. Spatial data mining has attracted attention in recent research (Miller and Han, 2001). It refers to the discovery and extraction of implicit information, spatial relationships, or spatial distribution patterns from spatial databases (Koperski et al., 1996). Spatial data mining can be used to understand spatial data and discover spatial relationships and relationships between spatial data and nonspatial data, etc. It has wide applications in geographical information systems (GISs), remote sensing, and many other areas related to spatial data.

Among different spatial data mining algorithms, spatial classification aims to assign an object to a class from a given set of classes based on spatial and nonspatial attribute values of the object. Decision tree classifiers (Ester et al., 1997) and neural networks (Gopal et al., 2001; Liu et al., 2001) have been widely used as base classification methods that can handle spatial and nonspatial data. Different from traditional classification methods, spatial classification will explicitly involve spatially related features or metrics (Koperski et al., 1996; Ester et al., 2001).

Urban expansion will be one of the biggest environmental challenges of the 21st century. Much research has been done to study the urban land-use change aiming at monitoring urban expansion and explaining the location and trajectories of urban land-use change across space and time (Seto and Liu, 2003; Seto et al., 2002). Recently, future urban land-use change has become a very common and geographic interest of diverse groups of scientists. Being able to predict the growth of cities will be of enormous benefit for urban planners and policymakers who must provide infrastructure services to a huge number of new urban residents.

As a complex system, urban land-use development hides spatial and temporal patterns that cannot be detected directly by human beings from huge amounts of historic urban growth data. However, through extracting informative spatial features over historic urban growth data, spatial data mining provides powerful tools for automatically detecting explicit urban development patterns across space and time. The urban growth prediction model based on spatial data mining techniques can apply hidden urban development patterns automatically to predict future urban expansion based on current land-use conditions. In this chapter, we present an ART-MMAP neural network-based spatial classification method to process multitemporal urban growth data for predicting future urban expansion.

8.2 URBAN ANALYSIS REVIEW

Urban growth and the resultant sprawling patterns of U.S. communities are causing social, economic, and environmental strain (Schmidt, 1998), which raise great concern from the federal government to local public sectors. However, urban growth is a complex issue that can best be understood through dynamic modeling. Although land-use change is generally considered as the signature of urban growth, land-use change modeling has been the focus of urban growth research. Very recently, computer-based urban system simulation models have been increasingly employed to forecast and evaluate land-use change (Batty and Xie, 1994; Engelen et al., 1995; Landis, 1994).

This spatial dynamic modeling approach enables planners to view and analyze the future of their decisions and policies even before they are put in action. Therefore, it can help improve fundamental understanding and communication of the dynamics of land-use transformation and the complex interactions between urban change and sustainable systems (Deal, 2001). Nowadays, spatial dynamic modeling techniques are considered essential to a planning support system (PSS) after being overshadowed by GIS applications since the 1980s (Hopkins, 1999; Kammeier, 1999).

To date, however, spatial dynamic urban modeling is still in its infancy. Due to the extreme complexity of an urban system, few, if any, models have been built that truly represent the dynamics of urban growth and can provide consistent results with what we know about such changes (Maria de Almeida et al., 2003). Consequently, such models are barely operational and rarely used to assist urban planning practice. Nevertheless, cellular automata (CA) and agent-based models representing a different approach to the traditional top-down approaches are emerging as promising model tools. Agent-based models possess characteristics analogous to those of cellular automata. Most agent-based models can be considered as extended CA models with "smart" cells, which can communicate with other cells and environments, make decisions based on information received, and sometimes move across the space. Therefore, no essential difference exists between CA models and agent-based models. Here, we focus our reviews on CA-based land-use change models.

Cellular automata are discrete dynamical systems whose behavior is completely specified in terms of a local relation. They are embedded with a spatial dynamic feature, which makes CA a natural tool for spatial modeling. CA application in geographic modeling dates back to the spatial diffusion model developed by Hagerstrand (1967), which was essentially a stochastic CA, although he did not use the term CA. Geographer Tobler (1979) first defined CA as geographical models, although he believed that some CA are too simple to be usefully applied. Later, the implications of CA for geographic modeling, including advantages and theoretical obstacles of applying CA to geographic modeling, were explored theoretically (Batty et al., 1997; Couclelis, 1985, 1987, 1997). CA are very appealing to geographic modelers because

The CA-based model is simple and intuitive, yet capable of simulating self-organizing complex systems.

The naturally born spatial dynamic feature enables modeling spatial dynamic systems in extreme spatial detail and spatial explicitness.

The cellular structure of CA has a natural affinity with the raster data format of remote sensing images and the GIS grid map. The CA model can be easily integrated with GIS through generalization of map algebra (Takeyama and Couclelis, 1997).

The bottom-up approach of CA provides a new strategy of geographic modeling.

The CA-based model is a computational model running in parallel, which fits the high-performance geocomputation.

Since its development, CA application in geography has been experiencing exponential growth, especially in urban land-use simulation. Batty was one of the earliest geographers to sketch the general framework of CA-based urban models (Batty and Xie, 1994). An integrated platform, named DUEM, designed for geographic CA exploration was also developed by Batty and his group (Batty et al., 1999). Engelen used CA to model urban land-use dynamics to forecast climate change on a small island setting (Engelen et al., 1995). Wu presented a model that also included user decisions to determine model outcomes (Wu and Webster, 1998). White's St. Lucia model (White and Engelen, 1997) is an example of high-resolution CA modeling of urban land-use dynamics and an attempt to use the standard nonspatial models of regional economics and demographics, as well as a simple model of environmental change for predicting demand for future agricultural, residential, and commercial/industrial land uses. An urban growth model of the San Francisco Bay Area (Clarke and Gaydos, 1998) is another example of using relatively simple rules in the CA environment to simulate urban growth patterns. Li and Yeh integrated neural network and CA in a GIS platform and successfully applied it to urban land-use change simulation in Guangdong, China (Li and Yeh, 2002).

Although a large number of models have been proposed and built over the last 20 years, the CA-based land-use modeling technique is still far from mature. Despite the flexibility of the CA approach, limitations remain (Torrens and O'Sullivan, 2001). The hypothetical urban forms emerging from CA models with surprisingly simple local transition rules are certainly plausible. However, an urban system evolves in a much more complex way in reality. The current CA-based urban models are just too simple to capture the richness of urban systems. Consequently, very few CA models are operational or are used as productive tools to support regional planning practice.

To build useful models, modelers try to extend the concept of CA and also integrate a diversity of models, such as traditional regional social-economic models (White and Engelen, 1997; Wu and Martin, 2002). Instead of using CA-based models in this chapter, we present a neural network-based model to learn urban growth patterns based on historical urban data and to predict future urban growth.

8.3 ART-MMAP NEURAL NETWORK-BASED PREDICTIVE MODEL

The adaptive resonance theory (ART) family of pattern recognition algorithms was developed by Carpenter and Grossberg (1991). ART is a match-based learning system, the major feature of which is its ability to solve the stability–plasticity dilemma or serial learning problem, in which successive training of a network interferes with previously acquired knowledge. Among the ART family models, fuzzy ARTMAP is a supervised learning system used widely in many fields. A comprehensive description of the model is detailed in Carpenter et al. (1992).

ART-MMAP, an extension of ARTMAP, decreases the effect of category proliferation in the testing process for mixture analysis (Liu et al., 2004). The ART-MMAP model keeps the learning process of the ARTMAP model and changes the testing process. During the testing process, ARTMAP selects a category to each test sample using the winner-take-all (WTA) rule. Instead of picking one winner, ART-MMAP selects winners (ART_a) based on one predefined threshold parameter: τ. Categories with activation value larger than the threshold value are selected. If none of them is selected, the WTA rule is activated. This winner selection strategy provides an enhanced interpolation function that is based on a weighted summation operator. The ART-MMAP model overcomes the limitation of class category of the ARTMAP model and increases prediction accuracy as well.

This study presents an ART-MMAP neural network-based spatial data mining method to learn urban growth patterns across space and time from historical urban growth data, and then to predict future urban expansion with the knowledge learned by this neural network. The model includes four main steps: spatial feature (predictor) extraction with GIS, neural network learning (model calibration), neural network testing (model validation), and future urban growth prediction.

Spatial feature extraction. Similar to other kinds of urban simulation or prediction models, the proposed model needs at least two time period (T_0 and T_1) land-use maps of the study area and relative land-use predictors (factors). All related spatial data are imported into and managed by GIS. Then the spatial features/predictors are extracted with GIS. Each cell of the study area is represented by a vector consisting of extracted spatial features/predictors and a class label.

Neural network learning. This procedure is also called model calibration and assigns values to each parameter of the model. For the purpose of learning, the proposed model selects a small subset of the whole data set. Each sample's spatial vector, consisting of land-use predictors and class label (nonurban or urbanization), is fed to the ART-MMAP neural network. Then this network automatically processes the sample data set, generalizes the urban growth knowledge, and implicitly stores the land-use type transition rules in its network architecture. Compared to CA models, the calibration process of the proposed model is simple, fast, and intervention free. Also, the functional relationship between the land-use change and its spatial predictor variables/features may not be a simple linear one. The ART-MMAP neural network-based model can capture complex nonlinear patterns. Once the learning is done, the learned neural network can be applied to any other time period land-use data (i.e., T_1) to predict future urban growth (T_2).

Neural network testing. This procedure is called model validation, which is an important step in every land-use model (Pontius et al., 2004). In land-use change modeling, there is a short of uniform method of model assessment. To quantify the predictive performance, Pontius and Malanson (2005) proposed several approaches, including separating calibration from validation and more useful accuracy assessment methods. The proposed model selects portion sample pixels of the whole study area for learning (calibration process) and the whole study area or another group of sample data can be used for testing (validation process). At space level, this method implements the separation of model validation from model calibration.

Pontius and Malanson (2005) strongly suggested using the entire landscape for model accuracy assessment. If accuracy assessment only considers the truly urbanized pixels, it may encourage the model to predict urban growth everywhere, which introduces a large amount of falsely predicted urban areas and thus causes misinterpretation of the model's predictive power. To avoid bias or misleading of prediction accuracy, this model uses the entire data set, including urbanized and nonchange pixels, to assess the model's accuracy at pixel level. A traditional pattern recognition accuracy assessment method, error matrix, is introduced in this model to measure the model's prediction performance quantitatively. The error matrix not only shows the correctly predicted urban pixels, but also lists the wrongly predicted urban pixels. This method fairly illustrates the model's prediction accuracy and is simple and straightforward.

Urban growth prediction. Unlike CA models, the ART-MMAP neural network-based predictive model does not need many iterations to calculate urban transition probability for each cell, although modelers can run as many iterations as they want to view different scenarios. With the learned neural network and spatial vector extracted based on the second time period (T_1) land-use map and land-use drivers, the ART-MMAP-based model directly generates a likelihood of urban land-use transition for the future (T_2) within one iteration. Thus, this model can predict the location and amount of future urban growth.

8.4 DATA

8.4.1 STUDY AREA DESCRIPTION

Spanning parts of the states of Missouri and Illinois on both sides of the Mississippi River, the great St. Louis metropolitan region (Figure 8.1) includes ten counties. This area is about 120 miles from east to west and about 90 miles from north to

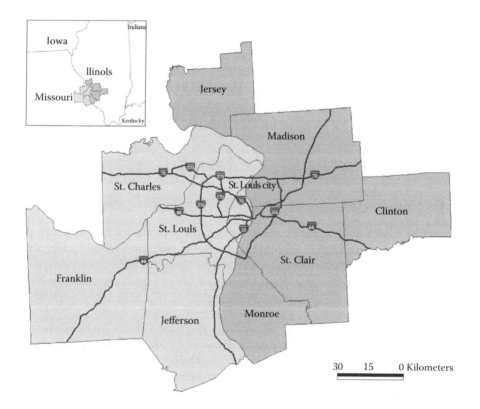

FIGURE 8.1 St. Louis metropolitan region.

south. It accounts for a little more than 30 million grid cells at 30- × 30-m spatial resolution. Like most other older metropolitan regions, St. Louis faces a great challenge of sustainable growth. With relatively slow population growth — even negative growth in its urban core — the city is continuing to sprawl. The St. Louis metropolitan region is already the third largest in the amount of land that it covers, but ranks 14th in terms of population. Under such conditions, prediction of and planning for urban growth become very important for the St. Louis region.

8.4.2 LAND-USE FACTORS

To simulate urban growth and then make a prediction, we need several important spatially related features extracted from land-use factors (also called land-use drivers) consisting of social, economic, transportation, and biophysical factors affecting land-use change. In reality, numerous factors affect urban land-use change more or less. Apparently, it is impossible to incorporate all these factors into one land-use model. Some of the most significant factors, which can be varied in different study areas, are selected. In addition to nonspatial factors, including most of the social-economic factors, which define the regional demand, the following factors are incorporated into our model to describe the land-use transition possibility of each cell:

City attractor: a gravity model is used to simulate the city's attractiveness. An attractiveness index of each cell is calculated as:

$$A_{cities} = \sum_i Pop_i \Big/ TT_i \tag{8.1}$$

where A_{cities} is the attractiveness index of a cell; Pop_i is the population of city i; and TT_i is the travel time from a cell to city i.

Employment attractor: similar to city attractor, the employment attractor represents the attractiveness impact of employment centers.

Neighbor: this describes the number of urbanized cells in the neighborhood (i.e., 3×3 window). The new growth more likely happens near developed or developing neighborhoods. In this study, we used three different window sizes (3×3, 5×5, and 7×7) to explore the effects of neighbors on land-use transition to urban area.

DEM/slope: suitability of development differs on various degrees of slope.

Water proximity: this is concerned with distance to lakes or rivers.

Forest proximity: this is concerned with distance to a forest area.

Transportation proximity factors: these are travel times to transportation facilities:

Proximity to interstate ramps
Proximity to state highways
Proximity to major roads
Proximity to major road intersections

With the extracted spatial features, each cell of the study area was assigned one vector consisting of the spatial value of each factor, land-use type in 1992, and land-use type in 2000 (urban or nonurban land-use type). The vectors will be used to describe the urban growth pattern through space and time. For the purpose of learning, we first selected 15,931 samples from the whole study area with a sample ratio of 900:1. Then the whole research area was used as testing data.

8.5 RESULTS AND DISCUSSION

With this data set, the parameters of the ART-MMAP model were set as: $\rho_a = 0.9$, $\rho_b = 1.0$, and $\tau = 0.98$ (threshold value for selecting winning F_2 nodes). After learning 15,931 training samples, the ART-MMAP network was applied to the whole research area for prediction accuracy validation.

8.5.1 MODEL PERFORMANCE AND VALIDATION

The ART-MMAP network-based predictive model assesses the likelihood of urbanization by assigning individual pixels a score ranging from 0 to 100. Higher scores correlate to an increased probability of changing from another land-use type to urban. The predicted class label can be assigned through setting a score threshold for the model. Scores equal to or greater than the score threshold are flagged as urbanized pixels. The choice of score threshold determines the number of pixels to be predicted as urban. As the score threshold is lowered, the total number of pixels predicted as urban and the number of urban pixels predicted correctly increase. The performance of a predictive model is characterized by plots of the percentage of urban pixels detected vs. the false positive ratio. Here, the false positive ratio is calculated as the ratio between the number of nonurbanized samples classified as urban and the number of detected urbanized samples at a certain score threshold.

Figure 8.2 shows the relationship between the cut-off score threshold and the percentage of urbanized samples detected. For example, if we set up the threshold score for urbanization as 12, approximately 50% of urbanized pixels are detected. Figure 8.3 shows the relationship between the percentage of detected urbanized samples and the false positive ratio.

Combining Figure 8.2 and Figure 8.3, we can determine, at a certain score threshold, the percentage of detected urbanized pixels and the number of falsely classified nonurbanized pixels. To view the performance in a traditional way, the classification error matrix built at score threshold 12 is shown in Table 8.1. With a total prediction accuracy of 94%, the ART-MMAP-based data mining model successfully predicted 50% of urbanized pixels with false predictions of a similar number of nonurbanized cells (437,027). Changing the cut-off score threshold, we can obtain different accuracy. To detect more urbanized pixels, we may reduce the cut-off threshold score. However, this will introduce more misclassified nonurbanized pixels as urban.

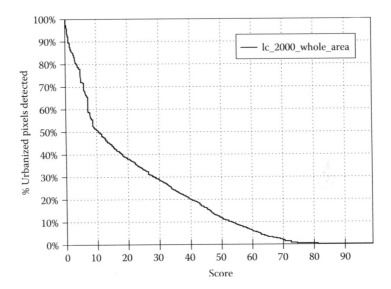

FIGURE 8.2 Relationship between cut-off score threshold and percentage of urbanized samples detected.

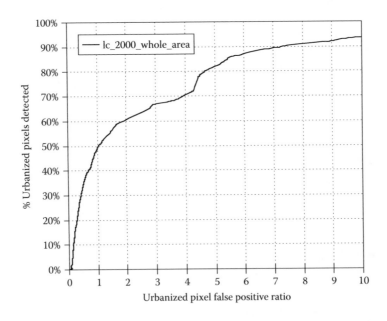

FIGURE 8.3 Relationship between percentage of detected urbanized samples and false positive ratio.

TABLE 8.1
Classification Error Matrix of Whole Study Area

Class	Urbanized	Nonurban	Total Pixels	Producer's Accuracy
Urbanized	402,520	396,410	798,930	0.504
Nonurban	437,027	12,632,090	13,069,117	0.967
Total pixels	839,547	13,028,500	Total accuracy: 0.94	
User's accuracy	0.479	0.970		

8.5.2 SIMULATION RESULTS OF YEAR 2008

Based on the land-use map of year 2000, we applied the trained ART-MMAP neural network model to predict urban growth of year 2008 to evaluate its predictive performance. Because transportation, social, and economic factors did not change much from 1992 to 2000, we kept the same value of these relative factors for each cell. The neighborhood value of urbanized pixels was recalculated with land use data of 2000.

Figure 8.4 displays the predicted urban growth area in 2008. The overall spatial pattern of the projected urban growth is quite reasonable. Most growth takes place around a city's peripheral region and is clustered around highway and major road intersections. Those areas are under growth pressure according to the local planners.

The transition statistics (in Table 8.2) show the number of pixels of each nonurban class changing to urban area from 2000 to 2008. Approximately 5% of herbaceous planted area changes into urban area, which is the majority of the urbanized pixels. The land-use types barren and forested upland are two other important types changing into urban.

TABLE 8.2
Land-Use Transition Statistics (2008)

Class Name	Class ID	Total Pixels	Pixels Changed	Change %
Water	1	404,567	3,141	0.78
Urban	2	2,650,997	0	0.00
Barren	3	102,742	19,161	18.65
Forested upland	4	4,151,988	87,335	2.10
Shrubland	5	11	0	0.00
Herbaceous upland natural/seminatural vegetation	6	161,269	7,809	4.84
Herbaceous planted/cultivated	7	7,690,045	378,031	4.92
Wetland	8	578,757	2,132	0.37

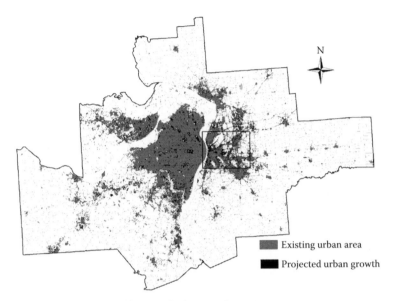

Existing urban area

Projected urban growth

(a) Regional urban growth overview

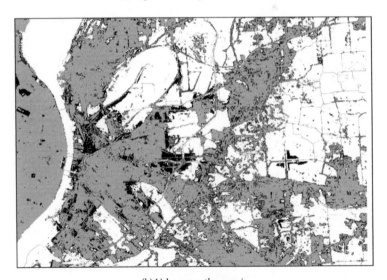

(b) Urban growth zoom in

FIGURE 8.4 Predicted urban growth area in 2008.

8.6 CONCLUSION

In the changing poor-data to rich-data environments in the field of geography, spatial data mining has become more interesting to many researchers. In this chapter, we presented an ART-MMAP neural network-based spatiotemporal data mining method to simulate future urban expansion and predict urban growth. With training based on multitemporal urban growth data, the ART-MMAP model can automatically predict the probability of urbanization of each pixel in the near future. Because the prediction is score based, we can get different urban expansion maps by setting different cut-off score thresholds.

Considering the goal of urban prediction, the prediction accuracy of the St. Louis data set is pretty good, although the model only detected 50% of urbanized pixels and also misidentified a similar amount of nonurban pixels. Realistically, no model will very accurately predict urban growth of the future. The predicted urbanized area with this model will provide a base probability map for urban planning. The goal of this model is to help planners answer "what if" questions by implementing planning scenarios. Exploring these scenarios will significantly enhance planners' insights into future land use and its impact. Although this model probably will not make decisions for planners or make them smarter, it certainly will help them make smarter decisions.

In this research, the training and testing data set were selected from the same time periods (1992 to 2000). To understand model performance more accurately, the trained ART-MMAP network with one time period of data (1992 to 2000) will be applied to a later time period (i.e., 2000 to 2004) for performance validation in future research. Also, integration of the prediction model and cellular automata-based models or agent-based models will be more interesting for realistic urban planning. It may also shed some light on urban simulation and urban modeling research.

ACKNOWLEDGMENT

We would like to thank the LEAM team at the University of Illinois at Urbana–Champaign for providing the St. Louis data set.

REFERENCES

Batty, M. and Xie, Y., 1994. From cells to cities, *Environ. Plann. B — Plann. Design*, 21, S31–S38.

Batty, M., Couclelis, H., and Eichen, M., 1997. Urban systems as cellular automata, *Environ. Plann. B — Plann. Design*, 24, 159–164.

Batty, M., Xie, Y., and Sun, Z., 1999. Modeling urban dynamics through GIS-based cellular automata, *Computer, Environ. Urban Syst.*, 23, 205–233.

Carpenter, G. and Grossberg, S., 1991. *Pattern Recognition by Self-Organizing Neural Networks*. Cambridge, MA: MIT Press.

Carpenter, G., Grossberg, S., Markuzon, S., Martens, N., Reynolds, J., and Rosen, D., 1992. Fuzzy ARTMAP: a neural network architecture for incremental supervised learning of analog multidimensional maps, *IEEE Trans. Neural Networks*, 3, 698–713.

Clarke, K.C. and Gaydos, L.J., 1998. Loose coupling a cellular automaton model and GIS: long-term urban growth prediction for San Francisco and Washington/Baltimore, *Int. J. Geogr. Inf. Sci.*, 12, 699–714.

Couclelis, H., 1985. Cellular worlds — a framework for modeling micro–macro dynamics, *Environ. Plann. A*, 17, 585–596.

Couclelis, H., 1987. Cellular dynamics — how individual decisions lead to global urban change, *Eur. J. Operational Res.*, 30, 344–346.

Couclelis, H., 1997. From cellular automata to urban models: new principles for model development and implementation, *Environ. Planning B-Planning Design*, 24, 165–174.

Deal, B., 2001. Ecological urban dynamics: the convergence of spatial modeling and sustainability, *Building Res. Inf.*, 29, 381–393.

Engelen, G., White, R., Uljee, I., and Drazan, P., 1995. Using cellular automata for integrated modeling of socio-environmental systems, *Environ. Monitoring Assessment*, 34, 203–214.

Ester, M., Kriegerl, H.P., and Sander, J., 1997. Spatial data mining: a database approach, *Proc. 5th Int. Symp. Large Spatial Databases*, Berlin, Germany, 47, 66.

Ester, M., Kriegerl, H.P., and Sander, J., 2001. Algorithms and applications for spatial data mining, in Miller, H. and Han, J. (Eds.), *Geographic Data Mining and Knowledge Discovery*. London: Taylor & Francis, 161–187.

Gopal, S., Liu, W., and Woodcock, C., 2001. Visualization based on the fuzzy ARTMAP neural network for mining remotely sensed data, in Miller, H. and Han, J. (Eds.), *Geographic Data Mining and Knowledge Discovery*. London: Taylor & Francis, 315–336.

Hagerstrand, T., 1967. *Innovation Diffusion as a Spatial Process*. Chicago: University of Chicago Press.

Hopkins, L.D., 1999. Structure of a planning support system for urban development, *Environ. Plann. B — Plann. Design*, 26, 333–343.

Kammeier, H.D., 1999. New tools for spatial analysis and planning as components of an incremental planning-support system, *Environ. Plann. B — Plann. Design*, 26, 365–380.

Klosgen, W. and Zytkow, J.M., 1996. Knowledge discovery in databases terminology, in Fayyar, E.M., Piatetsky–Shapiro, G., Smyth, P., and Ulthurusamy, R. (Eds.), *Advances in Knowledge Discovery and Data Mining*. Cambridge, MA: MIT Press, 573–592.

Koperski, K., Adhikary J., and Han, J., 1996. Knowledge discovery in spatial databases: progress and challenges. Proc. SIGMID Workshop Res. Issues in Data Mining Knowledge Discovery, technical report 96-08. University of British Columbia, Vancouver, Canada.

Landis, J.D., 1994. The California urban futures model — a new generation of metropolitan simulation models, *Environ. Plann. B — Plann. Design*, 21, 399–420.

Li, X. and Yeh, A.G., 2002. Neural-network-based cellular automata for simulating multiple land use changes using GIS, *Int. J. Geogr. Inf. Sci.*, 16, 323–343.

Liu, W., Gopal, S., and Woodcock, C., 2001. Spatial data mining for classification, visualization and interpretation with ARTMAP neural network, in Grossman, R. (Ed.), *Data Mining for Scientific and Engineering Applications*. The Netherlands: Kluwer Academic Publishers, 205–222.

Liu, W., Seto, K., Wu, E., Gopal, S., and Woodcock, C., 2004. ART-MMAP: a neural network approach to subpixel classification, *IEEE Trans. Geosci. Remote Sensing*, 42(9), 1976–1983.

Maria de Almeida, C., Batty, M., Vieira Monteiro, A.M., Camara, G., Soares–Filho, B.S., Cerqueira, G.C., and Pennachin, C.L., 2003. Stochastic cellular automata modeling of urban land use dynamics: empirical development and estimation, *Computers, Environ. Urban Syst.*, 27, 481–509.

Miller, H. and Han, J., (2001). *Geographic Data Mining and Knowledge Discovery*. London: Taylor & Francis, 1–31.

Pontius, R.G., Huffaker, D., and Denman, K., 2004. Useful techniques of validation for spatially explicit land-change models, *Ecol. Modeling*, 179, 445–461.

Pontius, G. and Malanson, J., 2005. Comparison of the structure and accuracy of two land change models, *Int. J. Geogr. Inf. Sci.*, 19, 243–265.

Schmidt, C.W., 1998. The specter of sprawl, *Environ. Health Perspect.*, 106, 274–279.

Seto, K.C., Woodcock, C.E., Huang, X., Lu, J., and Kaufmann, R.K., 2002. Monitoring land-use change in the Pearl River Delta using Landsat TM, *Int. J. Remote Sensing*, 23(10), 1985–2004.

Seto, K.C. and Liu, W., 2003. Comparing ARTMAP neural network with the maximum-likelihood classifier for detecting urban change, *Photogrammetric Eng. Remote Sensing*, 69(9), 981–990.

Takeyama, M. and Couclelis, H., 1997. Map dynamics: integrating cellular automata and GIS through geo-algebra, *Int. J. Geogr. Inf. Sci.*, 11, 73–91.

Tobler, W., 1979. Cellular geography, in Gale, G. and Olsson, S. (Eds.), *Philosophy in Geography*. Dordrecht: Reidel, 379–386.

Torrens, P.M. and O'Sullivan, D., 2001. Cellular automata and urban simulation: where do we go from here? *Environ. Plann. B — Plann. Design*, 28, 163–168.

White, R. and Engelen, G., 1997. Cellular automata as the basis of integrated dynamic regional modeling, *Environ. Plann. B — Plann. Design*, 24, 235–246.

Wu, F. and Webster, C.J., 1998. Simulation of land development through the integration of cellular automata and multicriteria evaluation, *Environ. Plann. B — Plann. Design*, 25, 103–126.

Wu, F.L. and Martin, D., 2002. Urban expansion simulation of Southeast England using population surface modeling and cellular automata, *Environ. Plann. A*, 34, 1855–1876.

9 Assessing Urban Growth with Subpixel Impervious Surface Coverage

George Xian

CONTENTS

9.1 INTRODUCTION

Urban development in the U. S. is increasingly using land for residential and commercial purposes. Over the past 100 years, cities have grown from small, isolated population centers to large, interconnected urban economic, physical, and cultural features of the landscape. Spatial distributions and patterns of urban land use and land cover (LULC) often affect socioeconomic (Douglass, 2000), environmental (Gillies et al., 2003; Yin et al., 2005), and climatic conditions (Arnfield, 2003; Kalnay and Cai, 2003; Voogt and Oke, 2003). Accurate and current information on the status and trends of urban LULC and ecosystems is needed to build strategies for sustainable development. The ability to assess urban LULC change is desirable for urban planners, local community decision makers, and urban environment researchers.

Considerable progress has been made in the development of monitoring and change detection methods using remotely sensed data (Jensen, 1995; Kam, 1995;

Ridd and Liu, 1998; Sohl, 1999; Weng, 2001; Seto and Liu, 2003). One approach designed for mapping impervious surface was also used for urban land-use estimation (Ward et al., 2000; Jantz et al., 2003; Xian and Crane, 2005a). Anthropogenic impervious surfaces can be generally defined as any materials, such as rooftops, parking lots, sidewalks, driveways, and roads, that are impervious to water (Arnold and Gibbons, 1996). Impervious surface area (ISA) is considered a key indicator of environmental quality and can be used to address complex urban environmental issues, particularly those related to the health of urban watersheds (Schueler, 1994), and is an indicator of non-point-source pollution or polluted runoff (Slonecker et al., 2001).

Most early remote sensed impervious surface studies, such as those by Jackson (1975), Plunk et al. (1990), Morgan et al. (1993), and Monday et al. (1994), used simple classification for streets, parking lots, and buildings to define ISA. More recent studies used basic land-use classification systems as a baseline for computing ISA as a function of land-use category, such as lot size (Slonecker et al., 2001; Hebble et al., 2001) and even the National Land Cover Dataset (Jennings et al., 2004). However, urban areas are highly heterogeneous, and most urban image pixels in remote sensed imagery, such as those from Landsat and other similar sensors, comprise a mixture of different surfaces. For example, pixels classified as residential may represent between 20 and 100% impervious surface, while also representing between 0 and 60% tree canopy coverage (Clapham, 2003).

Virtually every pixel in an urban area represents a mixture of different land-cover types, including grass, trees, sidewalks, driveways, roads, and buildings. Pixel-level analysis often represents a mixture of spectral features that may come from different types of LULC and may create considerable spectral confusion among urban land-cover types. This kind of confusion is especially relevant to urban area imperviousness because of the lack of spatial resolution to represent most impervious elements from satellite imagery-derived products. To quantify spatial extents and distribution patterns of urban LULC by using satellite remote sensing data, we need an approach meaningful to both — one that identifies the heterogeneity of urban LULC and provides large-area urban land-use information.

Remote sensing techniques have been developed to deal explicitly with this heterogeneity problem at a subpixel level (Ji and Jensen, 1999; Yang et al., 2003). Subpixel analysis breaks down the mixed pixel into percentages of its components based on spectral characteristics and provides quantifiable measurements for ISA. Subpixel percent ISA data obtained from historical Landsat imagery can be used to determine urban spatial extent and development density quantitatively by selecting different ISA threshold values. Subpixel imperviousness change detection (SICD) provides multitemporal urban LULC change information critical for urban dynamics research (Yang et al., 2003; Xian et al., 2005.).

Percent ISA data have been applied as useful sources of urban extent for urban growth modeling (Jantz et al., 2003; Xian and Crane, 2005a) and the assessment of storm water runoff and urban sprawl in a watershed scale (Carlson, 2004). Physical measurements of urban areas can provide a self-consistent metric for urban LULC analysis. Increased availability and improved quality of multiresolution and multitemporal remote sensing data, as well as other demographic or geographic information, make it possible to provide detailed urban LULC conditions in a timely and cost-effective way.

This chapter presents an integrated subpixel imperviousness assessment model (SIAM) approach improved from the original SICD to quantify multitemporal variations of urban spatial extents and development intensities in two geographic locations: Las Vegas, Nevada, and Tampa Bay, Florida. Subpixel ISAs were estimated from 1984 to 2002 for the Las Vegas Valley and from 1991 to 2002 for the Tampa Bay watershed. Landsat Thematic Mapper (TM) and Enhanced Thematic Mapper Plus (ETM+) images in combination with high-resolution aerial photographs were used as the primary source for estimating multitemporal, subpixel ISA distribution in the areas.

Data from the 2002 terra-advanced space-borne thermal emission and reflection radiometer (ASTER), which has a spatial resolution of 15 m in three visible and near-infrared (VNIR) bands, 30 m in six shortwave infrared (SWIR) bands, and 90 m in five thermal infrared (TIR) bands, were also applied to obtain ISA in a higher spatial resolution for the Las Vegas area. Urban spatial extents, determined by the ISA threshold value, encompassed a range of different definitions for urban areas. General ISA features and associated urban LULC characteristics and estimation accuracy were also analyzed.

9.2 METHODS AND DATA

9.2.1 DESCRIPTION OF STUDY AREAS

Two geographically diverse regions — Las Vegas, Nevada, and Tampa Bay, Florida — were selected as the study areas (Figure 9.1). The Las Vegas Valley is located in southern Nevada and encompasses about 1320 km^2, including the cities of Las Vegas, Henderson, North Las Vegas, and Boulder City. The area is characterized by a desert climate that is extremely hot and dry in the summer and relatively cold and wet in the winter. Natural vegetation contains a few desert floras in Las Vegas. However, landscaping with grass, shrubs, trees, gravel, and bare sandy soils that appear similar to concrete is found throughout the urban area (Xian and Crane, 2005b).

The region has experienced a remarkable increase in urban land use over the past 50 years. The population of Clark County increased from less than 50,000 in 1950 to slightly more than 740,000 in 1990 and more than 1.37 million in 2000, according to historical census data (Clark County, 2005). The population in the Las Vegas Valley urban area reached 1.36 million in 2000 and increased to more than 1.68 million in 2004. This increase indicates that Las Vegas became the fastest growing metro area in the United States (Frey, 2005). Associated with population growth was a tremendous surge in housing development in the region. Housing units reached approximately 540,000 in 2000. Single-family, detached housing and apartments made up 53.3 and 27.6% of total housing units, respectively. In other words, nearly 80% of housing units in the area are single-family units and apartments.

Tampa Bay is located on the Gulf Coast of west-central Florida and has an areal extent of approximately 1030 km^2, making it one of the largest open-water estuaries in the southeastern United States. Four major sources of surface water — the Hillsborough, Alafia, Little Manatee, and Manatee Rivers — flow into the bay. The watershed

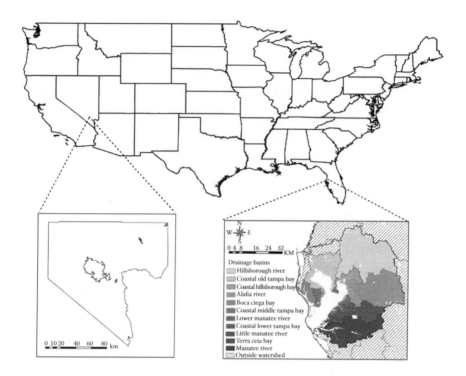

FIGURE 9.1 Las Vegas metropolitan area in the southern part of Nevada and the Tampa Bay watershed in west-central Florida.

around Tampa Bay covers approximately 6600 km² and encompasses most of Pinellas, Hillsborough, and Manatee Counties and portions of Pasco, Polk, and Sarasota Counties. The largest municipalities within the watershed are Tampa, St. Petersburg, Clearwater, and Bradenton.

A humid, subtropical climate with warm, wet summers and mild, dry winters characterizes and attracts many people to live in the region (Xian and Crane, 2003). Currently, more than 2 million people reside in the watershed. The Tampa Bay watershed features a high percentage of ISA mixed with vegetation and nonurban areas consisting of dense vegetation or water. Urbanization and suburban sprawl have changed the natural landscape into anthropogenic land and have significantly affected LULC conditions in the region.

9.2.2 SUBPIXEL IMPERVIOUSNESS ESTIMATION

To estimate spatial and temporal urban land use variation by remote sensing subpixel percent ISA, a SIAM approach was implemented. Generally, SIAM requires close-to-true ISA training data sets obtained from high-resolution imagery, such as 1-m digital orthophoto quarter quadrangles (DOQQ) from aerial photography, IKONOS, QuickBird, and 0.3-m orthoimagery. The high-resolution images are analyzed to classify urban and nonurban land use. The classification results are then rescaled to

1-m resolution for percent imperviousness calculation. Resulting percent ISA data are further scaled to 30-m resolution for development of training and validation data in regression tree modeling. The training data sets obtained from high-resolution imagery are converted to 15 m when implemented with ASTER data. Medium-resolution satellite imagery and derived information such as the normalized difference vegetation index (NDVI), together with geographic information such as slope, are then used to build regression tree models for extrapolating imperviousness over large spatial areas.

The regression tree is a machine-learning algorithm. It conducts a binary recursive partitioning and produces a set of rules for predicting a target variable (percent imperviousness) based on training data. Each rule set defines the condition under which a multivariate linear regression model is established for prediction (Breiman et al., 1984; Quinlan, 1993). The regression tree-based models provide a proposition logic representation of these conditions in the form of tree rules.

Regression trees are constructed using a recursive partitioning algorithm that builds a tree by recursively splitting the training sample into smaller subsets. In the partitioning process, each split is made so that the model's combined residual error for the two subsets is significantly lower than the residual error of the single best model. The main advantages of the regression tree algorithm are that it can account for a nonlinear relationship between predictive and target variables and allow continuous and discrete variables to be used as input data. The last step is to process accuracy assessment. Figure 9.2 illustrates detailed procedures for SIAM implementation. The regression tree models obtained from training data contain collections of rules in which each rule has an associated multivariate linear model, e.g.,

Rules m

if [conditions] are true then

$$y = f(x_1, x_2, \ldots, x_n)$$

where y is imperviousness and is defined as the dependent variable in the model, and $f(x_1, x_2, \ldots, x_n)$ is a linear combination of multiple independent variables including different spectral bands and other input geographic information. When a situation matches a rule's conditions, the associated rule-based model is used to calculate the dependent variable — the percent imperviousness — for the pixel.

9.2.3   Remote Sensing Data and Data Process

To estimate temporal and spatial urban development in Las Vegas and Tampa Bay through SIAM, high-resolution orthoimagery or DOQQs were used to build training data sets. The orthoimagery were a natural color with a spatial resolution of 0.3 m. The DOQQs were scanned from color infrared photographs acquired from the U.S. Geological Survey (USGS) National Aerial Photography Program. Each DOQQ comprised three colors — green, red, and near-infrared — with a nominal spatial resolution of 1 m. To estimate temporal and spatial urban development in the Las

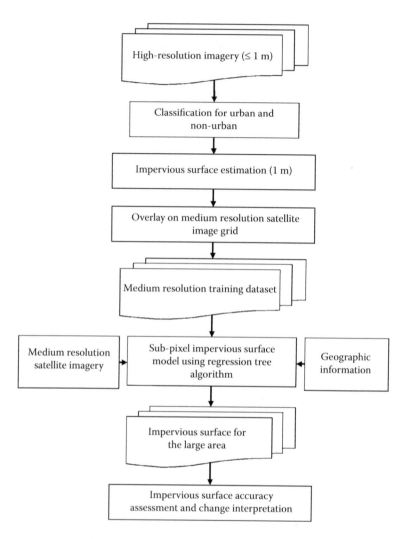

FIGURE 9.2 Implementation of SIAM for estimation of subpixel impervious surface using remote sensing data and regression tree algorithm.

Vegas Valley, eight orthoimages covering eight different locations in the region were selected to create the training data set. Eight DOQQs covering St. Petersburg, northern St. Petersburg, Tampa, and the southern and southeastern parts of Tampa Bay were utilized to develop training data sets for the Tampa Bay watershed.

One Landsat image for path 39, row 35 in 1984, 1986, 1992, and 1996, from Landsat TM and 2002 from Landsat ETM+ was selected for the Las Vegas area ISA estimation. Landsat TM images from 1990 and 1991, 1994 and 1995, and ETM+ images from 2000 and 2002 for paths 17 and 16, rows 40 and 41 were acquired for the Tampa Bay ISA estimation. All images were preprocessed by the USGS Center for Earth Resources Observation and Science (EROS) to correct radiometric and geometric distortions of the images. All images were rectified to a common universal transverse mercator (UTM) coordinate system.

Images for Las Vegas and Tampa Bay were georeferenced to UTM zones 11 and 17, respectively. Bands 1 through 5 and 7 were used at a spatial resolution of 30 m. The thermal bands had their original pixel sizes of 120 m for TM and 60 m for ETM+ images resampled to 30 m using the nearest neighbor algorithm to match the pixel size of the other spectral bands. These corrections resulted in digital number (DN) images that are measures of at-satellite radiance. DNs in each band were converted first to at-satellite radiance and then to at-satellite reflectance (Landsat Project Science Office, 2002). DNs were converted to at-satellite reflectance using the following equations:

$$L_\lambda = Gain_\lambda \cdot DN_\lambda + Bias_\lambda \qquad (9.1)$$

$$\rho_\lambda = (\pi L_\lambda d^2)/(ESUN_\lambda \cdot \sin(\theta)) \qquad (9.2)$$

where
L_λ is at-sensor radiance
$Gain_\lambda$ is the slope of radiance/DN conversion function
$Bias_\lambda$ is the intercept of the radiance/DN conversion function

Gain and bias values are provided in metadata that accompany each TM/ETM+ image (Landsat Project Science Office, 2002). ρ_λ is unitless at-satellite reflectance for TM/ETM+ bands (1 through 5; 7); θ is the solar elevation angle; and $ESUN_\lambda$ is mean solar exoatmospheric irradiance. ρ_λ values from six reflectance bands were implemented in regression modeling to obtain large-area subpixel percent ISA. Reflectance values from the visible (ρ_1) and near-infrared (ρ_2) bands of Landsat images were used to compute *NDVI* values using the formula:

$$NDVI = (\rho_2 - \rho_1)/(\rho_2 + \rho_1) \qquad (9.3)$$

The value of NDVI for each pixel was used as a variable in the regression models to eliminate confusion caused by vegetation canopy for ISA determination in urban areas, especially in medium-density urban and suburban areas.

The ASTER-registered radiance at the sensor level-1B product contains radiometrically calibrated and geometrically registered data for all bands. Data from the SWIR and TIR bands were rescaled to 15 m to be consistent with data from the VNIR. These are scaled or calibrated radiance DNs and were converted to at-sensor radiance using the following method:

$$R_\lambda = (DN_\lambda - 1) * UCC_\lambda \qquad (9.4)$$

where R_λ is the at-sensor radiance for different ASTER bands, and UCC_λ is the united conversion coefficient provided by Abrams et al. (2002) and listed in Table 9.1.

TABLE 9.1
Unit Conversion Coefficients of Each
ASTER Band

Band		Coefficient
VNIR (15 m)	1	1.688
	2	1.415
	3	0.862
SWIR (30 m)	4	0.2174
	5	0.0696
	6	0.0625
	7	0.0597
	8	0.0417
	9	0.0318
TIR (90 m)	10	6.882×10^{-3}
	11	6.780×10^{-3}
	12	6.590×10^{-3}
	13	6.693×10^{-3}
	14	5.225×10^{-3}

Large-area ISA was mapped using Landsat reflectance NDVI derived from reflectance bands for Tampa Bay, plus the thermal band from Landsat and ASTER and slope information for Las Vegas. Figure 9.3 shows the Landsat imagery used for ISA estimations in 2002 for Tampa and in 1984 and 2002 for Las Vegas. NDVI usually helped in identifying urban residential land use from rural land when housing was mixed with trees and other vegetation canopy. The Landsat TM thermal band or ETM+ band 6L, saturating at 347.5 K, was helpful in eliminating nonimpervious areas, especially at the urban fringe for the Las Vegas area. A slope layer eliminated the spectrally misclassified areas in the mountain ranges surrounding Las Vegas because most urban areas exist in valleys and on the lower alluvial flanks of the mountains.

9.3 ISA ESTIMATIONS FROM LANDSAT AND ASTER DATA

Using percent ISA as a threshold to define and quantify urban land use and extent could create a range of verifiable extent estimates that includes a range of different definitions of urban land-use densities. Urban extent defined using thresholds on the basis of spectral heterogeneity is far more consistent than that obtained from administrative maps alone (Small et al., 2005). This eliminates the considerable ambiguity that results from varying administrative and political definitions of urban areas.

The 10% threshold captured almost all developed land including low-, medium-, and high-density residential areas, as well as business areas (Xian and Crane, 2005a). Pixels with spatial resolution of 30 m were classified as urban when ISA was equal to or greater than 10%, whereas pixels of less than 10% were classified as nonurban.

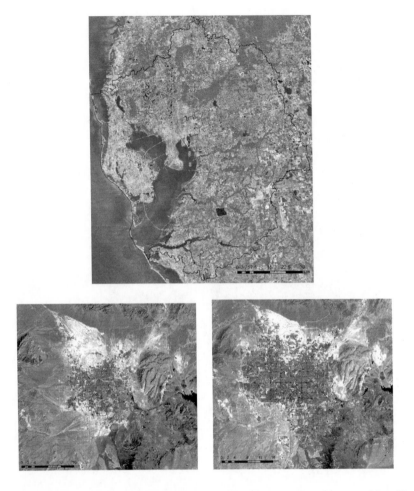

FIGURE 9.3 Landsat 7 image of Tampa Bay in 2002 (top); Landsat 5 image of Las Vegas in 1984 (bottom left) and 2002 (bottom right).

Furthermore, urban development density could also be defined by different ISA thresholds, such as 10 to 40% for low-density urban, 41 to 60% for medium-density urban, and above 60% for high-density urban.

9.3.1 ISA ESTIMATIONS FROM LANDSAT TM AND ETM+ FOR LAS VEGAS

Spatial and temporal variations in urban land use for the Las Vegas Valley from 1984 to 2002 are shown in Figure 9.4. Urban land use expanded in almost all directions in the valley. During the 1980s and early 1990s, most medium- to high-percent ISAs were located in the downtown and Las Vegas strip areas. More recently, high-percent ISA has expanded to the southeast and northwest portions of Las Vegas.

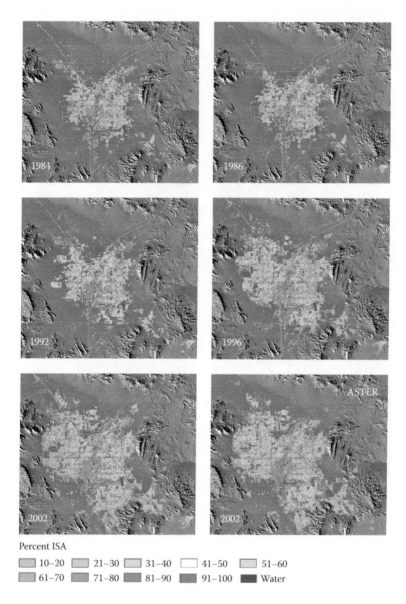

Percent ISA

| | 10–20 | | 21–30 | | 31–40 | | 41–50 | | 51–60 |
| | 61–70 | | 71–80 | | 81–90 | | 91–100 | | Water |

FIGURE 9.4 (Color Figure 9.4 follows page 240.) ISA distribution in Las Vegas from 1984 to 2002. All figures were obtained from Landsat imagery in 30-m resolution, except the lower right-hand one, which was mapped from ASTER 2002 imagery in 15-m resolution.

Consistent urban development accomplished with high- and medium-density residential areas, as well as commercial and business centers, enlarged the urban land-use boundary far from the 1984 urban boundary.

Multiyear spatial extent of urban area and the ratio of total nonurban land use to the entire mapping area for the Las Vegas Valley are displayed in Figure 9.5. The areal extent of urban land use was approximately 290 km² in 1984 and increased to

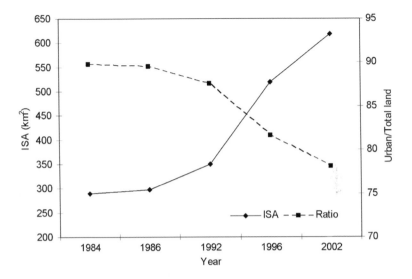

FIGURE 9.5 Urban extent delineated by 10% ISA threshold and the ratio of total nonurban land use to total area for the Las Vegas Valley.

about 620 km^2 in 2002, representing an increase of 113% during the period. Urban land use counted for approximately 10% of total land in the entire mapping area in 1984 and changed to 22% in 2002. To investigate growth rates for different urban development densities, pixels were regrouped into nine categories from 10 to 100% in every interval of 10% imperviousness.

Figure 9.6 presents variations of different ISA categories in 1984 and 2002, respectively. Categories 5 through 7 (50 to 70% ISA) contain the largest portion of

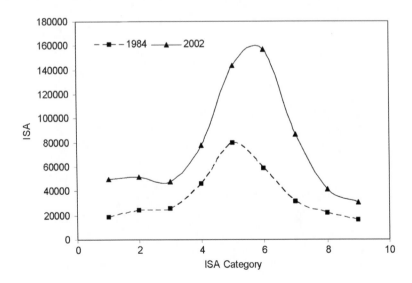

FIGURE 9.6 ISA category variation for 1984 and 2002 in Las Vegas.

total ISA on both dates. This indicates that the largest increase of urban land use is in medium to high urban development densities that include most single- to two-family housing units.

A naturally colored 0.3-m resolution orthoimagery acquired in the residential area in the northwest corner of Las Vegas was used to inspect percent ISA and its corresponding urban land use. A 5 × 5 30-m sampling unit registered to the Landsat pixel was defined to reduce the impact of geometric errors associated with orthoimagery and satellite images. In this sample unit, areas of interests (AOI) were outlined following the boundaries of interpreted impervious surfaces. The area of each AOI was determined by using AOI property functions.

The true fraction of ISA in each sample pixel was calculated by dividing the total area of AOI by each pixel area (Figure 9.7). The number on each sample pixel labeled the percent ISA. Single-family housing in one pixel was counted as 40 to 60% ISA. Two-family housing with relatively small lawns and few trees in one pixel was estimated as 60 to 70% (or more) ISA. A two-lane road in one sample pixel was also counted as approximately 60 to 70% ISA. These sample pixels with percent ISA greater than 40% could be categorized as medium- to high-density urban residential areas. Pixels located in commercial lands, freeways, and most business centers had higher percent ISA and therefore were classified as high-density urban lands. As urban land use defined by subpixel percent ISA indicated, most growth associated with urbanization in Las Vegas in the past 20 years was single- to two-family housing residential with a few lawns and trees.

FIGURE 9.7 Percent ISA measured from 1-ft DOQQ and displayed in 30-m pixels.

9.3.2 ISA Estimated from ASTER for Las Vegas

Using a similar regression tree model and the same training data set used for Landsat data, ASTER reflectance, and thermal bands, NDVI derived from reflectance bands and surface aspect data derived from digital elevation model (DEM) were used for ISA estimation. Theoretically, the 15-m resolution VNIR bands can identify urban features that cannot be discerned from 30-m Landsat imagery, especially for places where portions of buildings and streets are covered by trees or are adjacent to gravel. ASTER SWIR and TIR bands provide a wider range of electromagnetic spectrums that are useful in discriminating urban and rural areas in the Las Vegas environment.

One would expect that most pixels mixed with different urban LULC features, especially in high- to medium-density urban residential areas in 30-m resolution, would be discriminated as single urban LULC classes or higher ISA cover by ASTER imagery. Single-family housing estimated as 50% ISA by Landsat imagery would be determined as 90% or higher ISA by ASTER imagery. The spatial distribution of ISA estimated from ASTER data in 2002 in the Las Vegas Valley (Figure 9.4) indicated that the spatial extent of urban land use was almost the same as that obtained from Landsat imagery. However, fewer ISA pixels were found in low-density urban areas (ISA < 40%), and for medium- to high-density urban areas there were more pixels than those estimated from Landsat.

9.3.3 ISA in the Tampa Bay Watershed

Anthropogenic changes in the Tampa Bay landscape have significantly affected LULC conditions in the watershed. To estimate the change in urban extent for the Tampa Bay watershed, the subpixel percent imperviousness for the whole watershed was determined from Landsat imagery using a regression tree algorithm. TM and ETM+ images selected from early spring were implemented for ISA estimations from 1991 to 2002 using SIAM.

Many low-density urban areas were mixed with trees and grass in the region because its warm, moist year-round climate is conducive to lush vegetation. To apply SIAM for ISA estimation in the region, reflective bands and NDVI calculated from Landsat images were added to the model as independent variables. This resulted in an improved calculation that separated vegetated land-cover type from urban land cover. After water was masked out from the original maps, a unique temporal dimension documenting change was provided (Figure 9.8).

When ISA was regrouped into and displayed as different categories, the imperviousness distributions identified areas of residential growth and commercial and industrial development from 1991 to 2002. In comparison to the highly concentrated urban development pattern in the Las Vegas Valley, urban lands were widely spread in the watershed. Many new developments occurred in the northeastern and eastern portions of Tampa or the northern portion of Pinellas County. New developments along major transportation roads also suggested a transportation-orientated growth pattern in the region. The spatial extent of total urban land use and the ratio of nonurban land to total watershed area excluding water (Figure 9.9) indicated that urban area increased to

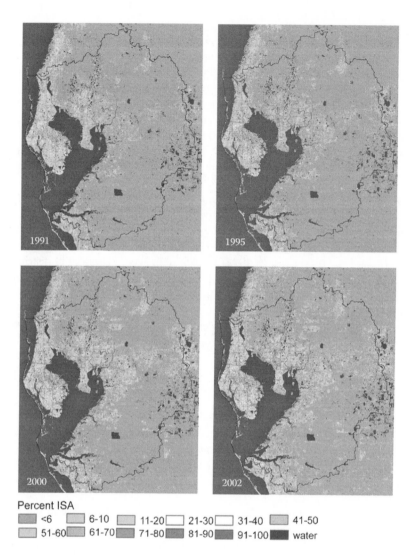

Percent ISA

| | <6 | | 6-10 | | 11-20 | | 21-30 | | 31-40 | | 41-50 |
| | 51-60 | | 61-70 | | 71-80 | | 81-90 | | 91-100 | | water |

FIGURE 9.8 (Color Figure 9.8 follows page 240.) ISA distribution in Tampa Bay. The black line is the watershed boundary.

approximately 1800 km², nearly threefold, from 1991 to 2002. The ratio of nonurban land to total watershed land was reduced to approximately 70% in 2002.

Impervious conditions vary throughout the watershed by different subdrainage basins. The variability of impervious surface from 1991 to 2002 in six major drainage basins (Figure 9.10) suggested that impervious categories 1 and 2 had the greatest extent in the watershed because urbanization densities were lower in suburban areas. Impervious distributions in the Alafia River, Hillsborough River, and Coastal Hillsborough Bay drainage basins showed similar patterns; apparent differences in different categories represented the density gradient between urban and suburban. New urban growth

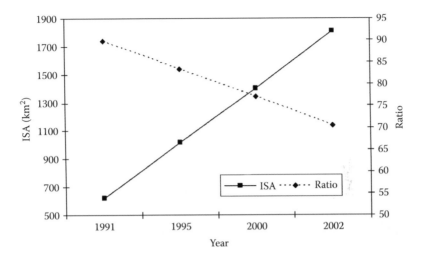

FIGURE 9.9 Total ISA change and the ratio of nonimpervious surface to total watershed land.

represented by 2002 impervious coverage occurred in suburban areas and accounted for most of the increase of imperviousness in these areas. In contrast with urban land-use characteristics in the Las Vegas area, most new residential areas were characterized by small lot sizes, with several buildings occupying a portion of a single TM/ETM+ pixel, resulting in a greater proportion of pixels in the lower category values. The impervious coverage and its category could be represented by a reverse relation as

$$\text{coverage} \propto 1/(\text{category})^n \qquad (9.5)$$

where n is a constant. However, this relation did not exist for the other three drainage basins.

In the Boca Ciega Bay drainage basin area, extents of impervious categories 3 through 5 were larger than any other categories. Temporal change of ISA indicated that impervious surface covered approximately 61% of the Boca Ciega Bay drainage basin, with impervious categories 3 through 5 accounting for 51% of total imperviousness in the area in 1991. The percentages increased to 86% for total impervious surface in the area in 2002 and 54% for impervious categories 3 through 5. Increases in impervious categories 3 to 5 in the Boca Ciega Bay drainage basin between 1991 and 2002 were associated with decreases in impervious categories 1 and 2 during the same period. This change pattern indicated that the basin was undergoing active urbanization and that new urban development occurred within existing low-density urban areas because few vacant patches were left for new development.

A similar pattern of change was seen in the Coastal Old Tampa Bay and Coastal Middle Tampa Bay drainage basins. Apparent increases from 1991 to 2002 for impervious categories 3 and 4 could be seen in areas where a remarkable degree of urbanization, including high-density urban and lower to medium-density residential areas,

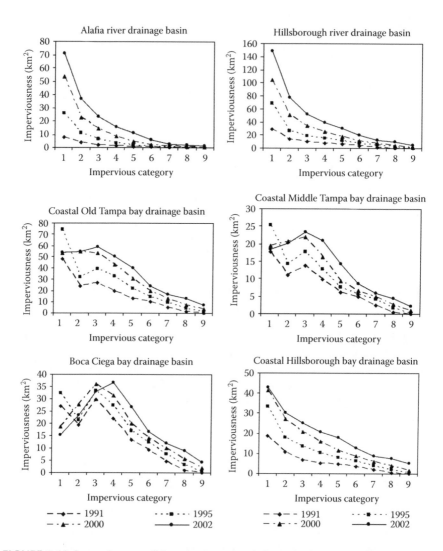

FIGURE 9.10 Impervious conditions in six major drainage basins and their variety from 1991 to 2002.

has been maintained. Analysis of the impervious distribution and change patterns for six drainage basins and the whole watershed suggests that impervious surface is a suitable surrogate for different urban densities and their changes; most new developments in the region are associated with low- to medium-density urban areas.

9.4 ACCURACY ASSESSMENT

The regression tree model implemented for large-area ISA estimation used several parameters to measure model prediction accuracy, including average error and correlation coefficient. Results for the Las Vegas 2002 ISA estimate using Landsat

ETM+ imagery indicated that the average error and correlation coefficient values were 0.34 and 0.85, respectively. Results derived from ASTER imagery, with average error and correlation coefficient values of 0.33 and 0.86, respectively, had the higher modeling accuracy. The 2002 Tampa Bay ISA estimation had average error and correlation coefficient values of 0.68 and 0.7, respectively. The model accuracy for Tampa Bay was lower than that for Las Vegas.

To perform accuracy assessment using true ISA data, root-mean-square error (RMSE) and systematic error (SE) were used. RMSE and SE are defined as:

$$RMSE = \sqrt{\frac{\sum_{i=1}^{N}(\hat{U}_i - U_i)^2}{N}} \tag{9.6}$$

$$SE = \frac{100}{N} \sum_{i=1}^{N}(\hat{U}_i - U_i) \tag{9.7}$$

where

$\hat{U}_i$ is the modeled ISA for sample i
U_i is the true ISA data for sample i
N is the total number of samples.

Six random sample locations were generated with the ERDAS imagine accuracy assessment module for the Las Vegas area. Corresponding to these sample locations, six naturally colored 0.3-m resolution orthoimages acquired in 2003 were used as ground truth reference for assessing accuracy of the model estimates. Using the sample procedures described in Section 9.3.1 to define a 5 × 5, 30-m sampling unit and to outline an AOI in each selected orthoimagery, the true fraction of ISA in each sample pixel was calculated by dividing the total area of the AOI by each pixel area: 30 m × 30 m for Landsat- and 15 m × 15 m for ASTER data-derived results.

To process the accuracy assessment for the Tampa Bay watershed, 16 naturally colored 0.3-m resolution orthoimages acquired in 2003 covering 16 sample locations in the Tampa Bay watershed were utilized. A 5 × 5 sampling unit was selected for each sample orthoimagery, and the true percent ISA for each sample pixel was determined by the same procedures used for the Las Vegas data. Mean RMSE and SE were calculated for all sample pixels distributed over the Tampa Bay region.

Table 9.2 presents average values of overall RMSE and SE from each sample site. Generally, the modeled ISAs had relatively lower accuracy in low-density urban areas than those in high-density urban areas. Mean SE and mean RMSE for ISA derived from Landsat imagery for Las Vegas were –6.0 and 16.1%, respectively. Lower accuracy for modeled ISA corresponded to the very low and very high imperviousness cover areas. The values of mean SE and RMSE for ISA derived

TABLE 9.2
Percent ISA for Each Sample Site and Accuracy

Sample site	Las Vegas	Tampa Bay
Mean ISA measured from DOQQ ISA	62	39
Modeling ISA % from Landsat	57	33
Mean SE	–6.0	–5.8
Mean RMSE	16.1	18.9
Modeling ISA% from ASTER	59	
Mean SE	–3.2	
Mean RMSE	14.7	

from ASTER data decreased to –3.2 and 14.7%, respectively, indicating apparent improvement because of the finer spatial resolution in VNIR bands and wider spectral ranges of TIR bands from ASTER. ISA obtained for the Tampa Bay watershed had mean SE and RMSE values of –5.8 and 18.9%, respectively. Inspection of ISA distribution indicated that the lowest accuracy came from medium- to low-residential areas where houses were usually surrounded by tall trees and other vegetation canopies.

9.5 DISCUSSION AND CONCLUSIONS

This study employed SIAM to quantify subpixel percent imperviousness over time for two metropolitan areas with distinct environs. The results provide a good estimation of the urban LULC change that occurred from 1984 to 2002 in Las Vegas, Nevada, and from 1991 to 2002 in Tampa Bay, Florida. The combination of Landsat and ASTER satellite data with high-resolution imagery provides the necessary spatial and spectral information needed to support subpixel impervious distribution estimation and change detection.

Subpixel analysis breaks down the mixed pixel into the percentage of its component parts based on spectral characteristics. The technique allows for characterization of spatial units at resolutions smaller than the size of a pixel and is effective for impervious features that are spatially small from a landscape perspective. Subpixel percent ISA provides a useful knowledge of urban LULC condition and information that can be used to represent the spatial extent and change intensity of urban LULC. Urban extent defined by percent ISA threshold values obtained on the basis of spectral heterogeneity of remote sensing imagery is more consistent than that obtained from administrative maps alone. This approach eliminates the considerable ambiguity resulting from varying administrative and political definitions of urban areas. SIAM also provides considerable flexibility in capturing the heterogenous characteristics of urban LULC.

Medium-resolution Landsat imagery has been widely implemented for extrapolating imperviousness in large areas. The growth patterns and detail structures of urban LULC can be successfully recognized using proper threshold values of percent subpixel ISA. The heterogeneous features of land-cover types in urban areas are effectively outlined at the subpixel level. However, the accuracy of spatial extents and temporal changes of urban LULC is highly dependent on spatial recognition accuracy of the imagery used for the large area. Considerable confusion can still result from use of such imagery for areas where urban LULC types are easily mixed with surrounding bare soils or vegetation canopy covers.

The accuracy of ISA estimation obtained from using ASTER imagery with relatively high spatial resolution for the Las Vegas region has apparently been improved. The wide spectral range of ASTER imagery also supplies more information for discriminating urban land-cover types from special landscapes in the region. Results also suggest that, for spectral complexity and heterogeneity of urban materials and land-cover types, more spectral details such as hyperspectral remote sensing data (Herold et al., 2004) could provide the capability to represent the spectral diversity of an urban environment and significantly improve spatial impervious surface mapping.

Quantification of urbanization through mapping impervious surface change also provides useful data for simulating and predicting growth using the urban growth model. The availability and consistency of satellite records make it possible to derive urban extent and development density and overcome the limitations or lack of multitemporal historical data sets for urban extent and LULC. Urban LULC information obtained through subpixel percent ISA estimation provides a consistent and realistic model calibration data source for future growth prediction and possible growth scenario investigation (Jantz et al., 2004; Xian and Crane, 2005). Detailed urban development information enables urban growth to be simulated by multiple categories of urban development intensity. The outcome provides information for potential applications related to urban planning and management of floods and urban water quality, as well as for urban ecosystem modeling.

ACKNOWLEDGMENTS

This research was performed under U.S. Geological Survey contact O3CRCN0001. I thank Drs. Darrel Napton and Dave Meyer for their suggestions and comments for this chapter. An anonymous reviewer is also thanked for providing several helpful suggestions.

REFERENCES

Abrams, M., Hook, S., and Ramachandran, B., *ASTER User Handbook*, Version 2.0, Pasadena: Jet Propulsion Laboratory, 2002. Accessed at http://asterweb.jpl.nasa.gov/content/03_data/04_Documents/aster_user_guide_v2.pdf.

Arnfield, A.J., Two decades of urban climate research: a review of turbulence, exchanges of energy and water, and the urban heat island, *Int. J. Climatol.*, 231, 1, 2003.

Arnold, C.A., Jr. and Gibbons, C.J., Impervious surface coverage: the emergence of a key urban environmental indicator, *J. Am. Planning Assoc.*, 62, 243, 1996.

Breiman, L., Friedman, J., Olshen, R., and Stone, C., *Classification and Regression Trees*, Wadsworth International Group, Belmont, CA, 1984, 358.

Carlson, T.N., Analysis and prediction of surface runoff in an urbanization watershed using satellite imagery, *J. Am. Water Resour. Assoc.*, 40, 1087, 2004.

Clapham, W.B., Jr., Continuum-based classification of remotely sensed imagery to describe urban sprawl on a watershed scale, *Remote Sensing Environ.*, 86, 322, 2003.

Clark County, Comprehensive Planning, Clark County, Nevada. Accessed on 12 May 2005 at http://www.co.clark.nv.us/Comprehensive-planning.

Douglass, M., Mega-urban regions and world city formations: globalization, the economic crisis and urban policy issues in Pacific Asia, *Urban Stud.*, 37, 2315, 2000.

Frey, W.H., Metro America in the new century: metropolitan and central city demographic shifts since 2000, Washington, D.C.: Brookings Institution, 2005. Accessed on 16 September 2005 at http://www.brookings.edu/metro/pubs/.

Gillies, R.R., Box, J.B., Symanzik, J., and Rodemaker, E.J., Effects of urbanization on the aquatic fauna of the Line Creek watershed, Atlanta — a satellite perspective, *Remote Sensing Environ.*, 86, 411, 2003.

Hebble, E.E., Carlson, T.N., and Daniel, K., Impervious surface area and residential housing density: a satellite perspective, *Geocarto Int.*, 16, 13, 2001.

Herold, M., Roberts, D.A., Gardner, M.E., and Dennison, P.E., Spectrometry for urban area remote sensing — development and analysis of a spectral library from 350 to 2400 nm, *Remote Sensing Environ.*, 91, 304, 2004.

Jackson, T.J., Computer-aided techniques for estimating the percent of impervious area from Landsat data, *Proceedings of Workshop on the Environmental Applications of Multispectral Imagery*, American Society of Photogrammetry, Fort Belvoir, Virginia, 140, 1975.

Jantz, C.A., Goetz, S.J., and Shelley, M.K., Using the SLEUTH urban growth model to simulate the impacts of future policy scenarios on urban land use in the Baltimore–Washington metropolitan area, *Environ. Plann., B, Plann. Design*, 31, 251, 2004.

Jennings, D.B., Jarnagin, S.T., and Ebert, D.W., A modeling approach for estimating watershed impervious surface area from National Land Cover Data 92, *Photogrammetric Eng. Remote Sensing*, 70, 1295, 2004.

Jensen, J.R., *Introductory Digital Image Processing: A Remote Sensing Perspective*, 2nd ed., Englewood Cliffs, NJ: Prentice Hall, 1995, 318.

Ji, M.H. and Jensen, J.R., Effectiveness of subpixel analysis in detecting and quantifying urban imperviousness from Landsat Thematic Mapper imagery, *Geocarto Int.*, 14, 31, 1999.

Kalnay, E. and Cai, M., Impact of urbanization and land-use change on climate, *Natural*, 423, 528, 2003.

Kam, T.S., Integrating GIS and remote sensing techniques for urban land-cover and land-use analysis, *Geocarto Int.*, 10, 39, 1995.

Landsat Project Science Office, 2002, Landsat 7 Science Data User's Handbook. URL: http://ltpwww.gsfc.nasa.gov/LAS/handbook/handbook_toc.html, Goddard Space Flight Center, NASA, Washington, DC.

Monday, H.M., Urban, J.S., Mulawa, D., and Benkelman, C.A., City of Irving utilizes high-resolution multispectral imagery for NPDES compliance, *Photogrammetric Eng. Remote Sensing*, 60, 411, 1994.

Morgan, K.M., Newland, L.W., Weber, E., and Busbey, A.B., Using spot satellite data to map impervious cover for urban runoff predictions, *Toxicol. Environ. Chemistry*, 40, 11, 1993.

Plunk, D.E., Morgan, K., and Newland, L., Mapping impervious cover using Landsat TM data, *J. Soil Water Conserv.*, 45, 589, 1990.

Quinlan, J.R., *C4.5: Programs for Machine Learning*, Morgan Kaufmann, San Mateo, CA, 1993, 302.

Ridd, M.K. and Liu, J., A comparison of four algorithms for change detection in an urban environment, *Remote Sensing Environ.*, 63, 95, 1998.

Schueler, T.R., The importance of imperviousness, *Watershed Prot. Tech.*, 1, 100, 1994.

Seto, K. and Liu, W., Comparing ARTMAP neural network with the maximum-likelihood classifier for detecting urban change, *Photogrammetric Eng. Remote Sensing*, 69, 981, 2003.

Slonecker, E.T., Jennings, D.B., and Garofalo, D., Remote sensing of impervious surfaces: a review, *Remote Sensing Rev.*, 20, 227, 2001.

Small, C., Pozzi, F., and Elvidge, C.D., Spatial analysis of global urban extent from DMSP-OLS night lights, *Remote Sensing Environ.*, 96, 277, 2005.

Sohl, T.L., Change analysis in the United Arab Emirates: an investigation of techniques, *Photogrammetric Eng. Remote Sensing*, 65, 475, 1999.

Voogt, J.A. and Oke, T.R., Thermal remote sensing of urban climates, *Remote Sensing Environ.*, 86, 370, 2003.

Ward, D., Phinn, S.R., and Murray, A.T., Monitoring growth in rapidly urbanized areas using remotely sensed data, *Prof. Geogr.*, 52, 371, 2000.

Weng, Q., A remote sensing-GIS evaluation of urban expansion and its impact on surface temperature in the Zhujiang Delta, China, *Int. J. Remote Sensing*, 22, 1999, 2001.

Xian, G. and Crane, M., Impacts of urban development on seagrass in Tampa Bay, *Proc. 19th Natl. Environ. Monitoring Conf.*, Arlington, Virginia, 23, 2003.

Xian, G. and Crane, M., Assessments of urban growth in the Tampa Bay watershed using remote sensing data, *Remote Sensing Environ.*, 97, 203, 2005a.

Xian, G. and Crane, M., Evaluation of urbanization influences on urban climate with remote sensing and climate observations, *Proc. 5th Int. Symp.: Remote Sensing Urban Area (URS2005)*, Tempe, Arizona, 2005b.

Xian, G., Yang, L., Klaver, J.M., and Hossain, N., Measuring urban sprawl and extent through multitemporal imperviousness mapping, Urban Dynamic Topic Report, USGS circular, 2005.

Yang, L., Xian, G., Klaver, J.M., and Deal, B., Urban land-cover change detection through subpixel imperviousness mapping using remotely sensed data, *Photogrammetric Eng. Remote Sensing*, 69, 1003, 2003.

Yin, Z.-Y., Walcott, S., Kaplan, B., Cao, J., Lin, W., Chen, M., Liu, D., and Ning, Y., An analysis of the relationship between spatial patterns of water quality and urban development in Shanghai, China, *Computers, Environ. Urban Syst.*, 29, 197, 2005.

10 Remote Sensing and Urban Growth Theory

Martin Herold, Jeff Hemphill, and Keith C. Clarke

CONTENTS

10.1 NEEDS AND CHALLENGES IN STUDYING URBAN DYNAMICS

Research regarding urban phenomena lacks common ground and testable concepts. Although this is to be expected to a certain degree in a multifaceted discipline, the discord among various perspectives and the lack of sound theories hampers progress that can lead to the application of new theoretically grounded understandings to contemporary issues. For those concerned with the management and planning of cities and developing urban regions, the potential rewards offered by remote sensing, urban models, and planning support systems as information aids in the decision-making process may seem dazzling.

When it comes down to the nuts and bolts of bringing these tools and the improved understanding they offer into operation, however, the glitter falls away. In the context of target end users, the confounding limitation of conflicting information drowns enthusiasm for actual use. This is the case even for remote sensing data. Practical usability suffers from the lack of a supporting framework with which to guide the application of these data for definitive purposes other than visualization.

Limitations also arise from questions of what, where, how, and why. What data are useful for what applications? Where are certain technologies and theoretical understandings applicable? How are data used to address particular problems? Why use these new tools? Answers for each of these questions can be expanded considerably depending on whether they are asked of an economist, a planner, a research scientist, an academic, or a member of management. The colloquial "big picture" rendered by the plethora of data, sophisticated analysis tools, and competing interests cannot be reconciled in the absence of a unifying conceptual framework within which questions and answers can be placed. The latter question of applicability is a major hurdle facing urban research that seeks to inform decision makers.

Common ground in urban geography and research that purports to address urban problems can be found by seeking a unifying framework within which applicable knowledge from different disciplines regarding particular issues can be placed. Enter remote sensing and the seemingly unambiguous nature of the information it contains. Pixels contain a wealth of data about virtually any topic to do with urban areas if their subject can be related to expression of natural or social systems. For the purposes of visualization, the allegorical image is said to be "worth a thousand words." However, given proper description, space — the urban built-up land — is the most unambiguous concept of all.

Beyond bickering over a technical definition of what is considered urban, the concept of urbanization as a process can be described from any number of perspectives with conceptual formulations and language that confuse and challenge, negating potentially useful or unique contributions. How can a particular pattern be described? What can remote sensing do to support such activities in a general sense, recognizing that many such projects are hobbled by narrowly defined scope or spatial extent? To answer this question, the most basic properties of spatiotemporal patterns must be studied, understood, and defined in a clear and concise manner.

Given these needs and challenges, this chapter attempts to outline ongoing research and how remote sensing can be used to assess the dynamics of the urban growth process. In the first part, a review of current issues in urban remote sensing and urban modeling will be presented. Different avenues taken in the study of urban dynamics are emphasized in order to outline the particular potentials of remote sensing as it may contribute to urban theory and modeling. In the second part, these points will be illustrated using an example of some research that makes use of spatial metrics for analysis of urban dynamics, outlining a generalized framework of urban growth, and developing the basis for a new contribution to urban theory.

10.2 PERSPECTIVES IN ANALYZING URBAN DYNAMICS

There are essentially two general perspectives from which to view spatiotemporal urban patterns (Figure 10.1). The traditional perspective follows a deductive top-down perspective: isolating urban structures as the outcomes of prespecified processes

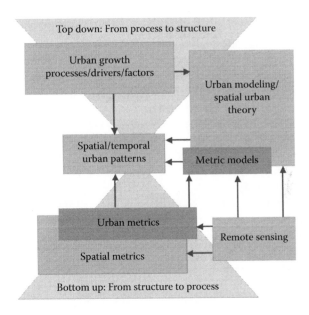

FIGURE 10.1 Conceptual approaches for studying spatial and temporal urban dynamics.

of urban change (from process to structure). This point of view is common in the fields of planning, geography, and economics. The main criticism of a reductionist perspective is that it is only marginally representative of the spatial and temporal complexities of urban change. Early demographic and socioeconomic research was limited by the ability to conduct detailed spatiotemporal pattern analysis at anything other than aggregate levels. These logistical constraints led to conclusions from most early investigations based on a top-down chain of causality. This era generated significant contributions and raised compelling questions regarding urban theory, but one unanswered question persists: how do cities form over time?

More recent studies from the various genres of urban research have started to address dynamics (White et al. 2001; Batty 2002). Research has become more focused on isolating the drivers of growth rather than solely the emerging geographic patterns. Although new urban models have provided insight into some aspects of urban dynamics, a deeper understanding of the spatial patterns and processes associated with urbanization is still limited by availability of suitable data and lack of compatible theory (Longley and Mesev 2000). It is well understood that good models and good theory necessitate reliable measurements that capture spatiotemporal dynamics. This need is emphasized in the inductive, bottom-up perspective. Consistent empirical observation of actual spatial structures in spatial and temporal detail is needed to link changes over time to specific hypotheses about the processes involved (from structure to process). A key source of such information is remotely sensed observations (Herold et al. 2003; Lo 2004).

10.3   A BRIEF HISTORY OF URBAN MODELING

Physicists developed models and theoretical frameworks around simple concepts and constructed experiments that were able to reproduce real observable actions precisely and accurately. From these elemental experiments came more complex models and improved understanding of systems. Looking back from the current state of research regarding urban systems, these early theories and concepts of urban spatial structure and urban dynamics (top-down) were not realistic representations of process. These models were not suitable for predictive applications or of use for understanding the spatial dynamic of urban growth processes. This prompted the need for more spatially explicit models and consequently led to development of the field of urban modeling.

The static ideal type of urban model that grew out of the social perspectives of the Chicago School in the 1960s sought only to represent visually the spatial arrangement of hypothetical urban socioeconomic systems. The famous concentric zone model published in *The City* was the spatial description of the outcomes of the ecological processes competition and succession, which were analogues with those of natural world systems (Park et al. 1925). The most successful group of models dealt with spatial interactions. Spatial interaction modeling tradition draws from the original efforts to model human activities (Reilly 1931; Zipf 1946). The model's formula, in its most basic form, is that of Newton's law of gravitation. Approaches and models included in this group are the well-known gravity-type models and their reincarnated formulations known more generally now as spatial interaction models. This class of models remains one of the most widely used.

Spatial interaction models are used to study a variety of interactions arising out of a host of human activities within the urban system, such as the journey to work, shopping, land-use transportation interactions, and urban change in general. The analogy of urban expansion with waves was described as a conceptual metaphor early on (Blumenfeld 1954; Boyce 1966). But beyond depicting urban expansion as an expanding wave of peak population density and drawing out the analogy to include tides and such, this work was only conceptual.

With the failures of large 1960s era urban modeling projects, the urban modeling movement slowly faded in the United States. Lee's "Requiem for Large Scale Models" (1973) revived the notion that urban models could be useful tools in planning, but pointed out that the failures of previous models were due to their general instead of specific purpose, command and control approach, and black-box modeling methods. Now, more than 30 years later, the urban modeling movement has generated successful efforts, much improved from the prematurely ambitious applications that led to Lee's "Requiem."

Especially during recent years, urban growth models have generated considerable renewed interest. Spatial modeling has matured and is now a viable tool for city planners, economists, ecologists, and resource managers oriented toward sustainable development of regions (Clarke et al. 2002). Model development has been driven by new data resources, computational processing capabilities, advanced image processing techniques, vastly improved GIS data integration capabilities, and the need for innovative planning tools for decision support (Wegener 1994; Klosterman 1999).

Models have demonstrated potential for supporting planning and management decisions; these include their ability to:

- Provide information and understanding of the dynamics (intuition structuring)
- Anticipate and forecast future changes or trends
- Describe and assess impacts of future development
- Explore different policies through scenario-based planning

Integration with GIS has helped move urban modeling to a morphology-oriented perspective. Prominent examples are the concepts of diffusion limited aggregation (Makse 1998; Andersson et al. 2002), which is based on a theoretical construct borrowed from physics — the concept of fractal cities (Batty and Longley 1994) — and cellular automata-based urban and regional models that developed from the realm of instructive metaphors to that of potentially useful qualitative forecasting tools (Couclelis 1997).

Cellular-based models have now come to the forefront in applied research and, although some research applications are computationally taxing, they are able to accurately generate growth dynamics that are predictive of evolving urban form (Clarke et al. 1997). Part of the recent progress in urban modeling is attributed to the availability of remote sensing. To date, however, remote sensing is seen more as a source of data for visualization rather than a tool directly applicable to the study of urban growth (Longley 2002; Lo 2004; Herold et al. 2005).

10.4 REMOTE SENSING OF URBAN DYNAMICS

Remote sensing, although challenged by the spatial and spectral heterogeneity of urban environments, may be used to delimit, classify, and measure a wide variety of urban phenomena at multiple resolutions (Jensen and Cowen 1999; Jensen 2000; Donnay et al. 2001; Herold et al. 2003). Remote sensing observations provide static freeze-frames of the spatiotemporal pattern associated with urban change (Batty and Howes 2001). Sequential snapshots can be used to generate quantitative descriptors of from–to spatial dynamics. Despite many recognized advantages, urban remote sensing has widely remained "blind to pattern and process" (Longley 2002). The spatial and temporal detail provided by space and airborne remote sensing platforms has yet to be broadly applied for purposes of developing understanding, representation, and modeling of the fundamental characteristics of spatial processes (Longley and Mesev 2000).

Only recently have researchers interested in understanding geographic phenomena combined remote sensing, spatial analysis, and spatial metrics to establish a link between urban form and process and tie empirical observation to theory (Dietzel et al. 2005; Herold et al. 2003, 2005). In that context, Longley and Tobon (2004) emphasize that extending the interests of urban geographers toward more direct, timely, spatially disaggregate urban indicators is key for developing the data foundations for a new, data-rich, and relevant urban geography.

10.5 REMOTE SENSING AND URBAN MODELING

Although it is necessary to advance theory before practice, it is the application of urban models that has shaped many recent advances. Good models and good theory (as the basis for good science) require the incorporation of multiscale dynamics. To this end, considerable progress has been made. However, building, calibrating, and applying models that accurately describe urban dynamics as they are revealed by time-series remote sensing data are quickly becoming an imperative for understanding the complexities of urban growth (Wegener 1994; Longley and Mesev 2000; Irwin and Geoghegan 2001).

Herold et al. (2005) described five different areas in which remote sensing combined with spatial analysis can support urban modeling. These are (1) basic mapping and data support, (2) model calibration and validation, (3) the interpretation, analysis and presentation of model results, (4) the representation of spatial heterogeneity in urban areas, and (5) the analysis of spatiotemporal urban growth patterns. With these areas in mind, remote sensing data have served the needs of many model applications as a source of spatially consistent and timely data. From a more theoretical perspective, remote sensing data provide a unique view of change and representation of spatial heterogeneity and the dynamics of urban structure. Using landscape features and socioeconomic phenomena captured with time-series remote sensing imagery is a significant improvement over traditional models that tend to reduce urban space to a unidimensional measure of distance (Irwin and Geoghegan 2001). Existing urban patterns, measurable with remote sensing, must be viewed as a growth process controlled and constrained by spatial factors. The simplest physical factors are those that are the primaries of the landscape within which cities are formed. For this influence to be quantifiable, there needs to be a link between evolving urban landscape configurations and the effective spatial distribution of underlying growth factors between different geographic scales.

10.6 TOWARDS IMPROVED URBAN THEORY

Thus far, the justifications for using remote sensing to support and improve urban theory and modeling have been presented. This section will present a specific empirical case study that is illustrative of a new concept for assessment of the spatial dynamics previously mentioned. The example will cover the *bottom-up* perspective (remote sensing-based empirical analysis), generalization of the findings, and resulting framework for a *top-down* urban modeling concept.

10.6.1 EMPIRICAL ANALYSIS OF URBAN DYNAMICS

The case study presented here focuses on California's Central Valley, encompassing the cities of Stockton–Modesto, Fresno, and Bakersfield (Dietzel et al. 2005). This study area was chosen because it contains one of the most rapidly urbanizing regions in the Western world. The time span of the data ranges from 1940 to 2040 with historical observations for 1940, 1954, 1962, 1974, 1984, 1992, 1996, and 2000.

The time span of the data series was extended using outputs from the SLEUTH urban growth model (Clarke et al. 1997) for 2010, 2020, 2030, and 2040. Buffers (2, 10, 30, and 60 miles) around the central urban cores of these cities, as defined by the Census 2000 urban areas data set, were used to conduct the multiscale analysis. Scaling was accomplished by the simplest means possible: by changing the radial spatial extent encompassed by buffers, but not changing the spatial resolution, which was fixed at 100 m × 100 m grid cell size.

Simple binary urban/nonurban categorization was used to represent urban extent for this study. Sequential snap-shots permit the application of quantitative descriptors of the geometric properties of urban form to be computed and compared over time. Analysis using multitemporal remote sensing and spatial metrics can provide a unique derived source of information regarding various spatial characteristics of cities changing over time. Spatiotemporal "signatures" generated using this combined approach capture dynamic processes that influence spatial urban structure (Herold et al. 2002). Spatial metrics for quantifying the structure and pattern of thematic maps (including those of urban areas) are commonly used in landscape ecology, where they are referred to as landscape metrics (O'Neill et al. 1988; Gustafson 1998). Calculation of spatial metrics is based on a categorical, patch-based representation of the landscape. The landscape perspective assumes abrupt transitions between individual patches that result in distinct edges. These measures provide a link with the detailed spatial structures captured by remote sensing (Herold et al. 2003). The metric calculations were performed using the public domain software FRAGSTATS, version 3.3 (McGarigal et al. 2002). Most metrics have fairly simple and intuitive values, such as the urban patch (PD) and edge density (ED), and the measures of mean Euclidean distance (ENN_MN) between individual urban areas. The contagion index (CONTAG) is a general measure of landscape heterogeneity and describes the extent to which landscapes are aggregated or clumped (O'Neill et al. 1988). Landscapes consisting of relatively large contiguous patches have a high contagion index. If a landscape is dominated by a relatively large number of small or highly fragmented patches, the contagion index is low. A detailed description of spatial metrics can be found in McGarigal et al. (2002).

Four spatial metrics were used in this study to compare growth signatures for three rapidly urbanizing areas (Figure 10.2). The different line styles in each graph represent the metric signatures derived for the four different spatial extents, plus the central core as defined by the 2000 Census. The contagion metric is a general measure of landscape heterogeneity and is lowest when the urban/rural configuration is most dispersed and fragmented. For the central area, the lowest contagion is found for 1974, when the landscape was most heterogeneous. With further expansion of the urban core, notice that contagion increases as the landscape homogenizes. The spatiotemporal signature of the contagion metric follows a pattern similar to a sine wave. The general wave shape is evident for all extents, but with varying wavelengths.

The wavelength represents the stage of urbanization for each scale and generally increases with distance from the central core. The average nearest neighbor distance shows a peak in the 1950s and 1960s for all scales. This time period represents the initial phase of diffuse allocation of new development units separated

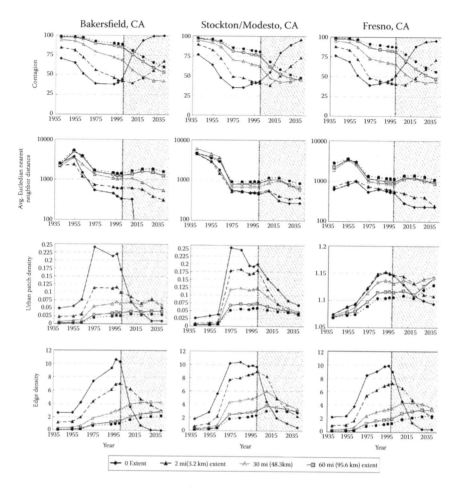

FIGURE 10.2 Spatial metric growth signatures for Bakersfield, Stockton–Modesto, and Fresno, California, for multiple concentric ring buffers. The metric values until 2000 were obtained from remote sensing; 2010 to 2040 is based on SLEUTH model outputs. (After Dietzel, C. et al., 2005, *Int. J. Geogr. Inf. Sci.*, 19, 2, 175–195.)

by large distances. With the major spread of distinct new urban development units in the late 1960s and 1970s, the Euclidian nearest neighbor distances show an accordant decrease. The system of urban areas grows increasingly dense until the year 2000. For the central urban area, the number of patches significantly increased between 1962 and 1974. This increase coincided with the highest rate of diffusive urban sprawl for these areas. The urban expansion is characterized by the diffuse allocation of new development units around the central core. The patch density metric decreases after 1974 as the new individual units grow together and become spatially connected to the urban center. This development results in larger, more heterogeneous and fragmented urban patches. The spatial process that generates this general fragmentation pattern is reflected by the edge density metric, which peaks in the mid-1990s.

The process of coalescence and expansion into open spaces continues toward the later stage of urbanization. This stage is indicated by decreasing patch density and edge density in later dates. Also observed in the contagion metric, patch density metric, and edge density metric is that they all appear to have similar wave-like shapes for all spatial extents. Except for the Euclidian nearest neighbor distance metric, the metric values peak first in the smallest scale, and in chronological order the larger scales respond as urbanization progresses outwards from the central core. In general, the sequence of metric development with an early peak of the nearest neighbor distance, followed by a peak in patch density and then in the edge density, is evident to different degrees in each of the metropolitan areas studied. Perhaps the use of SLEUTH, which is a cellular automata-based model (Clarke et al. 1997) to extend the mapped time series in this manner, allows for a full century of urban growth (Figure 10.2) and does reflect predictions rather than real measurements. This is to be considered in the interpretations, but the model outcomes generally confirm the trend derived from the actual observations for different cities (Figure 10.3).

10.6.2 FROM EMPIRICAL OBSERVATIONS TO A GENERALIZED FRAMEWORK

Figure 10.3 compares the metric growth signatures described for Fresno, which is located in the central valley region of California, with metric signatures for different cities in the California southern coast region (Santa Barbara and Carpinteria) using

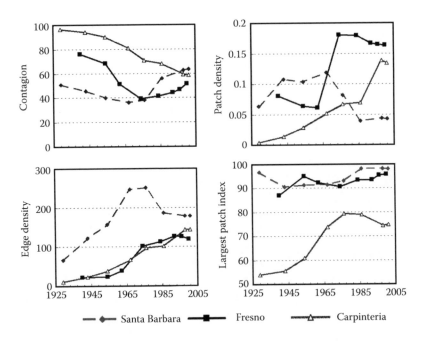

FIGURE 10.3 Temporal growth signatures of three different cities in California — Santa Barbara, Fresno, and Carpinteria — derived from remote sensing observations using spatial metrics.

other data sources derived from time-series remote sensing (Herold et al. 2003). These cities vary in size (50,000 to 1,000,000 inhabitants) and the regional growth histories have been quite different (California's central valley vs. southern coast regions). Santa Barbara showed major growth during the 1950s to 1960s and Fresno in the 1970s to 1990s; Carpinteria is just currently developing as a subsidiary center near Santa Barbara.

The temporal trajectories of the spatial metric signatures shown in Figure 10.3 reveal that comparable growth signatures exist for all three subregions of the California southern coast region, considering that the timing and absolute values are shifted due to specific local growth histories. There seem to be common growth patterns shown for each of the three cities. Dietzel et al. (2005) have distinguished these common growth patterns as phases of coalescence and diffusion and established an explanatory hypothetical framework for spatiotemporal urban expansion that theorizes dynamics in terms of alternating diffusion and coalescence processes. The framework hypothesizes that urban growth can be characterized as having two distinct processes and generally follows a harmonic oscillation (Figure 10.4).

As is shown with the contagion metric (Figure 10.2 and Figure 10.3), urbanization is reflected in a transformation from homogenous nonurban to a heterogeneous mix of urban and nonurban. At some time in the progression of the theoretical development trajectory depicted in Figure 10.4, there is a transition to a more homogenous urban landscape. The other three spatial metrics — average Euclidian nearest neighbor distance, urban patch density, and edge density — capture spatiotemporal phases. The phases can be differentiated into two diffusion phases and two coalescence phases. The first phase of diffusion represents the seeding of new development centers. This first diffusion is indicated by a peak in the nearest neighbor distance metric, which reflects the establishment of peripheral development centers around the original core. The second phase of diffusion is the allocation of a large number of new urban areas in the nascent urban system comprising the original core and peripheral development centers. The nearest neighbor distance metric drops and

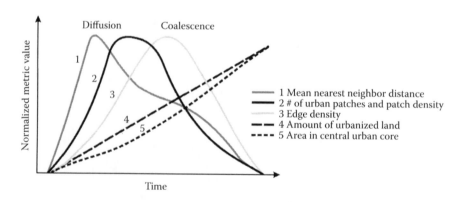

FIGURE 10.4 Theoretical spatial metric signatures for a full cycle of urbanization for uniform isotropic growth at a specific scale. (After Dietzel, C. et al., 2005, *Int. J. Geogr. Inf. Sci.*, 19, 2, 175–195.)

the patch density peaks during this second phase of diffusion. The amount of urban land in the largest patch is the lowest at this point. In Figure 10.3, the phase of diffusion corresponds to 1940s to 1960s for Santa Barbara and 1960s to 1970s for Fresno. Although Santa Barbara and Fresno are currently in the process of coalescence, Carpinteria is at the end of the first diffusion phase.

The low point in the contagion metric marks the transition from diffusion to coalescence. Coalescence starts as urbanized areas aggregate. This is reflected by a decrease in the patch density, edge density, and nearest neighbor distances metrics. The first phase of coalescence spatially aggregates individual urban centers close to each other first, indicated by a slight plateau in the nearest neighbor distance metric and a decrease in patch density. This increases the fragmentation of the urban area and the edge density metric peaks within the first phase of coalescence. The second phase of coalescence is characterized by a decrease in the patch density, edge density, and nearest neighbor distances, which ultimately results in complete urbanization. The terminal point of coalescence is complete urban build-out when all, or nearly all, of the available land has been urbanized.

This "final" stage can be seen as an initial urban core for further urbanization when viewed at a scaled-out extent. It is hypothesized that similar growth patterns — hence, a different temporal dimension — can be observed across scales with varying spatial extents (Dietzel et al. 2005). For example, if a city progressively expands from its central core, the described spatial processes will occur similarly across a ripple of concentric rings with increasing distance from the center (Figure 10.2). The periods of the phased related cycles were expected to be longer as a function of distance from the central core (hence at larger spatiotemporal scales).

The theoretical framework constructed around diffusion and coalescence is adaptable to the scaling of relationships concerned with changes in the spatial extent. This was investigated in this study — that is, the spatiotemporal characteristics of diffusing and coalescing processes (wavelength and amplitude of cycles) as a function of distance from the central urban core. In that context, empirical observations have revealed some interesting behavior with the spatial metrics. Most metrics follow the hypothesized pattern with longer cycles of urbanization with increasing distance from the central core (e.g., for contagion and edge density). On the other hand, the nearest neighbor standard distance-based measures and the urban patch density exhibit a parallel behavior among the multiple rings (Figure 10.4). This suggests that the diffusion process similarly affects all concentric rings, whereas coalescence propagates outwards with a distinct increase for larger concentric radial distances from the original core.

Urban modeling studies have shown (Torrens and O'Sullivan 2001; Wu 2002; Herold et al. 2003) that cellular automata-based urban models tend to generate too compact and overly aggregated patterns because of their dependence on local growth rules. In terms of spatial processes, the cellular automata model overemphasizes the coalescence of urban areas or does not allow for sufficient diffusion. In other words, the models do not sufficiently reflect the four stages of spatial evolution as described by empirical observations and the theoretical framework. Given sufficient information that describes the process of diffusion and coalescence for a particular area, the model outputs are able to serve as a guide or reference for potentially more accurate representations of dynamic spatial processes (Torrens and O'Sullivan 2001).

However, the link between empirical measurements (Figure 10.2 and Figure 10.3) and the theoretical concept (Figure 10.4) is, for now, only of a qualitative nature. A quantitative comparison reveals differences among metric signatures in amplitude, duration, location, and extent. These differences were anticipated in light of the fact that urban growth is not constant over time and among the different regions. Furthermore, the spatial configurations of these areas are not uniform and the initial conditions for each developing city system are not identical with regard to the starting point of the data time series. Local urban growth factors such as topography, transportation infrastructure, growth barriers or planning efforts affect the spatial growth pattern. However, local variations yield important information about the ongoing processes. They can be interpreted as "distortions" (i.e., amplifications, lagging, or damping the metric signatures). Such distortions can be thought of as residuals of an idealized baseline growth pattern under uniform, isotropic spatial and temporal conditions. These unique components could then be studied independently of those that are simply the geometries of the urban growth process that all growing cities could have in common.

10.6.3 Evolving a Novel Model of Urban Dynamics

We started to evolve the empirical observations and conceptual model thus far discussed into an urban modeling framework; the goal was to bridge the inductive, bottom-up, remote sensing observations with the top-down perspectives of urban theory and urban growth models. To further elaborate on this relationship, a simplified geometric model will be presented that has the potential of generating comparative baseline patterns. What is lacking in the framework as it has been developed is a means with which to isolate the components of observed urban growth patterns analytically. Circles have a long tradition in urban geographic research. Shown in Figure 10.5 is an illustration of interacting city systems. The two expanding radii are intended to be descriptive of the interacting scales of socioeconomic factors over time as they can be conceptualized for planning purposes. Using circles as a basic geometric shape to represent ranges of influence is also prevalent in contemporary urban geographic research.

The use of circles to represent the spatial evolution of urban systems, like the growth signatures of the hypothetical urbanizing area depicted in Figure 10.4, is an intuitive leap that superficially seems grossly oversimplified. Albeit abstract, a geographic model based on circles will enable further development of the theoretical framework by experimentally defining ranges of metric values for situations in which dynamics can be controlled.

Figure 10.6 shows a sequence of snapshots from a prototype model that generates patterns starting from a set of differently sized circles dispersed randomly and grown in a varied sequence on an isotropic surface with uniform growth characteristics. This sequence mimics the process of diffusion and coalescence found using the spatial metric analysis of time-series remote sensing data shown in Figure 10.3. This sequence can be seen as further support for the theoretical framework of urban growth harmonics (Dietzel et al. 2005). Outputs of this experimental model can be used to scale the empirically derived spatial metric signatures against a common baseline expectation. Values for the number of circles, circle areas, overlap areas,

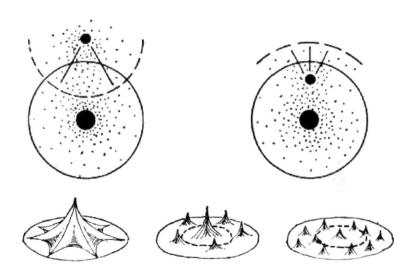

FIGURE 10.5 This illustration shows theoretical interactions of two major city systems. (From Guttenburg, E.Z., 1964, in Webber, M.M. (Ed.). *Explorations into Urban Structure*, University of Pennsylvania Press, Philadelphia, 197–219.)

separation distance, edge distance, and other metrics can be iteratively generated and analyzed.

This model will be used in future research to observe the pattern signature that results from differing arrangements of seed locations, circle sizes, and origin locations in simulated urban systems. An improved version will contribute a simplified case as a baseline with which indicators can be derived that describe changing landscape patterns. Beyond this overly simplified model, more complex geometric structures can be incorporated by integrating specific distortions (roads, topography, water, etc.) that constrain or contort patterns in urban development. The complexity of the experimental environment will be increased through the introduction of distorting factors, which will alter the simplified growth pattern and permit a more analytically based under-standing of the role that these factors play in shaping urban form.

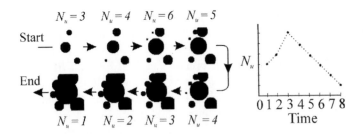

FIGURE 10.6 Sequential frames from the simple circle model. The graph on the right shows $N\mu$, number of agglomerations, through a sequence of time steps.

10.7 THINKING AHEAD: ECOLOGY CONCEPTS AND SCALING IN URBAN SYSTEMS

Underlying any geographical framework, such as diffusion and coalescence, is the concept of scaling. Dynamics can be measured with spatial metrics, as has been shown by Herold et al. (2002, 2005) and Dietzel et al. (2005); however, to model the changing spatial character of urban growth, a much simpler concept is needed. Although it may not be comparable directly, the simple models developed in ecology and population biology for describing living systems provide some foundation for describing social systems — the systems of cities.

Thinking imaginatively of a city as a living organism, or of a system of cities as an ecosystem, allows for translation of some of the quantitative theoretical models developed to describe them. The early work of sociologists seeking to understand urban structure in socioeconomic terms (Park et al. 1925) adapted the concepts of competition and succession from ecology. This translation is, of course, not direct because social systems are not controlled in the same manner as natural systems are by environment constraints. What do cities and living things have in common? Without delving into drawing out analogies, it can be pointed out for the purposes of the following description that cities and living systems consume, grow, and disperse.

The ultimate purpose of ecology is the understanding of consequences (Levin 1992). To this end, numerous methods have been discovered based on controlled experiments and empirical investigations. Metabolism is essentially the processes of living; it involves inputs and outputs, growth, action, and energy. Living systems as well as social systems, cities have a characteristic metabolism. This is a simple concept to grasp in basic terms. In detail, however, the complexity of metabolic processes and the analogous processes of the functioning and growth of cities is extremely complex. Metabolic theory, in short, is a causal explanation for the simple but universal relation of size, temperature, and metabolic rate. A theoretical framework explaining the simple but ubiquitous scaling relationship of body mass and metabolic rate found in the natural world was first introduced by West et al. (1997). In 2001 and 2002, two significant articles along this same line of inquiry were published (Gillooly et al. 2001, 2002). These publications have had a wide-ranging influence on the fields of biology and ecology. The ideas put forth are powerful in terms of shaping new perspectives regarding the understanding of living systems.

The variability of the natural world leads one to appreciate what is distinct, but without a baseline, measuring distinctiveness is almost impossible. Having a valid theoretical framework allows for formulation of a null hypothesis or, at the very least, a means of quantitatively interpreting variability (Whitfield 2004). Complex systems can be described with simple power law relationships. With regard to urban geography, Zipf's law is a prominent example. Cities scale according to their population sizes as a power law distribution (Zipf 1949). With a few exceptions, that the scaling exponent is approximately three fourths for size vs. metabolic rate of living systems is not in doubt. But both of these scaling "laws" were empirical oddities, not laws in a theoretical sense, until relatively recently.

Cross-pollination of concepts, theories, and methods across disciplines encourages the adoption of new perspectives. Understanding systems requires knowledge of scaling, the causes and consequences that can be inferred from observable patterns. In addition to being an analysis tool, spatial metrics have enabled a new perspective for study of spatial heterogeneity. This new concept from ecology has been adopted wholeheartedly by a large number of scientists. The core utility of these new theoretical frameworks within ecology and biology is that they can be used to relate measurements across scales. The tantalizing implications of scaling relationships in ecology, like landscape metrics, have motivated considerable research.

Pattern depends on scale; the ability to observe pattern is one thing but the ability to quantify it accurately is another. Levin describes a limited scale of observation as a "low-dimensional slice through a high-dimensional cake." Scale is imposed by a deliberately chosen extent or resolution often to contain variability within a manageable range. Levin (1992) and Chave and Levin (2003) relate the theoretical and empirical understandings directly and indirectly associated with describing system dynamics with models and the modeling of multiscale processes.

For most geographers, the concepts of detail and resolution are linked inextricably to space. Successful models that are developed empirically or experimentally describe system behavior across a range of scales in space, time, and organizational complexity. The important point is that models that incorporate multiscale variability need to retain only enough detail to capture the essence of the pattern under consideration (Levin 1992). Simplicity in the context of models refers to the level of detail required to reproduce essential components of the patterns of interest; individual level quantification is irrelevant for modeling of planet scale variability. But what patterns are relevant for studies of urbanization? It is exactly this question that the concepts presented in this chapter and the ongoing research that has thus far been presented hope to answer.

Urban modeling has necessitated certain perspectives and scales that are chosen based on conventional as well as logistical considerations. Also important are mechanistic translations of dominant scale patterns and how they change across spatial and temporal scales. Response is observable as a spatial pattern, but lacking understanding of the most essential components of that pattern limits which responses can transfuse to finer scales. Macro- to microscaling is often critically important for the applications of predictive models, or at least it is the most demanding translation that is often requested. To this end there are the top-down vs. bottom-up perspectives for urban growth modeling, neither of which is universally more correct that the other. Detail in such models only need be enough to capture the variability relevant to a highly aggregated and simplified system, but much of the observed responses are microscale in origin. In fact, the temporal and spatial mismatch of relevant scales limits insights about the system as a whole, as does limited understanding of observations of the whole being transferable to the parts.

The individual cases in a system are in and of themselves unpredictable and random; however, with knowledge of their regular behavior as a collection of cases, useful generalizations permit inference about process (Levin 1992). Detailed knowledge is given up for reliable generalizations because, by aggregating, local variability is reduced and system predictability is improved. O'Neill et al. (1988) recognized this spatially with the purpose behind the advent of landscape metrics — the spatial description of changing patterns occurring at multiple scales simultaneously.

Identification of pattern is a necessary precursor to description of variation, but quantification of variation first requires understanding of the relationship of process and scale. Thus, pattern and scale are inextricably linked (Levin 1992). To this end, considerable effort has been devoted to the quantitative description of pattern (e.g., Burrough 1981; Gardener et al. 1987). But, as Levin (1992) points out, "There are many roads to Rome … there will be many conceivable mechanisms that could give rise to any set of patterns. All that theory alone can do is to create a catalogue of possible mechanisms." Thus, the purpose of an experiment is to distinguish among candidate mechanisms, but first the components of pattern that are extrinsically determined must be identified and factored out. According to Levin, this is accomplished by understanding how or what information is transferred across scales, what information is lost, and what is preserved.

The superficial simplicity of the theoretical model of urban growth proposed in the previous section is misleading. To address the lack of dynamics in urban growth models, what is first necessary are adequately testable hypotheses regarding the spatiotemporal patterns of urban growth. A whole host of philosophical and methodological research questions surrounds the development of such an experiment as we have devised it. In addressing the relation of multiscale urban growth dynamics using the combined approach of spatial metrics, time-series remote sensing data, and a theoretical baseline as an ideal case, it will be possible to identify components of urbanization that are the spatial responses we need to understand if urban growth modeling is ever to address dynamics adequately.

10.8 CONCLUSIONS

Understanding urban dynamics based on actual measurements from remote sensing and in the context of existing and evolving urban theories remains challenging. However, through the use of remote sensing, the spatial evolution of urban systems can be better described, measured, and modeled. We have outlined an integrative approach whereby empirical observations can be used for comparative analysis based on time-series remote sensing data and an experimental model. Also incorporated is a bridge connecting theoretical understandings of the spatial evolution of urban areas, the analytical modeling of systems, and the role that urban remote sensing can play. Contributions in the form of understandings regarding spatial components of urban growth dynamics with this approach are potentially rewarding and uniquely insightful.

The main objectives of empirically based research and the conceptual development of our theoretical model discussed in this chapter rely on remote sensing for: (1) historical time-series of urban extent that is our means of quantitatively assessing the spatial evolution of the urban system; (2) ability to observe patterns in spatiotemporal metric signatures; and (3) development and testing of a theoretical framework that will help explain the dynamic evolution of cities through time. An integral part of developing a method for assessing theoretical phase-related patterns (Figure 10.4) will be experimenting with the manipulation of a simplified growth model (Figure 10.6) and introducing perturbations. Future research will be testing the hypothesized metric signatures shown in Figure 10.5 against multiple urban areas and developing an analytically solid means of diagnosing the phases of diffusion and coalescence, thus using remote sensing to validate urban growth theory.

With a controlled experiment, it may be possible to characterize spatial responses to factors that distort overall patterns. Such a finding may lead to identification and characterization of commonalities that stem from the geometric properties of the space filling spatial behaviors of the urbanization process as well as a means of isolating unique and potentially interesting patterns. Studying the dynamic nature of the urbanization process as it is captured by data sources that are static snapshots (such as from time-series remote sensing-derived land-cover products) involves a difficult set of assumptions — difficult because they do not lend themselves well to unambiguous identification and description. However, some assumptions are necessary to address spatiotemporal dynamics. First, the process of urbanization is in reality continuous and nonuniform, and the temporal scale of analysis is fixed by the dates of the data sets used. Second, the spatial resolution of data sources used for the comparative part of the analysis will also be fixed and thus impart uncertainty regarding any findings and most certainly will influence calculation of spatial metrics.

The results of this research and the preliminary development of the theoretical framework based on urban growth phases provide encouragement for future research. Remote sensing delivers accurate urban mapping capabilities and a wide range of temporal and spatial scales, which are necessary for validation of urban theory. What is necessary is the articulation of a theoretically sound approach with which to address cross-scale urban dynamics. The results presented make it clear that the combination of remotely sensed data and spatial measurements (metrics) has the potential to aid in development and validation of new urban theory.

REFERENCES

Andersson, C., Lindgren, K., Rasmussen, S., and White, R. 2002. Urban growth simulation from "first principles." *Phys. Rev. E* 66(02620).

Batty, M. 2002. Thinking about cities as spatial events. *Environ. Plann. B* 29, 1–2.

Batty, M. and D. Howes, 2001. Predicting temporal patterns in urban development from remote imagery, in Donnay, J.P., Batty, M., and Longley, P.A. 1994. *Fractal Cities: A Geometry of Form and Function*, Academic Press, London.

Batty, M. and P.A. Longley. 1994. *Fractal Cities: A Geometry of Form and Function*, Academic Press, London.

Blumenfeld, H. 1954. The tidal wave of metropolitan expansion. *J. Am. Instit. Planners* 20, 3–14.

Boyce, R.R. 1966. The edge of the metropolis: the wave theory analog approach. *Br. Columb. Geogr. Ser.* 7, 31–40.

Burrough, P.A. 1981. Fractal dimension of landscapes and other environmental data. *Nature* 294: 240–242.

Chave, J. and S. Levin, 2003. Scale and scaling in ecological and economic systems. *Environ. Resour. Econ.* 26, 527–557.

Clarke, K.C., Parks, B.O., and Crane, M.P. 2002. *Geographic Information Systems and Environmental Modeling*, Prentice Hall, Englewood Cliffs, NJ.

Clarke, K.C., Hoppen, S., and Gaydos, L. 1997. A self-modifying cellular automata model of historical urbanization in the San Francisco Bay area. *Environ. Plann. B* 24, 247–261.

Couclelis, H. 1997. From cellular automata to urban models: new principles for model development and implementation. *Environ. Plann. B – Plann. Design* 24, 165–174.

Dietzel, C., Herold, M., Hemphill, J.J., and Clarke, K.C. 2005. Spatio-temporal dynamics in California's Central Valley: empirical links urban theory. *Int. J. Geogr. Inf. Sci.* 19, 2, 175–195.

Donnay, J.P., Barnsley, M.J., and Longley, P.A. 2001. Remote sensing and urban analysis, in Donnay, J.P., Barnsley, M.J., and Longley, P.A. (Eds.). *Remote Sensing and Urban Analysis*, Taylor & Francis, London, 3–18.

Gardner, R.H., Milne, B.T., Turner, M.G., and O'Neill, R.V. 1987. Neutral models for the analysis of broad-scale landscape pattern. *Landscape Ecology*, 1: 19–28.

Gillooly, J.F., Brown, J.H., West, G.B., Savage, V.M., and Charnov, E.L. 2001. Effects of size and temperature on metabolic rate. *Science* 293(5538), 2248–2251.

Gillooly, J.F., Charnov, E.L., West, G.B., Savage, V.M., and Brown, J.H. 2002. Effects of size and Temperature on Developmental Time. *Nature* 417: 70–73.

Gustafson, E.J. 1998. Quantifying landscape spatial pattern: What is the state of the art? *Ecosystems* 1, 143–156.

Guttenburg, A.Z. 1964. The tactical plan, in Webber, M.M. (Ed.). *Explorations into Urban Structure*, University of Pennsylvania Press, Philadelphia, 197–219.

Herold, M., Couclelis, H., and Clarke, K.C. 2005. The role of spatial metrics in the analysis and modeling of land use change. *Computers, Environ. Urban Syst.* 29(4), 369–399.

Herold, M., Goldstein, N.C., and Clarke, K.C. 2003. The spatiotemporal form of urban growth: measurement, analysis and modeling. *Remote Sensing Environ.* 86, 286–302.

Herold, M., Clarke, K.C., and Scepan, J. 2002. Remote sensing and landscape metrics to describe structures and changes in urban land use. *Environ. Plann. A* 34, 1443–1458.

Irwin, E.G. and J. Geoghegan, 2001. Theory, data, methods: developing spatially explicit economic models of land use change. *Agric., Ecosyst. Environ.* 85, 7–23.

Jensen, J.R. 2000. Remote sensing in the urban landscape, in *Remote Sensing of the Environment: An Earth Resource Perspective*, Prentice Hall, Upper Saddle River, NJ, 407–470, 544.

Jensen, J.R. and D.C. Cowen, 1999. Remote sensing of urban/suburban infrastructure and socio-economic attributes. *Photogrammetric Eng. Remote Sensing* 65, 611–622.

Klosterman, R.E. 1999. The what if? Collaborative planning support system. *Environ. Plann. B: Plann. Design* 26, 393–408.

Lee, D.B. 1973. Requiem fro large-scale models. *Journal of the American Institute of Planners* 39, 163–178.

Levin, S.A. 1992. The problem of pattern and scale in ecology. *Ecology* 73: 1943–1967.

Lo, C.P. 2004. Testing urban theories using remote sensing. *GISci. Remote Sensing* 41(2), 95–115.

Longley, P.A. and C. Tobon, 2004. Spatial dependence and heterogeneity in patterns of hardship: an intra-urban analysis. *Ann. Assoc. Am. Geogr.* 94(3), 503–519.

Longley, P.A. 2002. Geographical information systems: will developments in urban remote sensing and GIS lead to "better" urban geography? *Prog. Hum. Geogr.* 26, 231–239.

Longley, P.A. and V. Mesev, 2000. On the measurement of urban form. *Environ. Plann. A* 32, 473–488.

Makse, H., Andrade, J., Batty, M., Havlin, S., and Stanley, E. 1998. Modeling urban growth patterns with correlated percolation. *Phys. Rev. E*, 58, 6, 7054–7062.

McGarigal, K., Cushman, S.A., Neel, M.C., and Ene, E. 2002. FRAGSTATS: spatial pattern analysis program for categorical maps, www.umass.edu/landeco/research/fragstats/ fragstats.html, (accessed 30 Jan 2005).

O'Neill, R.V., Krummel, J.R., Gardner, R.H., Sugihara, G., Jackson, B., Deangelis, D.L., Milne, B.T., Turner, M.G., Zygmunt, B., Christensen, S.W., Dale, V.H., and Graham, R.L. 1988. Indices of landscape pattern. *Landscape Ecol.* 1, 153–162.

Park, R., Burgess, E., and McKenzie, R. 1925. *The City.* University of Chicago Press, Chicago.

Reilly, W. 1931. *The Law of Retail Gravitation.* The Knickerbocker Press, New York.

Torrens P. and D., O'Sullivan, 2001. Cellular automata and urban simulation: where do we go from here? *Environ. Plann. B* 28, 163–168.

Wegener, M. 1994. Operational urban models: state of the art. *J. Am. Plann. Assoc.* 60(1), 17–30.

West, B., Brown, J., and Enquist, B. 1997. A general model for the origin of allometric scaling models in biology. *Science* 276, 122–126.

White, R., Luo, W., and Hatna, E. 2001. Fractal structures in land use patterns of European cities: form and process. 12th European Colloquium on Quantitative and Theoretical Geography, September 7–11 2001. St-Valery-en-Caux, France.

Whitfield, J. 2004. Ecology's big, hot idea. *PLoS Biol.* 2:1–10.

Wu, F. 2002. Calibration of stochastic cellular automata: the application to rural–urban land conversions. *Int. J. Geogr. Inf. Sci.* 16, 795–818.

Zipf, G. 1949. *Human Behavior and the Principles of Least Effort.* Addison Wesley, Cambridge, MA.

Zipf, G. 1946. The P1P2/D hypothesis: on the intercity movement of persons. *Am. Sociol. Rev.* December 1946, 677–686.

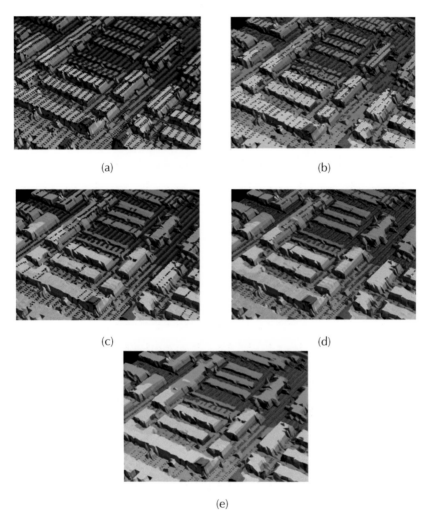

(a)

(b)

(c)

(d)

(e)

COLOR FIGURE 2.4 Stepwise three-dimensional illustration of the filtering process (Baltimore suburb). a: Raw data; b: ground; c: ground points from backward filtering points from forward filtering; d: ground points after forward and backward filtering; e: final ground points after regression.

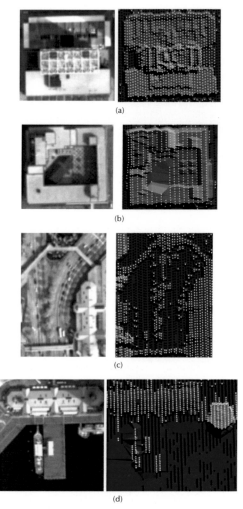

(a)

(b)

(c)

(d)

COLOR FIGURE 2.5 Filtering results of urban features. a: Building 1; b: building 2; c: bridges, trees; d: vehicles; e: water.

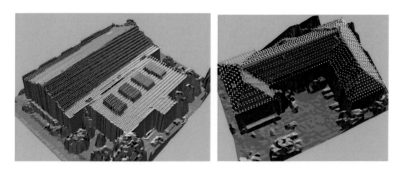

COLOR FIGURE 2.10 Color-coded planar roof points via clustering (Purdue campus).

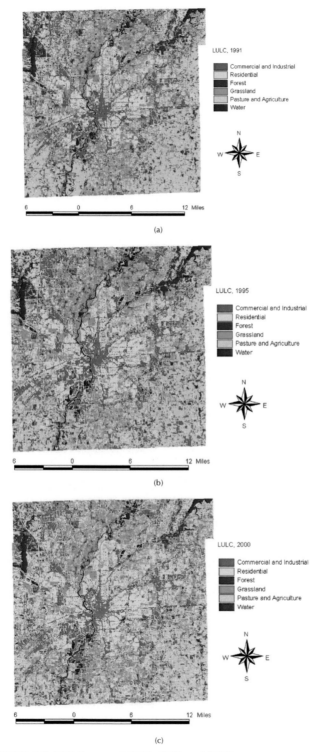

COLOR FIGURE 4.6 LULC maps of 1991, 1995, and 2000.

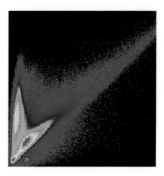

COLOR FIGURE 5.1 The red/near-infrared feature space (horizontal = red, vertical = near-infrared) for an Ikonos image collected over Bangkok, Thailand, on November 27, 2002. At 4-m spatial resolution, most of the pixels are pure pixels. Vegetation and other urban components separate well in the feature space, but each varies continuously in the spectral space.

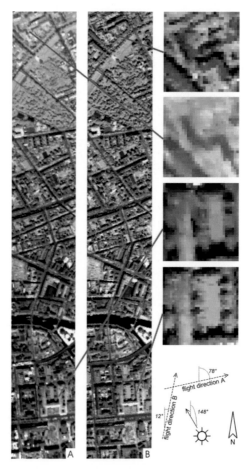

COLOR FIGURE 7.4 Subsets from two HyMap images of the same area in Berlin, Germany, flown at 70° (A) and 44° (B) to the principal plane (red: 828 nm; green: 1647 nm; blue: 661 nm). A severe brightness gradient can be observed in A, whereas the corresponding nadir region of B is free of such phenomena. The zoom areas show that the effect is greatest for vegetated surfaces.

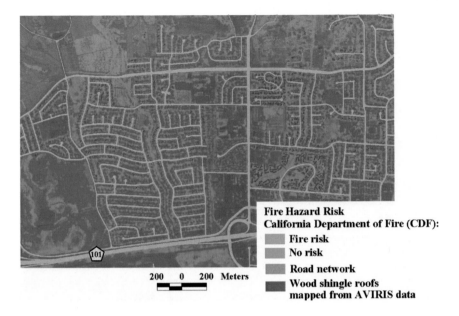

Fire Hazard Risk
California Department of Fire (CDF):

▢ Fire risk
▢ No risk
▢ Road network
■ Wood shingle roofs
  mapped from AVIRIS data

200   0   200   Meters

COLOR FIGURE 7.7 Distribution of wood shingle roofs in the Fairview region of the Santa Barbara urban area derived from AVIRIS high-resolution data with 4-m spatial resolution compared to the California Department of Fire risk map.

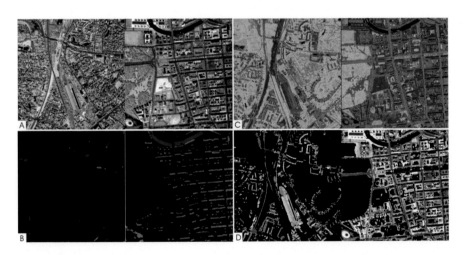

COLOR FIGURE 7.8 Processing steps of impervious surface mapping based on HyMap data from two areas in Berlin, Germany. A: original image data (red: 828 nm; green: 1647 nm; blue: 661 nm); B: separation of shaded areas (red) and water (blue); C: results from SMA (red: asphalt; green: vegetation; blue: soil); D:- mask of impervious surfaces with same band combination as in A.

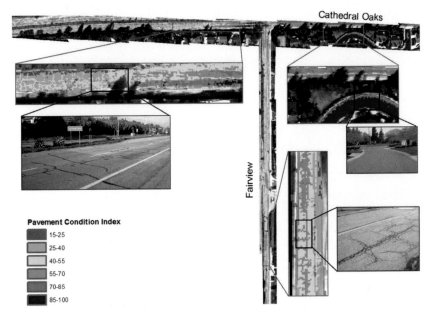

COLOR FIGURE 7.10 Spatial distribution of pavement condition index for selected asphalt roads in the Goleta, California, area derived from hyperspectral HyperSpectir data with 0.5-m spatial resolution using band ratios.

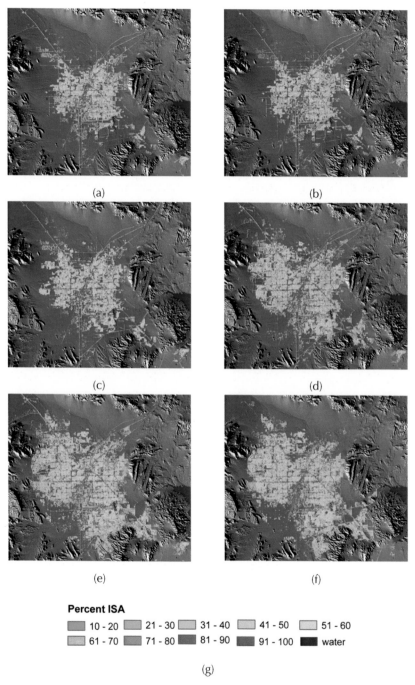

(a)

(b)

(c)

(d)

(e)

(f)

Percent ISA

■ 10 - 20	□ 21 - 30	□ 31 - 40	□ 41 - 50	□ 51 - 60
□ 61 - 70	□ 71 - 80	■ 81 - 90	■ 91 - 100	■ water

(g)

COLOR FIGURE 9.4 ISA distribution in Las Vegas from 1984 to 2002. All figures were obtained from Landsat imagery in 30-m resolution, except the lower right-hand one, which was mapped from ASTER 2002 imagery in 15-m resolution.

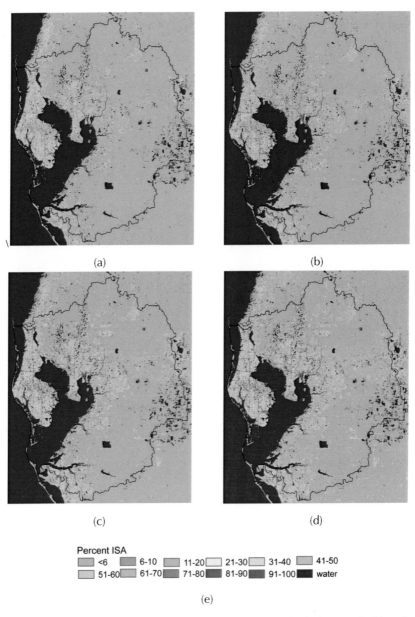

Percent ISA

<6	6-10	11-20	21-30	31-40	41-50
51-60	61-70	71-80	81-90	91-100	water

(e)

COLOR FIGURE 9.8 ISA distribution in Tampa Bay. The black line is the watershed boundary.

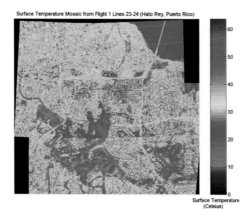

COLOR FIGURE 11.9 Mosaic of Hato Rey flight 1, lines 23 and 24. Surface temperatures are measured in degrees Celsius.

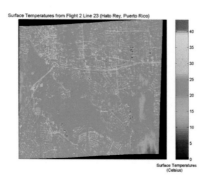

COLOR FIGURE 11.13 Hato Rey flight 2, line 23. Surface temperatures measured in degrees Celsius.

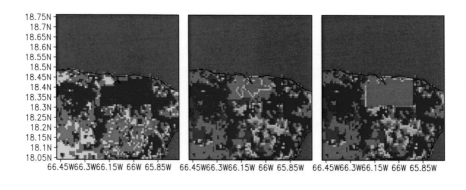

COLOR FIGURE 11.19 Specification of the surface characteristics used in the three runs of the atmospheric model, simulating, from left to right, natural, present, and urban scenarios.

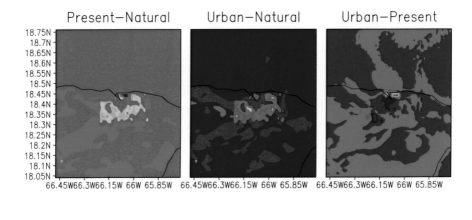

COLOR FIGURE 11.21 Spatial distribution of the air temperature difference (degrees Celsius) at 2 m AGL among the three scenarios simulated for the analysis. The interval of the contours is 0.5°C for the first two panels and of 0.2°C for the third panel.

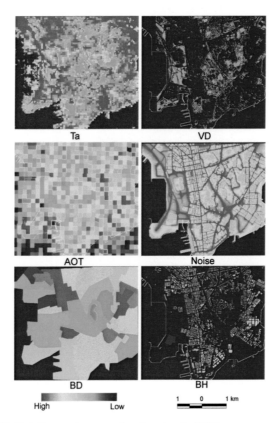

COLOR FIGURE 12.1 Raster data sets showing the six UEQ parameters.

COLOR FIGURE 12.9 Terrain model looking north across Victoria Harbor to the shoreline at Causeway Bay. Upper and lower left-hand figures show selected buildings extruded to actual height, colored according to image-derived Ts. Lower right hand represents model overlain with IKONOS image. Buildings blocking fresh-air corridors are labeled "A."

COLOR FIGURE 14.6 DMSP OLS image and Google Earth image of Mexico City, Mexico.

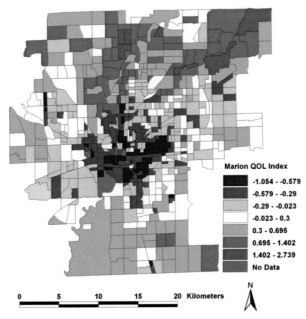

COLOR FIGURE 15.6 Synthetic QOL index in Marion County.

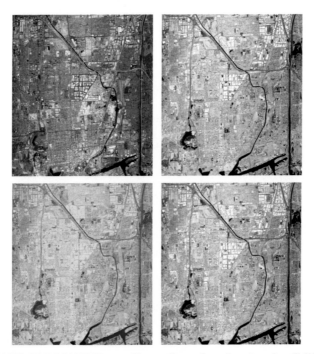

COLOR FIGURE 16.5 MASTER upwelling radiance from Los Angeles, California, on 06 June 2000: (upper left) reflected shortwave VNIR (R = 0.910, G = 0.658, B = 0.542 μm); (upper right) emitted longwave TIR (R = 10.6, G = 11.2, B = 12.1 μm); (lower left) reflected + emitted MIR 3 to 5 μm. (R = avg, G = stdev, B = skewness); (lower right) MIR + TIR (R = avg MIR, G = 11.2, B = 12.1 μm). The lower panels demonstrate that, within this scene, there is more information content in the MIR than in the TIR.

COLOR FIGURE 16.7 False color composites from a MASTER scene of Albuquerque, New Mexico, acquired June 3, 1999: (left) conventional RGB composite displaying near infrared/red/green bands; (right) the experimental composite normalized difference MIR indices displayed as red = organic matter index (OMI), green = quartz surface index (QSI), and blue = carbonate surface index (CSI). Magenta tones indicate high levels of dried plant materials at the surface. Green tones indicate concrete surfaces. Bright blue tones reveal exposed subsoils (caliche).

(a)

COLOR FIGURE 17.1 Ikonos image of Bremen, Germany. The panchromatic resolution of 1 m offers suitability for many urban applications (a). The multispectral bands allow true-color (b) and false-color (c) infrared information at a resolution of 4 m. (Images courtesy of OHB Systems, Bremen, Germany.)

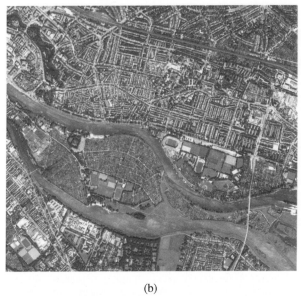

(b)

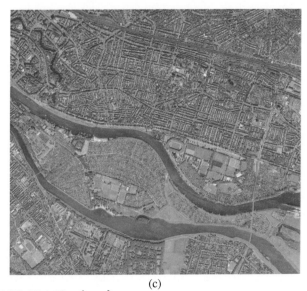

(c)

COLOR FIGURE 17.1 (Continued).

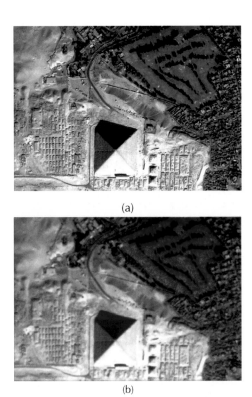

(a)

(b)

COLOR FIGURE 17.2 Quickbird image of the pyramids of Giza, Egypt. The panchromatic resolution is 0.70 cm due to the camera tilt for the acquisition (maximum resolution for nadir view is 0.61 cm). The information content challenges aerial photography (a). The Quickbird multispectral CIR bands were resampled to the panchromatic pixel size. Even the resolution of 2.80 m (maximum resolution for nadir view is 2.44 m) can prove sufficient for many urban remote sensing applications (b). (Images courtesy of Globe View, Longmont, Colorado, and distributed by Leica Geosystems Geospatial Imaging, Norcross, Georgia.)

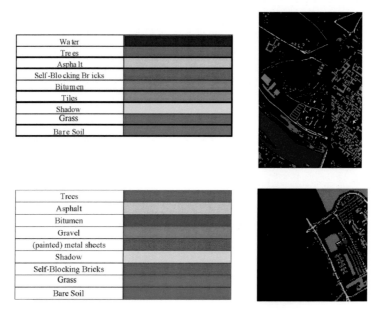

Water	
Trees	
Asphalt	
Self-Blocking Bricks	
Bitumen	
Tiles	
Shadow	
Grass	
Bare Soil	

Trees	
Asphalt	
Bitumen	
Gravel	
(painted) metal sheets	
Shadow	
Self-Blocking Bricks	
Grass	
Bare Soil	

COLOR FIGURE 18.2 Test sets for the two validation areas (sampled to match ROSIS spatial resolution), together with corresponding legends.

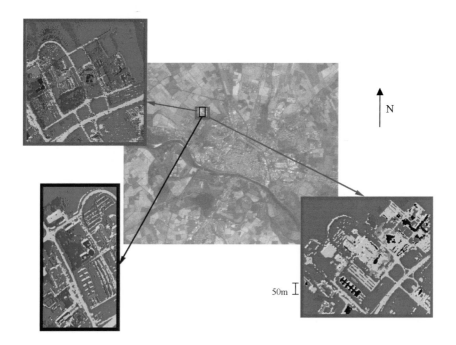

N

50m

COLOR FIGURE 18.3 Second test area (university campus): classification maps obtained using the DAIS (red border), ROSIS (blue border), and Quickbird-1 (green border) data.

Part IV

---

*Urban Planning
and Socioeconomic Applications*

11 Urban Heat Island Identification and Climatological Analysis in a Coastal, Tropical City: San Juan, Puerto Rico

Jorge E. González, Jeffrey C. Luvall, Douglas L. Rickman, Daniel Comarazamy, and Ana J. Picón

CONTENTS

11.1 INTRODUCTION

It is difficult to imagine that cities on small tropical islands show local climate change effects similar to those in great continental cities. Sea breeze usually controls urban climates of topical coastal cities, overcoming any local effect.

However, this may change as significant population growth in tropical coastal locations is expected (IPCC, 2001). It was recently discovered that this might be the case for the city of San Juan, Puerto Rico, a relatively affluent coastal tropical city of nearly 2 million inhabitants, which is growing rapidly and occupying areas around it that were once rural. Until recent times, the forest area around the metropolitan area of San Juan recovered from deforestation of the past. Nevertheless, the fast expansion of urban areas has given rise to loss of much of the reforested area in recent years. A climatological analysis of the surface temperature has revealed that the surface temperature of San Juan is increasing when compared to the surrounding vegetated areas at a rate of 0.06°C every year for the past 30 years, a tendency comparable with other regional climate changes induced by global warming.

These results encouraged planning and execution of an intense field campaign in February 2004 that was denominated the San Juan ATLAS Mission. The main scientific objective of this field campaign was to investigate the impact of fast urbanization on the local climate. The campaign included observations of remote sensing from an airplane, launchings of weather balloons, and data from a network of weather stations and temperature sensors deployed in the urban area and its surroundings. This chapter is centered in the information provided by remote and surface sensors during the San Juan ATLAS Mission and in results of simulations made with an atmospheric numerical model that demonstrate the impact of the metropolitan area of San Juan on different local climatic variables.

11.2   URBANIZATION AND THE URBAN HEAT ISLAND (UHI)

Urbanization is an extreme case of land-cover or land-use (LCLU) change. Although only 1.2% of the Earth's surface is at the moment urban, it contains half of the population, with a tendency of increasing spatial cover and density. It is estimated that in the year 2025, 60% of the world's population will live in cities (UNFP, 1999). Human activity in urban environments has impacts on a local scale, including changes in atmospheric composition, impact on the water cycle, and modifying ecosystems. Our understanding of the role of urbanization in the Earth–climate system is incomplete; nevertheless, it is critical to determine how Earth's components — atmosphere, ocean, land, and biosphere — act reciprocally in a connected system. The clearest local indicator of climate changes due to urbanization is a well-known urban/rural convective circulation known as the urban heat island (UHI).

The UHI is defined as a dome of high temperatures observed over urban centers as compared to the relatively low temperatures of rural surroundings (see Figure 11.1). These temperature contrasts are greater in clear and calm conditions and tend to disappear in cloudy and windy weather by effects of thermal and mechanical mixture. Among the factors that cause formation of a heat island is the replacement of natural vegetation with man-made materials that have significantly different energy and water balance. The partitioning of the sun's energy by these materials results in a change from mostly latent heat fluxes to sensible heat flux and into storage.

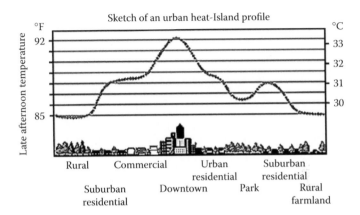

FIGURE 11.1 Schematic of the profile of temperatures typically observed in the great urban centers of the planet. (Courtesy of the Lawrence–Berkeley National Laboratory.)

This change leads to the great temperature differences observed during the early hours of the night when the energy stored during the daytime is released to the low atmosphere over the cities.

Typical urban surfaces do not allow infiltration of precipitation into the soil, thereby increasing storm water runoff and the likelihood of flash flooding. Another microclimatological effect caused by urbanization that is poorly understood is the "urban canyon effect" (Oke, 1987). In the spaces between buildings, the long wave radiation emitted by the surface at night is absorbed by the walls, resulting in trapped energy and higher temperatures. The urban topography also interrupts the natural flow of the wind, generating losses in mechanical energy and trapping the heat of the city.

Consequences induced by the UHI include poor air quality, mainly produced by industrial and residual gases of combustion; changes in the urban boundary layer and in the atmospheric stability (Bornstein and Lin, 2000); and prolongation and intensification of heat waves in cities. High energy consumption is also a direct consequence of UHI caused by increase in the demand for air conditioners. It has been estimated that the increase in electricity in southern California exceeds 200 MW per degree of air temperature in the city (Akbari et al., 1998). Another not so obvious consequence is a regional weather modification reflected in increased precipitation due to enhanced convection in the low atmosphere as reported by Shepard and coworkers.

UHI effects of diverse magnitude have been reported for various cities (Inoue and Kimura, 2004; Landsberg, 1981; Tso, 1995; Jauregui, 1997; Gallo et al., 1993; Lo et al., 1997; Bornstein and Lin, 2000; Noto, 1996; Poreh, 1996). Because each city is exposed to diverse local and synoptic factors, the study of the UHI is complex and specific to the locality. The general patterns, however, are very similar. Several climatological and observational studies have concluded that the UHI can have a significant influence in circulations of mesoscale and the resulting convection.

An important effect of large cities around the world recently studied is that of precipitation induced by the heat island. Past studies have suggested three main factors as possible causes of anomalies in precipitation patterns induced by urban centers: mixing and mechanical turbulence as a result of increased surface roughness due to the urban topography; sensible heat addition by warm air that ascends from the city; and the anthropogenic cloud condensation nuclei floating in the urban air. It has also been observed that cities, depending on characteristics of the storms that come near them, tend to branch off the precipitation systems due to the barrier effect that the buildings exert. This type of investigation has been carried out in a few cities, including Houston (Shepherd and Burian, 2003), Mexico City (Jauregui and Romales, 1996), and Atlanta and New York (Bornstein and Lin, 2000). Research on the impact of LCLU changes in tropical regions has been very limited and the San Juan field campaign was designed to fill this knowledge gap.

As was mentioned previously, the case study of the UHI in a coastal tropical city was developed in San Juan, Puerto Rico. Analysis of characteristics and patterns of this phenomenon was divided into two areas of study: analysis of long-term and recent observations, and the numerical analysis. The analysis of observations consists of gathering data of existing surface stations in the San Juan metropolitan area (SJMA) during last the 40 years and deployment of new stations and sensors during the period of the ATLAS Mission campaign. For the numerical analysis, an atmospheric model of regional scale was configured for the characteristics of the area of interest, validated against the experimental data to thus design several experiments of simulations around the SJMA.

11.3 CLIMATOLOGICAL ANALYSIS

The primary objectives of the climatological analysis are to determine the geographic extension of the SJMA and its effects in the atmosphere, to corroborate the existence of a UHI centered in the city of San Juan, and to recognize the typical and determining characteristics of the UHI in the SJMA.

To fulfill the first objective, an aerial image of the SJMA taken by an airborne camera in 1993 was analyzed (Figure 11.2). The photograph was generated by the NASA Ames Research Center in California. From this image, a total extension of San Juan of 310 km^2 was determined. This is the urban area used for design of the experiments and numerical simulations that will be presented. It is estimated that the area of urban coverage has grown significantly during the past 10 years, and part of the work of the ATLAS Mission will serve to determine its present extent.

The climatological analysis of air temperatures at 2 m above the surface was made taking into consideration four cooperative stations: two located in the urban area and two in the rural surroundings to the west of the city. Two scenarios were analyzed — a coastal scenario (ITL-DO) and an inland scenario (RP-MT) — using the four stations located along the north coast of Puerto Rico and shown in Figure 11.3. The cooperative stations included a complete database that dates 1900 for some stations with minimum missing data. For this study, continuous monthly data were available for all stations except for the Dorado station (station 1 in the map), which had missing data between 1980 and 1990. The temperature variable was analyzed

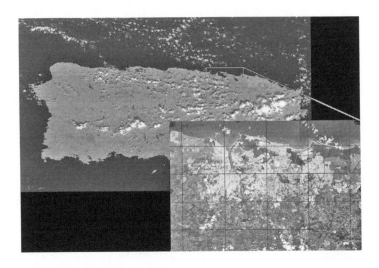

FIGURE 11.2 Aerial image of the metropolitan area of San Juan, Puerto Rico, obtained on December 11, 1993, by the NASA Ames Research Center. From this image a geographic extension of the SJMA of 310 km² was determined.

by calculating the differences between daily averages of minimum, maximum, and average values of the urban stations and the values of the rural stations. The results of this exercise are presented in Figure 11.4, along with a linear projection toward the middle of the present century.

These results show statistically significant average temperature increases in the urbanized areas over the vegetated areas ranging from 0.025°C/year inland to 0.05°C/year in the coastal areas. The increases in the differences of the minimum temperatures are larger, reaching values of 0.15°C/year in the coastal areas.

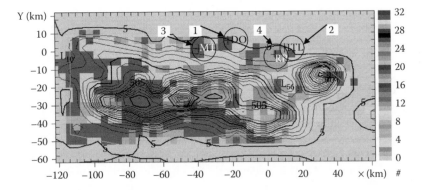

FIGURE 11.3 COOP weather stations' locations at the north of the island of Puerto Rico: station 1 (Dorado-Do); station 2 (International Airport-ITL); station 3 (Manatí-MT); station 4 (Río Piedras-RP).

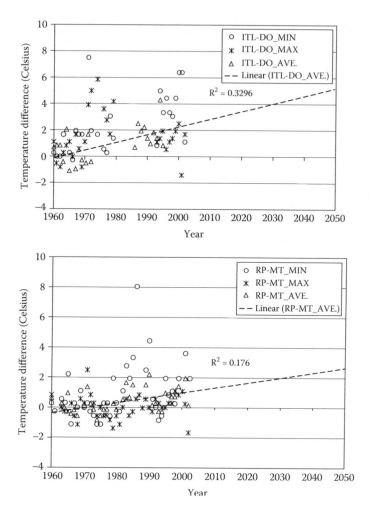

FIGURE 11.4 Daily averages per year for last the 40 years for the values of minimum, maximum, and average temperatures in urban and rural localities for (a) coastal areas and (b) inland areas on the north coast of the island of Puerto Rico. The solid, dashed, and dotted lines represent the projections of minimum, maximum, and average temperatures, respectively.

The projected average temperature difference trend can be estimated in 0.06°C/year. The higher temperature difference values for inland areas may be a consequence of larger development near the coastlines, as shown in Figure 11.2. Similar results as those presented here were reported by Duchon (1986) by climatological analysis for surface temperatures in San Juan, where a temperature increase of 2.5°C was found between 1956 and 1983. This earlier work did not distinguish between temperatures in urbanized and vegetated areas; however, the temperature increase was attributed to heat island effects.

The consistent presence of positive values found in the climatological analysis here is a clear indication that the temperatures in the city are greater than the temperatures in the country; the positive slope of the linear regression indicates that this difference could be intensified if present and past conditions prevail without mitigation and suitable policies for urban development. It is also evident that the effects of urbanization are larger on the coast than inland.

Figure 11.5 shows a recent annual pattern of the calculated temperature differences. It can be observed that the greater differences are located during the first months of the year, which are less humid, in what is known as the early rainfall season in the Caribbean when the convective activity is much smaller than during the late rainfall season (Daly et al., 2003; Malmgren and Winter, 1999; Taylor et al., 2002). These results agree with other studies that stipulate that the UHI is more intense

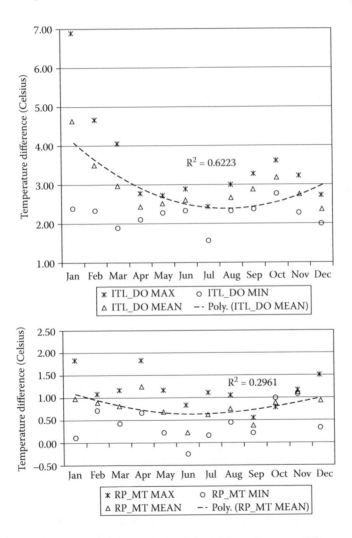

FIGURE 11.5 Recent annual climatological patterns of the temperature differences between the urban and rural stations for the north coast of the island of Puerto Rico for (a) coastal areas and (b) inland areas.

during stable and calm atmospheric conditions. These results of the climatological analysis suggested the existence of a UHI in the SJMA and motivated development of more extensive and wide-ranging experimental campaigns to determine particular characteristics and patterns of the UHI in San Juan.

11.4 DESCRIPTION OF THE ATLAS MISSION

The airborne thermal and land applications sensor (ATLAS) of NASA/Stennis operates in the visual and infrared bands. The ATLAS can detect 15 multispectral channels of the radiation through the visible, near-infrared, and thermal spectra (see Figure 11.6). The sensor also incorporates the active sources of calibration needed for all bands. Data are corrected for atmospheric radiation and georectified before analysis is performed. The ATLAS sensor has been used in other field campaigns to investigate the UHI in Atlanta, Salt Lake City, Baton Rouge, and Sacramento in the continental mass of the United States (Luvall et al., 2005).

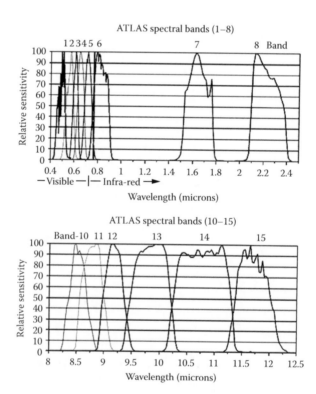

FIGURE 11.6 Spectral resolution of the airborne thermal and land applications sensor (ATLAS) from NASA/Stennis. It flies on a Lear 23 jet for flexibility and produces high-resolution images of value to observe UHI.

TABLE 11.1
Execution of Flight Plan

Day	Lines	Flight	Time (LST)	Resolution
Feb. 11	18–22	Mayaguez	09:27–10:03	10 m
Feb. 11	23–24	Hato Rey (day)	10:22–10:36	5 m
Feb. 13	1–7	El Yunque	13:54–14:49	10 m
Feb. 13	23–24	Hato Rey and Arecibo (night)	19:59–20:29	5 m
Feb. 16	8–17	San Juan	10:03–11:29	10 m

The ATLAS Mission of San Juan, Puerto Rico, was conducted during February 2004 to investigate the impact of urban growth and landscape in the climate of this tropical city. The flight plan of the mission covered the metropolitan area within San Juan, the national forest of El Yunque to the east of San Juan, the city of Mayagüez on the west coast of Puerto Rico, and the Arecibo Observatory located on the north-central coast, for a total of 25 flight lines (see Table 11.1 and Figure 11.7). This reports information used for San Juan only.

The downtown area of San Juan was covered in a horizontal resolution of 5 m in flights during the day and the night. The remaining areas of the city were covered in 10 m of resolution. The flights were executed between February 11 and 16, 2004. To analyze the existence of an urban heat island in San Juan and support the data of the ATLAS sensor, several experimental campaigns for data collection were designed and conducted by different teams. In addition, diverse numerical experiments were performed that helped us to understand the phenomenon and its characteristics.

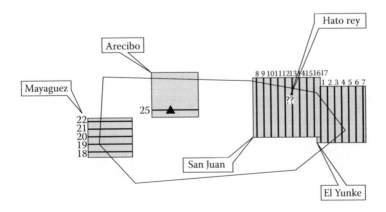

FIGURE 11.7 San Juan ATLAS Mission flight plan.

11.4.1 ATLAS DATA PROCESSING

The atmospheric corrections needed to produce calibrated data sets from ATLAS involve a complex procedure. They require direct measurements of the atmosphere extinction coefficients by wavelength and profiles of atmospheric temperatures and water vapor. ATLAS instrument characteristics and calibration are also required. Figure 11.8 details the process flow followed for this project, including resultant images from every relevant routine. A combination of software was used for processing, including the public domain image processing/remote sensing package ELAS (Beverley and Penton, 1989; Graham et al., 1986) and a series of custom programs, Watts and Energy from ELASII (Rickman et al., 2000).

MODTRAN4 (Berk et al., 1999) was used to model the atmospheric radiance and transmittance using input from radiosonde data and shadow band radiometers. Rickman et al. (2000) detail the procedure for calibrating the ATLAS sensor to produce the system transfer function to convert digital values (DV) into radiance measurements. These procedures produce ATLAS data files in physical units of energy. The files are used for generation of files that derive albedo and surface temperature.

Surface temperature is a major component of the surface energy budget. Use of energy terms in modeling surface energy budgets allows direct comparison of various

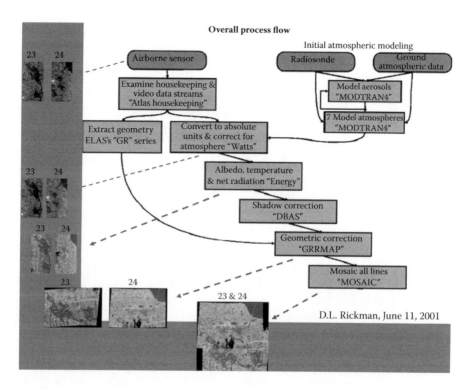

FIGURE 11.8 ATLAS overall data-processing flow to produce atmospheric, radiometric, and geometrically corrected data (Rickman et al. 2000).

land surfaces encountered in a landscape, from vegetated (forest and herbaceous) to nonvegetated (bare soil, roads, and buildings) (Oke, 1987). Partitioning of energy budget terms depends on the surface type. In natural landscapes, the partitioning depends on canopy biomass, leaf area index, aerodynamic roughness, and moisture status, all of which are influenced by the development stage of the ecosystem. In urban landscapes, coverage by man-made materials substantially alters the surface energy budget. The net all-wave radiation balance (W/m^2) of landscape canopies can be determined following Oke (1987).

The net solar radiation, K^*, is given by

$$K^* = (1 - a)(K \downarrow) \qquad (11.1)$$

where

a = site albedo
$K\downarrow$ = incoming solar radiation

The albedo is defined as

$$a = \frac{K\uparrow}{K\downarrow} \qquad (11.2)$$

where
$K\uparrow$ = reflected solar radiation

The long wave energy emitted from a surface ($L(\uparrow)$ is dependent on surface temperature:

$$L\uparrow = \varepsilon[\sigma T^4] \qquad (11.3)$$

where
ε = emissivity
σ = Stefan–Boltzman constant (5.7×10^{-8} W/m^2-K^4)
T = land surface temperature (Kelvin)

The net long wave radiation at the surface, L^*, is given by

$$L^* = L\downarrow - L\uparrow \qquad (11.4)$$

where
$L\downarrow$ = long wave radiation from the atmosphere

The net all-wave radiation, Q^*, can be given as:

$$Q^* = K^* + L^* \qquad (11.5)$$

Net radiation, under most conditions, represents the total amount of energy available to the land surface for partitioning into nonradiative processes (mass heating, biological synthesis, etc.) at the surface. It is the amount of energy the system holds on to and degrades. In vegetated areas, the amount of net radiation depends upon vegetation type and varies with canopy leaf area and structure. The net radiation may be expressed as the sum of these nonradiative fluxes:

$$Q^* = \lambda E + H + G \qquad (11.6)$$

where

H = sensible heat flux
λ = latent heat of vaporization of water
E = transpiration flux
G = energy flux into or out of storage (both canopy and soil)

The partitioning λE, H, and G also depend on the makeup of the surface. Physiological control of moisture loss (stomatal resistance) and leaf/canopy morphology for vegetation determine how Q^* is partitioned among λE, H, and G. For urban surfaces, the coverage of man-made materials and vegetation results in a heterogeneous mixture of surfaces that determine the partitioning of energy. The ATLAS remotely sensed data allow measurement of important terms in the radiative surface energy budget: $K\uparrow$ and $L\uparrow$ on an urban landscape scale. When combined with output from MODTRAN4 (Berk et al., 1999) atmospheric radiance models, the remaining terms $K\downarrow$ and Q^* can be determined.

11.4.2 REMOTE SENSING RESULTS FOR HATO REY, PUERTO RICO, DURING DAYTIME

During flight 1 of the mission, Hato Rey surface temperatures were obtained from the ATLAS sensor. With the use of ENVI™ and Matlab™ software, energy files from Hato Rey were manipulated to visualize and determine frequency distributions from the observed surface data. Flight lines 23 and 24 covered the area of Hato Rey at a resolution of 5 m during the daytime. By looking at the surface temperatures from the mosaic of flight line 23 and flight line 24 (see Figure 11.9), urbanized areas can easily be identified. Lakes and vegetated areas are the coldest compared to the warmest — roof tops and paved areas. Notice the large temperature differences between northeast Hato Rey and northwest Hato Rey. The former shows the coldest temperatures over the entire Hato Rey area. A significant urbanized area around northwest Hato Rey can be seen in the thermal mosaic with warmer temperatures.

The temperature frequency distribution for flight line 24 (Figure 11.10) identifies important land surface features that make up the thermal fabric of the city. First, the Teodoro Moscoso Lake is identifiable by its peak as the coldest surface of 24.7°C. The next coldest peak identifies the vegetated component near the experimental station at around 27.2°C. The hottest surface is associated with roofs and asphalt pavements and is represented by the "tail" of the distribution between 49.9 and 63.7°C.

Surface temperature mosaic from flight 1
lines 23–24 (Hato Rey. Puerto Rico)

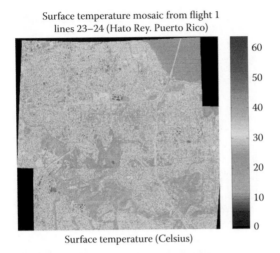

Surface temperature (Celsius)

FIGURE 11.9 (Color Figure 11.9 follows page 240.) Mosaic of Hato Rey flight 1, lines 23 and 24. Surface temperatures are measured in degrees Celsius.

Albedo alone does not truly reflect how the land's surface partitions energy. A good example is the comparison between vegetated and asphalt surfaces. Both surfaces have a low albedo, but the asphalt surface temperature can be over 34°C greater than that of the vegetated surface. If the surface temperature is included, the additional information needed to asses the "urban fabric" of the city is provided.

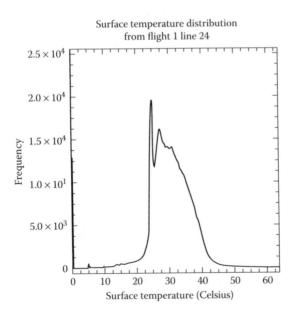

FIGURE 11.10 Frequency distribution for flight 1, line 24.

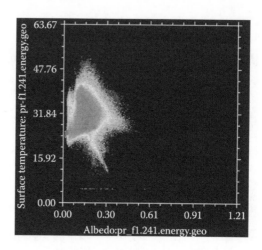

FIGURE 11.11 Scatter plot of surface temperature vs. albedo: Hato Rey flight 1, line 24.

The surface temperature and albedo classifications represent a functional classification of that surface that can readily be incorporated into the surface parameterization of meteorological and air quality models. Within each city, each land use has a unique "energy print" that is directly physically related to how that surface is processing energy. These energy prints of land use are unique for each city. Figure 11.11 shows a scatter plot of flight line 24.

As can be seen, the relation between surface temperature and albedo indicates the energy print of that portion of Hato Rey. Red pixels show the highest frequency of occurrences for a given pixel value of temperature and albedo, whereas the blue pixels show the lowest frequency. The lower left-hand corner in the scatter plot identifies the dark and cool water bodies in flight line 24. The upper side of the scatter plot identifies the light and hot buildings and paved areas. Because these energy prints of land use are unique for each city, a scatter plot from one city will differ significantly from the scatter plot of another city. Results from the scatter plot emphasize that classifications based on cover type/land use cannot be applied across a variety of cities because they cannot represent true energy partitioning of that surface.

Figure 11.12 shows the mosaic for Sacramento, California, taken with the same ATLAS sensor in a previous mission (Luvall et al., 2001) along with the mosaic of Hato Rey, San Juan, Puerto Rico. Hato Rey skattergrams for residential and forest areas reflect a higher energy contribution to the whole mosaic than residential and forest-related skattergrams from Sacramento. In the case of Sacramento of the skatterg rams shown, the forest emitted the lowest amount of energy. Water bodies reflect the lower amount of energy in the case of Hato Rey. It seems that urbanized and industrial areas contribute the largest amount of energy for healing the air.

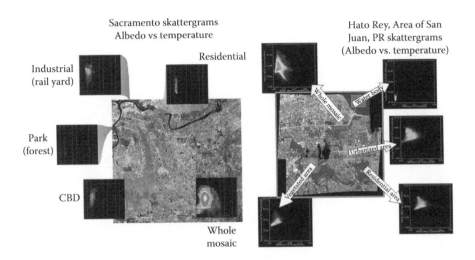

Sacramento skattergrams
Albedo vs temperature

Industrial
(rail yard)

Residential

Park
(forest)

CBD

Whole
mosaic

Hato Rey, Area of San
Juan, PR skattergrams
(Albedo vs. temperature)

Whole mosaic

Water body

Urbanized area

Vegetated area

Residential area

FIGURE 11.12 ATLAS mosaic for Sacramento, California, and Hato Rey, San Juan, Puerto Rico, with their respective energy prints.

11.4.3 REMOTE SENSING RESULTS FROM HATO REY, PUERTO RICO, DURING NIGHTTIME

During flight 2 of the mission, Hato Rey surface nighttime temperatures were obtained from the ATLAS sensor. Flight lines 23 and 24 correspond to the area of Hato Rey (Figure 11.7). By looking at the surface temperatures from flight line 23 (Figure 11.13), most rooftops and paved areas are 27.2 through 34.6°C.

Surface temperatures from flight 2,
line 23 (Hato Rey. Puerto Rico)

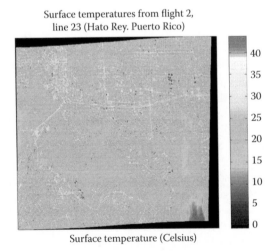

Surface temperature (Celsius)

FIGURE 11.13 (Color Figure 11.13 follows page 240.) Hato Rey flight 2, line 23. Surface temperatures measured in degrees Celsius.

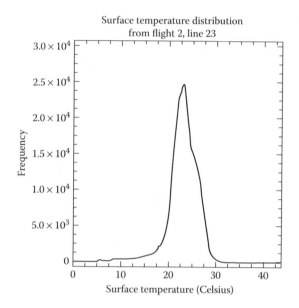

FIGURE 11.14 Frequency distribution for flight 2, line 23.

Vegetated areas are 23.0 through 26.9°C. It can be noted that even if there is not a large temperature difference between urban and vegetated areas, there is a signature temperature range between both data sets. The temperature frequency distribution for flight line 23 (Figure 11.14) better identifies the signature temperature range that separates urbanized areas from vegetated areas. The highest temperature is 43.7°C, which represents just one rooftop. Most paved areas can be as high as 34.6°C.

11.4.4 SURFACE STATION AND SENSOR DATA

The information provided by the weather balloon launchings and the synoptic information provided by the National Center for Environmental Prediction (NCEP) demonstrate that during the days of the mission, the middle and upper atmosphere in the Caribbean were relatively dry and highly stable (not shown), an ideal condition to conduct UHI studies.

The weather stations and the temperature sensors were placed in strategic locations throughout lines that follow the climatological pattern of the northeastern trade winds, to analyze possible differences in the temperature among commercial, residential, and rural areas (see Table 11.2 and Table 11.3). The data of these weather stations and temperature sensors provide further validation of a strong UHI in the San Juan metropolitan area. Figure 11.15 shows the results of the surface stations' noon-averaged data interpolated on a regular grid in the locations selected through the SJMA and the neighboring rural municipalities.

It is clearly seen that the noon average temperatures during the period of the ATLAS mission demonstrate the existence of a pronounced UHI, with the tip of the

TABLE 11.2
Surface Stations in San Juan Metropolitan Area and Rural Surroundings

Station	Location	Geographical Location		Variables						
		Latitude	Longitude	Temp.	HR	Wind	Precipitation	Solar Radiation	Pressure	Soil Moisture
Bayamón station	Sci. Park	18°24'41"	66°09'37"	X	X	X	X	X	X	
Polytechnic University	Hato Rey	18°25'19"	66°03'19"	X	X	X	X	X	X	
Dorado station	Dorado environmental house	18°27'55"	66°19'37"	X	X	X		X		
UPR stations	Rio Piedras 1	18°24'08"	66°03'04"	X			X			X
	Rio Piedras 2	18°24'12"	66°02'52"	X			X			X
	Rio Piedras 3	18°24'14"	66°02'52"	X			X			X
Rio Grande station	Rio Mar Beach resort	18°22'44"	65°45'22"	X			X			

TABLE 11.3
HOBO Sensors

		Latitude	Longitude	Temp.
Bayamón sensor	Science Park, North Bayamón	18°24′41″	66°09′37″	X
Cupey sensor	South Guaynabo	18°21′12″	66°05′13″	X
CUSC sensor	Santurce, downtown SJU	18°26′29″	66°03′31″	X
Guaynabo sensor	North Guaynabo	18°24′23″	66°06′07″	X
Interamericana sensor	South Bayamón	18°21′06″	66°11′00″	X
NWS sensor	North Carolina	18°25′53″	65°59′29″	X
Toa Baja sensor	Sabana Seca Naval Base	18°27′28″	66°11′47″	X
UPR Bayamón sensor	East Bayamón	18°22′14″	66°08′36″	X

dome of high temperature exactly over the commercial area of downtown San Juan, represented by the stations within the areas shaded in red. Most of the suburban areas are located to the west of the center of San Juan and are represented by shades of light green and yellow. The tropical forest of El Yunque, to the east of San Juan, as well as the central mountains to the south of San Juan, appears to be well vegetated and supplied with water. A cross-section in the east–west direction, following the trade wind direction, indicates the dome of elevated temperatures on the commercial

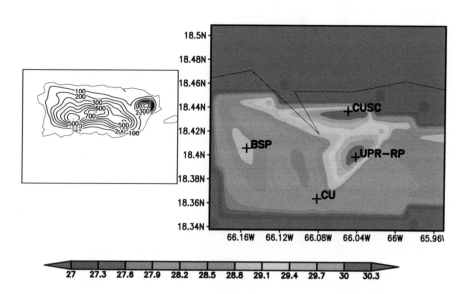

FIGURE 11.15 Observed air temperatures averaged at noon during the ATLAS field campaign period in the stations and sensors located in the San Juan metropolitan area, and rural and suburban areas (identified points: UPR-RP = University of Puerto Rico-Rio Piedras; CUSC = Catholic University of the Sacred Heart; BSP = Bayamón Sciences Park; CU = Cupey). The contours in the left panel represent elevations.

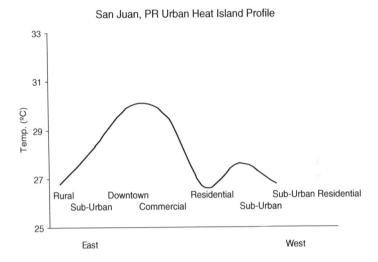

San Juan, PR Urban Heat Island Profile

FIGURE 11.16 Cross-section of average temperatures during the period of February 13 to 16, 2004, in the east–west gradient line for the selected stations and sensors.

center of San Juan (Figure 11.16), showing a profile quite similar to the definition of the UHI presented in Figure 11.1.

An analysis of the data collected shows that the peaks of the temperature readings occurred in the middle of the afternoon, ranging between 33 and 35°C in urban areas and between 26 and 28°C in suburban and rural areas. Low temperatures recorded by the stations and sensors in urban, suburban, and rural locations constantly reached 20 to 22°C in the late evening (around 10:00 p.m., local standard time). The high temperatures in rural areas were substantially lower than those in urban areas between midmorning and the late hours of the morning, more specifically between 9:00 a.m. and noon. Average temperature differences of 4.5°C were found in such a time frame, a temporal pattern not observed in previous studies of UHI in big continental cities. Still, it is not clear why the San Juan UHI peaks in the last hours of the morning and not in the early evening, a pattern shown in previous experiments of UHI conducted in continental cities (http://www.ghcc.msfc. nasa.gov/atlanta/). The nonexistence of a cool island — a negative immersion of the temperature difference between the urban and rural areas dT(U-R) — opposite to the heat island caused by thermal heat storage was also observed. These results are summarized in Figure 11.17.

The high concrete density covering the surface and the importance of the volumetric soil moisture content and the evapotranspiration in controlling the urban tropical climate are demonstrated by the variability of the UHI pattern with respect to precipitation. Even for relatively small and short precipitation events on the order of only 8.4 mm during the period of study, registered by the station of the National Weather Service of San Juan, temperatures in the commercial center of San Juan and surrounding areas approached a very similar value, demonstrated by a low dT(U-R).

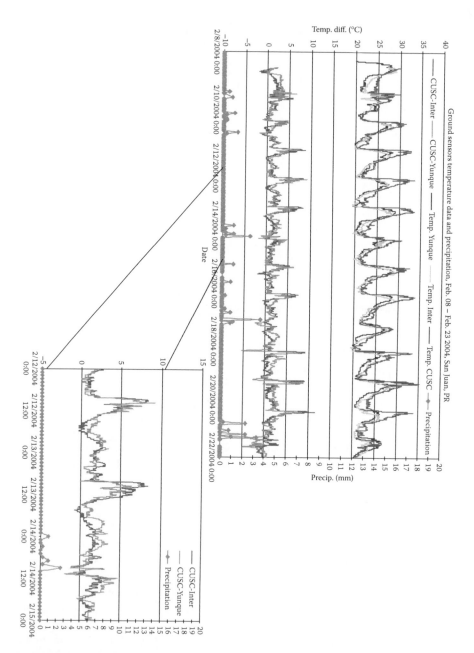

FIGURE 11.17 Time series of observed temperatures (top series), temperature differences between the commercial center and rural areas (intermediate series), and precipitation in the area of study (bottom series) for the urban (CSCU), suburban (Inter-American), and rural (El Yunque) reference stations.

11.5 NUMERICAL ANALYSIS

The main objective of the numerical analysis is to investigate the impact of the land use for urbanization in regional environmental variables. The approach used includes configuration of an atmospheric mesoscale model for Puerto Rico and San Juan, validation of control simulations with the observations presented in Section 11.3.2, and quantification of the impact of land cover/land use by the cities. The regional model used for the study presented in this document is the regional atmospheric modeling system (RAMS) developed at the Colorado State University (Pielke et al., 1992; Cotton et al., 2003).

11.5.1 DESCRIPTION OF THE MODEL AND EXPERIMENTAL DISPOSITION

RAMS is a highly versatile numerical code developed to simulate and forecast meteorological phenomena. The atmospheric model is constructed around the complete system of nonhydrostatic dynamic equations that govern atmospheric dynamics and thermodynamics and the conservation equations for scalar quantities such as mass and humidity. These equations are complemented by a wide selection of parameterizations available in the model. The version of RAMS used for this document contains an upgraded cloud microphysics module described by Saleeby and Cotton (2004), an advance over the original package available in the current model release (Meyers et al., 1997; Walko et al., 1995). This new cloud microphysics module includes activation of the cloud condensation nuclei by means of the use of a Lagrangian parcel model that directly considers environmental conditions of the cloud for the initial formation of the cloud droplets on the aerosol.

The simulations were conducted with three grids making use of the grid nesting capabilities of the model used. Grid 1 covered a great part of the Caribbean basin with a horizontal resolution of 25 km. Grid 2, which nested within grid 1, covered the island of Puerto Rico in 5 km of horizontal resolution. Grid 3 nested within grid 2 and centered in the city of San Juan with a resolution of 1 km (see Figure 11.18). For the vertical coordinate, all the grids had the same specification. A vertical grid spacing of 100 m was used near the surface and stretched at a constant ratio of 1.1 until a Δz of 1000 m was reached. The depth of the model was approximately 22.83 km with 40 vertical levels. Time-variable boundary conditions were used. The cloud microphysics humidity complexity was set at the highest level. This level incorporated all the categories of the water in the atmosphere (cloud water, rainwater, pristine ice crystals, snow, aggregates, graupel, and hail) and included the precipitation process. Marine atmosphere clouds have low concentrations of large drops and a wide concentration spectrum.

All the simulations were forced with the same initial conditions and variable lateral conditions for the period of February 10 to February 20, 2004, given by the National Centers for Environmental Prediction (NCEP) atmospheric fields. The use of this regional atmospheric model already has been demonstrated to produce satisfactory results in the Caribbean basin simulating the precipitation pattern on the island of Puerto Rico in months of the early rainy season (Comarazamy, 2001).

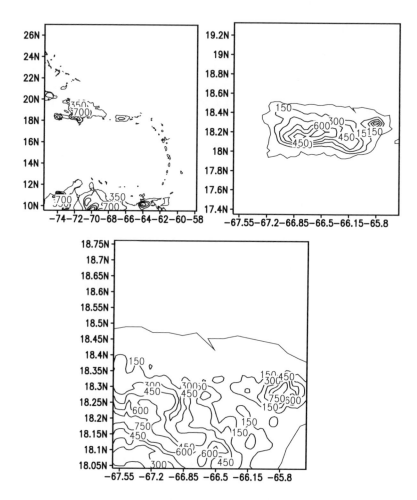

FIGURE 11.18 Model grids used in numerical simulations to investigate the impact of urban LCLU on the local climate. The topography contours have an interval of 150 m in the three panels.

To quantify the impact of the LCLU change in the metropolitan area of San Juan through the time period, three different scenarios were configured. First, the standard specification of the surface characteristics used in regional atmospheric models was specified. Then, one of the model subroutines was modified to represent the urban extension and configuration of San Juan as it was observed in the aerial photography shown in Figure 11.2. The third configuration was designed to represent the possible natural vegetation of the zone occupied by the city, interpolating the surrounding vegetation covering all the area. The runs were denominated "present," "urban," and "natural," respectively. The variable modified for these numerical simulations was the denominated vegetation index defined by the biosphere–atmosphere transfer scheme (Dickinson et al., 1986). This index includes physical parameters of albedo, emissivity, leaf area index, vegetation cover percentage, surface roughness,

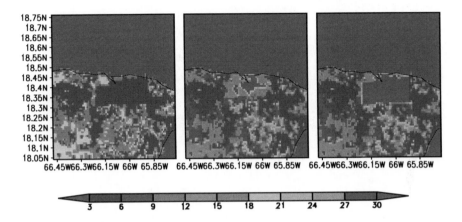

FIGURE 11.19 (Color Figure 11.19 follows page 240.) Specification of the surface characteristics used in the three runs of the atmospheric model, simulating, from left to right, natural, present, and urban scenarios.

and root depth. The configuration of the LCLU index used in the three simulations is presented in Figure 11.19.

11.5.2 MODEL RESULTS: AIR TEMPERATURE AND VERTICAL PROFILES

Validation of the mesoscale model was performed by comparing the air temperatures at 2 m above ground level produced by the simulation with the values recorded by the stations, as presented in Figure 11.15. The daily cycle presented in Figure 11.20 was obtained by averaging the temperature values predicted by the model over the entire area represented by the city at each hour for the duration of the ATLAS Mission and comparing them with the stations and sensors averaged over the same geographical area and time span (February 11 through February 16, 2004). This comparison shows that the model performed satisfactorily even though it produced temperatures higher during the heating hours. This overprediction could be explained by the model's use of a homogenous urban LCLU, and therefore not capturing the different microclimates present in the metropolitan area and producing a more uniform spatial temperature distribution.

To study the impact of the urban LCLU on the temperatures of the San Juan metropolitan area, an analysis of the air temperatures at 2 m above ground level (AGL) was performed with the results produced by the three simulated scenarios. The analysis consisted of calculating the difference of the values averaged during the period of greater heating (considered to be 3 p.m.) in this case, with the following combinations: urban–natural, present–natural, and urban–present. To visualize the effect of the slab of concrete that represents the city of San Juan on the wind pattern, a similar procedure was followed with the diurnal cycle of the marine/land breeze circulation.

The results of the analysis of air temperatures averaged in midafternoon through the complete period of simulation are shown in Figure 11.21. Here it is shown that the

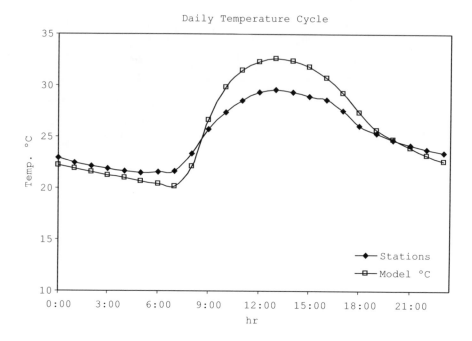

FIGURE 11.20 Comparison of averaged air temperatures between the regional model results and the stations and sensors deployed in the San Juan metropolitan area during the San Juan ATLAS Mission (February 11 to 16, 2004).

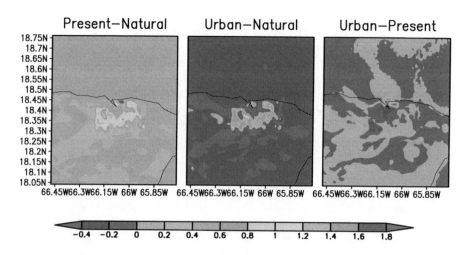

FIGURE 11.21 (Color Figure 11.21 follows page 240.) Spatial distribution of the air temperature difference (degrees Celsius) at 2 m AGL among the three scenarios simulated for the analysis. The interval of the contours is 0.5°C for the first two panels and of 0.2°C for the third panel.

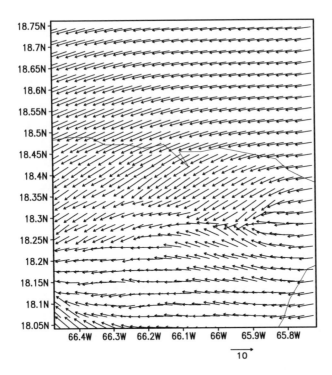

FIGURE 11.22 Wind field averaged at 3 p.m., local time, during the complete period of simulation of the present run.

atmospheric model predicts that the presence of a large urban area in San Juan has an impact on the low atmosphere of the area occupied by the city; this impact is reflected in higher temperatures for simulations that have a specified urban LCLU in the inferior border. The temperature difference is more pronounced between the urban and natural runs, with positive values of up to 2.5°C, mainly downwind of the city. The spatial pattern of the temperature differences on the zone represented by the SJMA can be explained by the presence of a sustained wind from the northeastern direction during a great part of the afternoon (see Figure 11.22). The three simulations produced the same daily cycle for wind pattern, characterized by a strong influence of the northeasterly trade winds. Nevertheless, differential heating between the Atlantic Ocean and the north coast of Puerto Rico induced an inland circulation during the day, as can be observed in Figure 11.22, and a wind direction inversion at night. Both circulations showed a slanted pattern of approximately 45° due to the synoptic influence.

The impact of the presence of model grid cells that specify an urban LCLU is also significant. In Figure 11.23, it can be observed that the difference of the wind field among the three scenarios lies essentially over the area covered by the city and in the direction of the prevailing winds. The effect is of an accelerated wind reflecting an increment of the wind vectors represented in the panels that show the differences urban–natural and present–natural, in the order of 3 m/s. It is worth noting that the

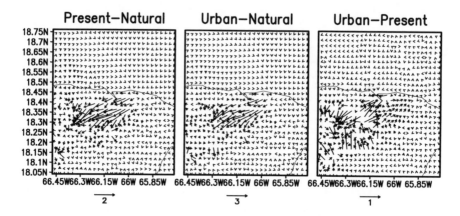

FIGURE 11.23 Average differences in the modeled wind fields calculated at 3 p.m. for the three LCLU scenarios.

differences in the temperature and wind were found very close to the surface, especially in the case of the vertical wind profile where differences were limited to a height of below 1 km (see Figure 11.24) and below 3 km when comparing the temperature vertical profiles (see Figure 11.25).

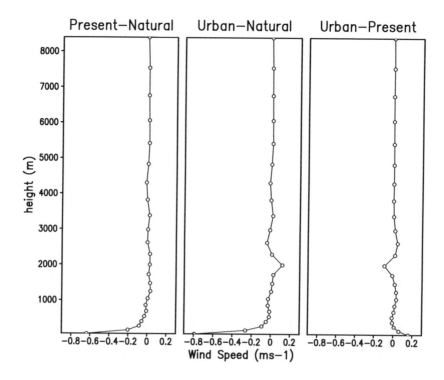

FIGURE 11.24 Average differences in the modeled wind vertical profiles (ms^{-1}) calculated at 3 p.m. at the location of the San Juan National Weather Office.

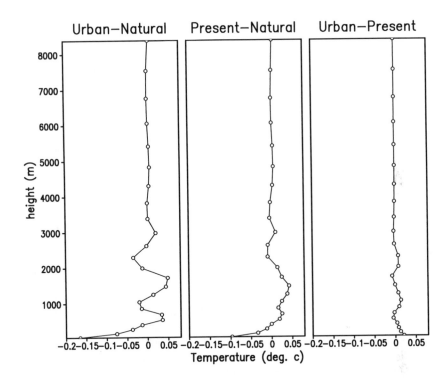

FIGURE 11.25 This figure is the same as Figure 11.24, except for temperature (degrees Celsius).

11.6 SUMMARY AND CONCLUSIONS

The work presented here is a comprehensive investigation of the impact of land use for urbanization on the environment of a city located on a small tropical island — in this case, San Juan, Puerto Rico. The findings can be summarized as follows.

- The empirical analysis revealed a prevailing UHI in the metropolitan area of San Juan that is growing together with the urban development.
- The UHI in San Juan has greater intensity during the early rainy season, when the atmospheric conditions are more stable than during other times of the year.
- The ATLAS field campaign conducted in February 2004 validates the development of this UHI showing temperatures as high as 60°C with temperature differences between the developed and vegetated areas of more than 30°C during daytime. This translates into temperature differences of less than 10°C during nighttime. The thermal fingerprint of a large urban development is clearly observed from the high-resolution remote sensing images.
- Surface temperatures were collected for the same period of the San Juan ATLAS Mission. Several surface stations were deployed throughout San Juan, revealing hot spots on the commercial areas of downtown San Juan

and a cross-sectional temperature profile in the east–west direction that shows a dome of high temperatures over the center of the SJMA, with consistent values of UHI between 5 and 10°C during the period of observation.

- The occurrence of rain strongly affects the pattern of the San Juan UHI, even for weak precipitation events.
- The atmospheric numerical model RAMS was validated to capture the impact of the urban LCLU of San Juan on different atmospheric variables.
- The analysis of three simulated land-use scenarios leads to the conclusion that the urban LCLU has an impact on the general atmospheric dynamics of the north coast of the island of Puerto Rico.
- Model results demonstrate that the influence of San Juan on the local air temperature is to produce higher temperatures in the region where the urban area was represented. This influence could be quantified in air temperature increases between 2.5 and 3°C and a local low pressure that accelerates the local sea breeze in the area of study.

After these characteristics of the urban atmosphere in the San Juan metropolitan area were investigated, several points were left as research questions that should be approached in future studies. With the presence of a UHI in San Juan, what would be the effect on the predominant winds and the overall impact on the sea/land breeze circulation? Is the UHI inducing a convective precipitation effect in San Juan? What is its influence on cloud formation and height? How does the UHI affect the vertical structure of the atmosphere over the city? What is the effect on the hydrological cycle of the region?

In order to try to respond to these questions, future experiments and numerical simulations should be designed. Observations should include long-term remote sensing analysis of cloud height and precipitation and its possible correlation with land use. The simulations should include a heterogeneous surface with specifications of albedos and extracted temperatures from the data of the ATLAS sensor, and the urban topography that is of extreme importance for the dynamic and heat storage terms and thus a more realistic way to approach the problem. Finally, further comprehensive studies should be conducted in other major tropical coastal cities.

ACKNOWLEDGMENTS

This investigation was made under the sponsorship of the NASA-EPSCoR program of the University of Puerto Rico and carried out at the University of Puerto Rico, Mayagüez, the University of Santa Clara, Santa Clara, California, and the Global Hydrological and Climate Center of Huntsville, Alabama (GHCC). The atmospheric simulations were performed in the High Performance Computational Facilities of the University of Puerto Rico — Rio Piedras (UPR). The data of the cooperative stations were collected from the Regional Climatic Center of the southeastern part of the United States. Parts of the climatological empirical analysis were conducted by Alexander Velázquez.

REFERENCES

Akbari, H., Pomerantz, M., and Taha, H., 1998. Cool surfaces and shade trees to reduce energy use and improve air quality in urban areas, report no. LBNL-42637, Lawrence Berkeley National Laboratory, Berkeley, CA.

Berk, A., Anderson, G.P., Acharya, P.K., Chetwynd, J.H., Bernstein, L.S., Shettle, E.P., Matthew, M.W., and Adler–Golden, S.M., 1999. MODTRAN4 Users Manual. Air Force Research Laboratory, Space Vehicles Directorate, Air Force Material Command, Anscom AFB, MA 01731-3010.

Beverley, A.M. and Penton, P.G., 1989. ELAS, Earth resources laboratory applications software, vol. II, user reference, Earth Resources Lab, report no. 183; NSTL, Mississippi.

Bornstein, R. and Lin, Q., 2000. Urban heat islands and summertime convective thunderstorm in Atlanta: three case studies, *Atmos. Environ.*, 34, 507–516.

Comarazamy, D.E., 2001. Atmospheric modeling of the Caribbean region: precipitation and wind analysis in Puerto Rico for April 1998. Master's thesis, Department of Mechanical Engineering, University of Puerto Rico-Mayagüez, 95 pp.

Cotton, W.R., Pielke, R.A., Sr., Walko, R.L., Liston, G.E., Tremback, C.J., Jiang, H., McAnelly, R.L., Harrington, J.Y., Nicholls, M.E., Carrio, G.G., and McFadden, J.P., 2003. RAMS 2001: current status and future directions, *Meteor. Atmos. Phys.*, 82, 5–29.

Daly, C., Helmer, E.H., and Quiñones, M., 2003. Mapping the climate of Puerto Rico, Vieques and Culebra, *Int. J. Climatol.*, 23, 1359–1381.

Dickinson, R.E., Sellers, A.H., Kennedy, P.J., and Wilson, M.F., 1986. Biosphere–atmosphere transfer scheme (BATS) for the NCAR climate community model, technical note NCAR/TN-275+STR. 69 pp.

Duchon, C.E., 1986. Temperature trends at San Juan, Puerto Rico, *Bull. Am. Meteorol. Soc.* (*BAMS*), 67, 1370–1377.

Gallo, K.P., McNab, A.L., Karl, T.R., Brown, J.F., Hood, J.J., and Tarpley, J.D., 1993. The use of NOAWA AVHRR data foe the assessment of the urban heat island effect, *J. Appl. Meteorol.*, 32, 899–908.

Graham, M.H., Junkin, B.G., Kalcic, M.T., Pearson, R.W., and Seyfarth, B.R., 1986. ELAS — Earth Resources Laboratory applications software. Revised Jan 1986, NASA/NSTL/ERL, report no. 183.

Inoue, T., and Kimura, F., 2004. Urban effects on low-level clouds around the Tokyo metropolitan area on clear summer days, *Geophys. Res. Lett.*, 31, L05103, doi: 10.1029/2003GL018908.

IPCC, 2001. Third assessment report of working group I of the IPCC — climate change 2001, summary for policy makers. 1(1): 1–16. Retrieved January 15, 2003 from http:// www.grida.no/climate/ipcc_tar/vol4/spanish/pdf/sum.pdf.

Jauregui, E., 1997. Heat island development in Mexico City, *Atmos. Environ.*, 31, 3821–3831.

Jauregui, E. and Romales, E., 1996. Urban effects of convective precipitation in Mexico City, *Atmos. Environ.*, 30, 3383–3389.

Landsberg, H.E., 1981. *The Urban Climate*, Academic Press, New York, 275 pp.

Lo, C.P., Quattrochi D.A., and Luvall, J.C., 1997. Applications of high-resolution thermal infrared remote sensing and GIS to asses the urban heat island effect, *Int. J. Remote Sensing*, 18(2), 287–304.

Luvall, J.C., Rickman, D., Quattrochi, D., and Estes, M., 2005. Aircraft-based remotely sensed albedo and surface temperatures for three U.S. cities: color roofing: cutting through the glare, Roof Consultants Institute Foundation, May 12–13, 2005, Atlanta, GA.; 8 pp.

Luvall, J.C., Morris, L., Stewart, F., Thretheway, R., Gartland, L., Russell, C., and Reddish, M., 2001. EPA/NASA Urban Heat Island Mitigation Pilot Cities Project. http://www.epa.gov/heatisland/pilot/index.html.

Malmgren, B.A. and Winter, A., 1999. Climate zonation in Puerto Rico based on principal components analysis and an artificial neural network, *J. Climate*, 12(4), 977–985.

Meyers, M.P., Walko, R.L., Harrington, J.Y., and Cotton, W.R., 1997. New RAMS cloud microphysics parameterization. Part II: the two-moment scheme, *Atmos. Res.*, 45, 3–39.

Noto, K., 1996. Dependence of heat island phenomena on stable stratification and heat quantity in calm environment, *Atmos. Environ.*, 30, 475–485.

Oke, T.R., 1987. *Boundary Layer Climates*. Cambridge University Press, New York, 450 pp.

Pielke, R.A., Cotton, W.R., Walko, R.L., Tremback, C.J., Lyons, W.A., Grasso, L.D., Nicholls, M.E., Moran, M.D., Wesley, D.A., Lee, T.J., and Copeland, J.H., 1992. A comprehensive meteorological modeling system-RAMS, *Meteor. Atmos. Phys.*, 49, 69–91.

Poreh, M., 1996. Investigation of heat islands using small scale models, *Atmos. Environ.*, 30, 467–474.

Rickman, D.L., Luvall, J.C., and Schiller, S., 2000. An algorithm to atmospherically correct visible and thermal airborne imagery. Workshop on Multi/Hyperspectral Technology and Applications, Redstone Arsenal, Alabama.

Saleeby, M.S. and Cotton, W.R., 2004. A large-droplet mode and prognostic number concentration of cloud droplets in the Colorado State University Regional Atmospheric Modeling System (RAMS). Part I: Module descriptions and supercell test simulations, *J. Appl. Meteorol.*, 43, 182–195.

Shepherd, J M. and Burian, S.J., 2003. Detection of urban-induced rainfall anomalies in a major coastal city, *Earth Interaction*, 7, 1–17.

Shepherd, J.M., Pierce, H.F., and Negri, A.J., 2002. Rainfall modification by major urban areas: observations from space-borne rain radar on the TRMM satellite, *J. A ppl. Meteorol.*, 41, 689–701.

Taylor, M.A., Enfield, D.B., and Chen, A.A., 2002. Influence of the tropical Atlantic vs. the tropical Pacific on Caribbean rainfall, *J. Geophys. Res.*, 107, 10.1–10.14.

Tso, C.P., 1995. A survey of urban heat island studies in two tropical cities, *Atmos. Environ.*, 30, 507–519.

UNFP, 1999. The state of world population 1999. United Nations Population Fund, United Nations Publications, New York, 76 pp. (http://www.unfpa.org/swp/1999/index.htm).

Walko, R.L., Cotton, W.R., Meyers, M.P., and Harrington, J.Y., 1995. New RAMS cloud microphysics parameterization: Part I: the single-moment scheme, *Atmos. Res.*, 38, 29–62.

12 Assessing Urban Environmental Quality with Multiple Parameters

Janet Elizabeth Nichol and Man Sing Wong

CONTENTS

12.1 INTRODUCTION

Urban environmental quality is an abstract, dynamic, and multifaceted concept comprising human and natural factors operating at different spatial scales. To capture the continuous variability of the urban environment over whole cities, satellite sensor images are the only suitable data source. However, until recently, satellite sensors were unable to capture data detailed enough to represent the fragmented land cover of urban areas. Indeed, a comparative study of environmental quality in Hong Kong (Fung and Siu, 2001) used medium resolution SPOT images to estimate green space at the generalized level of tertiary planning units. A similar scale study to assess the quality of life in Athens, Georgia (Lo and Faber, 1997), used the NDVI to estimate greenness and temperature from Landsat TM images at 30- and 60-m resolution, respectively, and data were averaged at the census district level.

Surface temperature derived from satellite thermal sensors has been used to characterize urban heat islands (Roth et al., 1989), but thermal image data are of low resolution (60 m for Landsat ETM and 90 m for ASTER). Therefore, the majority of satellite-based studies of the UHI have been confined to climatologic assessments of broad land-cover types at regional (city) scale (Lo et al., 1997; Weng, 2001). Maps derived from these studies are at electoral or administrative district level; thus, their utility for urban and environmental planning is limited because landscaping and redevelopment in cities takes place at the local scale of individual buildings, blocks, or streets. A new generation of very high resolution (VHR) satellite sensors such as IKONOS and Quickbird enables detailed mapping of vegetation (Nichol and Lee, 2005) and detailed temperature products can be derived by adding land-cover information (Nichol, 1994; Nichol and Wong, 2005).

Apart from deficiencies in spatial resolution, the mapping of UEQ has been constrained because UEQ is a holistic concept, comprising numerous parameters that affect urbanites synergistically but are measured on different scales. Therefore, previous attempts to use remote sensing for urban quality analysis have involved the use of factor analysis to identify a relationship between one (Fung and Siu, 2000) or two (Lo and Faber, 1997) image-derived parameters such as vegetation or temperature and socioeconomic data such as population density and income levels. Fung and Siu's 2000 study of environmental quality in Hong Kong was based on the premise that the NDVI from Landsat data is closely related to other environmental indicators.

However, urban environmental quality as a holistic concept cannot be considered in the absence of the two factors that most people (Tzeng et al., 2002) regard as the most important: air quality and noise. Tzeng et al (2002) observe that these two parameters have the same sources — primarily motor vehicles, factories, and construction works. Although methodologies for deriving air-quality parameters from remote sensing systems such as MODIS (Kaufman, 1998), Landsat (Sifakis et al., 1998), and SPOT (Retalis et al., 1999) are now available, the spatial resolution of the derived products (approximately 250 to 450 m) is still too low for application at city street level. Because it is an invisible parameter, noise is not amenable to remote sensing and therefore must be derived from other data sources or by establishing a significant relationship with remotely sensed indices.

This chapter demonstrates how high-resolution satellite sensor data can be utilized in combination with other parameters of UEQ to derive an integrated UEQ (IUEQ) index permitting detailed spatial analysis of environmental quality as it varies over whole cities. It also demonstrates methodologies for mapping and data output in two and three dimensions, using animated VRML modeling, to give more complete visualizations.

12.2 BUILDING THE DATABASE FOR UEQ MAPPING

Because environmental quality may be thought of as continuously variable over an urban area, the use of VHR sensors such as IKONOS and Quickbird provides a level of detail down to the individual building (Nichol, 2005) or tree on a street (Nichol and Lee, 2005). Furthermore, increasing availability of digital map data representing urban infrastructure down to building level, as well as the hybrid, raster–vector data-handling ability of modern GIS, permits other environmental data to be stored in the same database as the image data. Except for discrete data such as building density or socioeconomic data measured within predetermined areal units, analysis may be done at pixel level. The two major constraints in such an integrated analysis are the different units of measurement of the parameters and their different spatial scales of operation; these are discussed later.

12.3 METHODS FOR DATA INTEGRATION

Due to the different scales of measurement as well as different measurement units, approaches to integrating the parameters are limited. Two general approaches have been demonstrated in previous studies: principal component analysis (PCA) (Lo and Faber, 1997; 15] and the GIS overlay approach (Lo and Faber, 1997; Nichol and Wong, 2005a, b, in press). In a quality of life assessment in Athens, Georgia, Lo and Faber (1997) aggregated satellite-derived parameters at the census district level for integration with socioeconomic data. Nichol and Wong (2005a) used pixels as the mapping unit for UEQ in Hong Kong while testing for the optimum resolution for aggregating the data. Both methodologies resulted in relative UEQ indices that are scene dependent and map UEQ on a graduated scale of lowest to highest within the study area.

12.3.1 PCA METHOD

This is a procedure for compressing multidimensional data into fewer representative dimensions or scales. It is commonly used in remote sensing to reduce the dimensionality of a multispectral data set by compressing the data from all wavebands onto a few components that comprise most of the overall variance in the data. Standard PCA (which uses a covariance matrix) is, however, unsuitable for data in different measurement units (Mather, 1999). Therefore, the PCA method using correlation matrix, which standardizes the variables, should be used.

In the case of UEQ mapping, in which the objective is to obtain a single representative index, if a large proportion of the variance among all parameters is found to be represented by one or more PCs, this may provide a single integrated UEQ index. This is a likely situation because the parameters are selected initially as indicators of UEQ, and they are expected to be correlated to some degree. However, in their 1997 study, Lo and Faber found that only 54% of the total variance was represented by the first PC. According to the modifiable areal unit problem (MAUP) (Fotheringham and Wong, 1991), the degree of correlation between parameters is highly scale dependent (see also Openshaw and Taylor, 1979).

Thus, one level of aggregation of pixel data may produce higher component loadings than another, according to differences in synergism between parameters at each spatial scale. This may also depend on the parameters involved as well as the nature of the study area. In a case in which no resolution was able to represent a significantly large proportion of the total variability, one possible conclusion would be that PCA was not suitable for UEQ mapping in that study area because some parameters were not represented due to low or zero loading on the UEQ index (e.g., PC 1 or PCs 1 through 3). However, such a conclusion would be subject to more detailed analysis of the parameters involved in the factor loadings, as well as their relative importance, if known.

12.3.2 GIS OVERLAY METHOD

Because correlation is not involved, this method does not rely so strictly on the scale dependence between parameters. Stacked GIS layers are combined by overlay techniques, but require subjective determination of thresholds. The range of values for each parameter may be ranked on a scale such as 1 to 10 or quantiles, and the resulting UEQ values would correspond to the sum of the data layers. An alternative approach to GIS overlay is the application of multicriteria queries to the data layers, using specific environmental thresholds, if known, or a more generic threshold such as standard deviation. The GIS overlay method can more readily accommodate data originating from different resolutions or units of measurement.

12.4 CASE STUDY: ANALYSIS OF URBAN ENVIRONMENTAL QUALITY IN HONG KONG

This study describes the mapping and analysis of UEQ in the Kowloon Peninsula, the most continuous, densely built urban area in Hong Kong. The GIS overlay and the PCA approaches are demonstrated for the integration of two satellite-derived parameters — biomass and temperature — with four other GIS-derived parameters: noise, air quality, building density, and building height. The mapping units correspond to 35 electoral subdistricts, as well as image pixels and raster data at a base mapping resolution of 4 m.

12.4.1 PARAMETERS

The individual mapped parameters are shown in Figure 12.1. Their derivation is:

Air temperature (Ta) is derived from surface temperature (Ts) from the thermal image waveband of Landsat ETM+. Ts is first obtained by conversion of the image DN values to black body temperature in Celsius using standard conversion factors. This is then adjusted to account for emissivity differences between vegetated and nonvegetated areas by fusion with an IKONOS image having 4-m resolution (see Nichol, 1994, 2005). The corrected surface temperatures are thus detailed enough to be related to individual buildings, streets, or even single trees. Air temperature is then

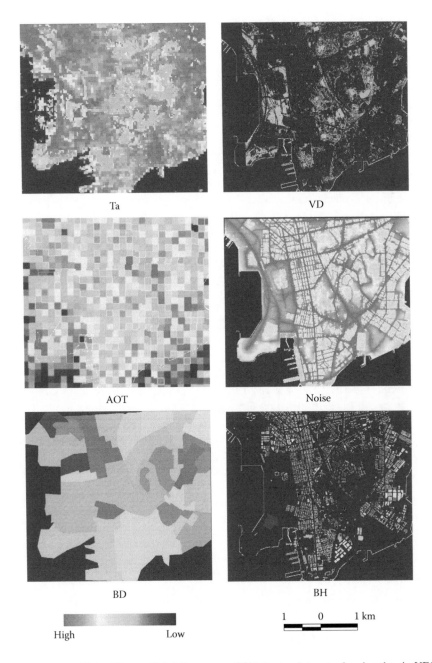

FIGURE 12.1 (Color Figure 12.1 follows page 240.) Raster data sets showing the six UEQ parameters.

obtained by regression of 400 simultaneous Ts and Ta measurements of a variety of horizontal surfaces in the study area, with Ta measured approximately 1 m above the surface.

Vegetation density (VD) is a measure of total biomass (see Nichol and Lee, 2005) and is derived from an IKONOS multispectral image of November 23, 2000, with 4-m resolution by regression of image wavebands against field data.

Air quality (AOT) is represented in this study by aerosol optical thickness (AOT). This is a dimensionless number representing the extent to which aerosols in the atmospheric column scatter or absorb energy of a particular wavelength. A method for estimating AOT using the difference between a reference image representing "nonpolluted" conditions and a "polluted" image of a different date is described by Retalis et al. (1999), using the 0.45 to 0.52 (blue) waveband of Landsat. This technique uses the difference between standard deviation of pixels within a moving window. Therefore, the resulting resolution corresponds to the window size: in this case, 5*5 Landsat pixels giving the AOT image a resolution of 150 m. Landsat ETM+ images of September 14, 2000, and September 17, 2001, were used for the polluted and nonpolluted cases.

Noise (N) is derived at 10-m resolution from data on traffic flow volumes and building locations using LIMA environmental noise calculation and mapping software. Generated in decibels, the data represent the noise levels at 4 m above ground.

Building density (BD) is derived from digital map data of building outlines from the Land Information Center, Lands Department, Hong Kong. The building areas are summed, and the percentage of built area per district and subdistrict is obtained.

Building height (BH) uses the same vector map database as BD. The BH value for each spatial unit is the mean height of all buildings within that unit.

For pixel-based PCA and GIS overlay operations, AOT, N, BD, and BH data are resampled to a base resolution of 4 m in order to correspond to the image-based parameters.

12.4.2 TECHNIQUES OF DATA INTEGRATION

The methods and procedures for data integration are summarized in Figure 12.2.

12.4.2.1 PCA: Pixel Based

PCA was performed on the six project parameters, from 4 to 80 m in increments of 4 m, to discover the optimum scale at which PC 1 as a single index could most represent the combined parameters. This was done by resampling each parameter at each resolution. Figure 12.3 shows that the resolution at which PC 1 holds the largest percentage of the variance occurs at 64 m, with 66% of the variance; at 4-m resolution, PC 1 contains only 27% of the variance. Therefore, to maximize data

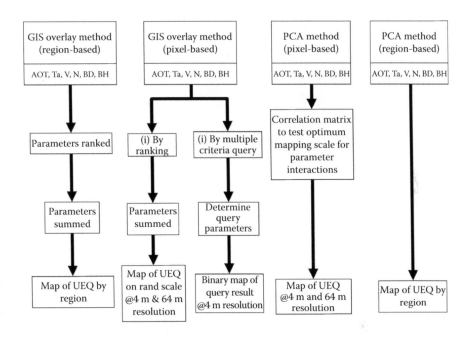

FIGURE 12.2 Methodologies for integration of UEQ parameters.

variability contained in the IUEQ index at 4-m resolution and because the first three PCs combined contain 61% of the total variability, these were normalized and weighted by their respective eigenvalues. The final map (Figure 12.4a) was produced using the sum of weighted PCs 1 through 3. Figure 12.4b similarly shows the weighted sum of the first three PCs at 64-m resolution that, together, contain 91% of the variability. Light-toned areas represent high UEQ.

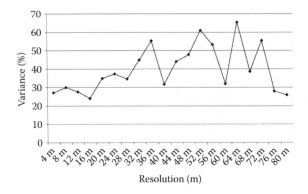

FIGURE 12.3 Total percentage loadings of parameters on PC 1 at different resolutions.

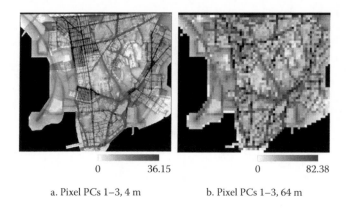

| 0 | 36.15 | 0 | 82.38 |

a. Pixel PCs 1–3, 4 m   b. Pixel PCs 1–3, 64 m

FIGURE 12.4 Map produced from pixel-based PCAs 1 through 3: (a) 4-m resolution; (b) 64-m resolution.

12.4.2.2 PCA: Region Based

The data for all parameters were averaged for the 35 electoral subdistricts prior to PCA analysis and a map of UEQ produced from PC 1 (Figure 12.5). However, only 41% of the variability was represented: an even lower finding than that of Lo and Faber (1997), where PC 1 contained 54% of the variability within census districts. The relatively low factor loading is not surprising due to the likely irrelevance of administrative boundaries to environmental factors whose variability is lost when the detailed distributions within each district are averaged to a single value. Therefore, no further processing was done at regional levels.

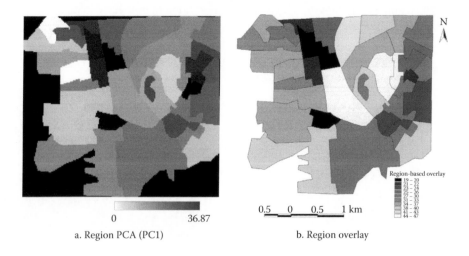

a. Region PCA (PC1)   b. Region overlay

FIGURE 12.5 a. Region-based map from PCA (PC 1); b. region-based map from overlay of ranked parameters.

12.4.2.3   GIS Overlay: Region-Based Overlay of Ranked Parameters

As with region-based PCA, the data were averaged by electoral subdistrict and ranked by dividing into ten equal classes; the higher values represented the most desirable environmental quality. The aggregate score for each subdistrict was obtained by summing the ranked scores for each variable, giving possible values from 6 to 60. The resulting map, with a range of values from 18 to 47, is shown in Figure 12.5.

12.4.2.4   GIS Overlay: Pixel-Based Overlay of Ranked Parameters

The pixel data at resolutions of 4 and 64 m were ranked as for region-based GIS overlay (discussed previously), and the ranked scores were summed (Figure 12.6a and b). The 4- and 64-m resolutions were selected in order to compare results with PCA mapping at the same resolution. The base mapping resolution of 4 m gave a range of data values of 14 to 53; at 64-m resolution, a smaller range of 18 to 51 was obtained, due to the spatial averaging process.

12.4.3   ACCURACY ASSESSMENT

Map accuracy is usually tested against a reference data set of known accuracy, with the results presented in the form of a contingency table. However, because UEQ is a subjective criterion, it is more appropriate to test the mapped values using sociological research methods (Tzeng, 2002; Bonaiuto et al., 2003; Wong and Domroes, 2005). Thus, reference values may be derived from interviews at sample field locations, with respondents who live and work in the study area, to obtain their perceptions of UEQ. They may be asked to rank the place at which they are standing, photographs of urban scenes, and districts that they know. Values from each UEQ map can be correlated with the questionnaire rankings. Respondents may additionally be asked to rank the

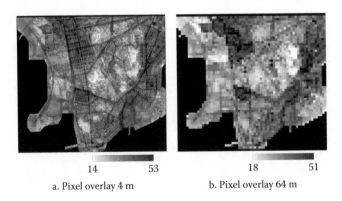

| 14 | 53 | | 18 | 51 |
| a. Pixel overlay 4 m | | | b. Pixel overlay 64 m | |

FIGURE 12.6 Pixel-based map GIS overlay of ranked parameters: (a) 4-m resolution; (b) 64-m resolution

importance of individual UEQ parameters, and the results of this may feed back into the mapping process to allocate weightings to each parameter.

No such data were collected for the present study or are known for other similar studies. However, a test of consistency between the methodologies was performed by correlating the mapped outputs. Correlations between the PCA method using the first three PCs and the pixel overlay method at 4- and 64-m resolution obtained significant correlations at the 99% confidence level ($R = 0.57$ and $R = 0.54$, respectively), and a higher correlation of $R = 0.77$ was obtained between the region-based PCA (which used PC 1) and region-based overlay methods (Table 12.1).

TABLE 12.1
Correlation of UEQ Values between Mapping Methods

Mapping Method	Mapping Method	R	Deg. Freedom	% Variance Represented by PCs
PC 1, 4 m	Pixel overlay 4 m	0.38[a]	104,000	27
PC 1–3, 4 m	Pixel overlay 4 m	0.57[a]	104,000	61
PC 1, 64 m	Pixel overlay 64 m	0.37[a]	4,020	66
PC 1–3, 64 m	Pixel overlay 64 m	0.54[a]	4,020	91
Region PCA	Region overlay	0.77[a]	35	41

[a] Significant at 99% confidence level.

12.4.4 Pixel-Based GIS Overlay: Multiple Criteria Query

An additional method of data integration using the GIS overlay approach is by multiple-criteria query. If environmental thresholds for certain parameters are known, queries may be formulated to return pixels satisfying the query parameters in the form of a binary map. Figure 12.7 shows the result of such a query on the two image-based parameters, VD and Ta, mapped at 4-m resolution. Areas of high environmental quality are defined as greater than one standard deviation (SD) above mean VD and greater than one standard deviation below mean Ta — that is, well-vegetated and cool areas.

Although the two parameters are often claimed to be highly correlated (Gallo et al., 1993; Weng et al., 2004), such claims are unsubstantiated at the microscale of the present study; this, as well as the multiple benefits of vegetation in addition to temperature control, suggests that both of these image-derived parameters should be present to designate areas of high UEQ. Other such queries may return areas that are:

Warm and unvegetated
Warm but vegetated
Cool but unvegetated

All of these offer insights into the physical controls on the urban microclimate, especially when overlaid with infrastructural details such as roads and building outlines in either two or three dimensions.

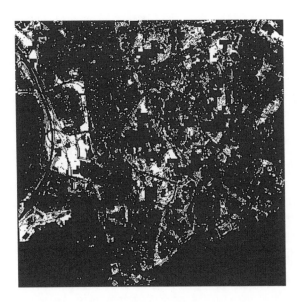

FIGURE 12.7 Multiple-criteria query of Kowloon based on greater than one standard deviation below mean Ts and greater than one standard deviation above mean VD.

Figure 12.8 shows the preceding query at a more detailed level, with the example of a small public park of only 0.15 ha within the densely urbanized and congested streets of Mongkok. Figure 12.8a is a binary image of pixels representing low temperature and high biomass. Surrounding buildings approximately 30 m high cast shadows over street canyons at the image time, in a similar direction as those on the digital color air photo (Figure 12.8b). However, Figure 12.8c shows that the dominant cooling effect is not shadow but vegetation because the surface temperature of the park is 6 to 7°C cooler than surrounding areas, which include shady street canyons.

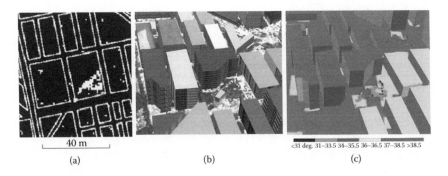

FIGURE 12.8 a. Result of multicriteria query showing pixels of high UEQ as white, overlaid with street outlines; b. three-dimensional model with texture derived from color orthophoto showing that the area is a small park within the densely built Mongkok district of Kowloon; c. three-dimensional thermal model showing the park 6 to 7°C cooler than surrounding areas.

Such small islands of greenery are especially important for their higher environmental quality. When these are combined with three-dimensional representation (see Section 12.4.5), a better understanding of environmental relationships at local and regional levels can be obtained.

12.4.5 VISUALIZATION OF UEQ

Three-dimensional models were constructed from digital data of building outlines from the Hong Kong Lands Department using three-dimensional Studio Max software and ArcGIS. This permits a more representative visualization of UEQ because most parameters operate not just at ground level, but within the depth of the urban canopy layer (Oke, 1976) and are three dimensional. For example, conventional satellite-derived heat islands measure the temperature of only the horizontal "seen" surface, which includes the ground and building roofs; however, in high-rise areas, this is considerably smaller than the complete active surface (Weng, 2001; Voogt and Oke, 1996; Nichol, 1998).

Yet all natural and man-made urban surfaces are active radiating surfaces as well as heat sources and sinks for the adjacent atmosphere. They also play a role in providing shade and act as barriers and funnels for fresh rural air, on the one hand, or stale urban air on the other. Thus, all surfaces collectively influence urban air quality, including the dynamics of the urban heat island. The under-representation of the active surface by satellite image data is accentuated in tropical cities such as Hong Kong due to the high sun angle. Thus, horizontal surfaces "seen" by the satellite, such as roofs and tree canopies, may be significantly hotter than the mean temperature of the active surface. Noise and aerosol models also require representation in three dimensions because noise is modified by barriers such as vertical building walls and, like aerosols, street level measurements may not resemble those at higher levels or across barriers.

To visualize the complete active surface, it was necessary to add vertical surfaces as three-dimensional facets whose temperature could ideally be varied according to the effects of sun angle and azimuth at the image time. For horizontal surfaces, temperatures were derived by overlaying the image data with building outlines after the method of Nichol (1998). The temperature of each building was based on the weighted mean temperature of image pixels intersecting the outline. Temperatures for vertical facets were determined according to relationships between horizontal and vertical surfaces from fieldwork conducted at the same season and time of day as those for the image. The resulting models provide three-dimensional visualization of the urban microclimate in relation to urban morphology and regional topography at the image time.

For example, Figure 12.9 shows Causeway Bay, which generally has the highest air pollution levels in Hong Kong due to extremely high rise building geometry and the mountain backdrop in relation to the northerly, pollution-bearing winds from mainland China. Despite the tall buildings' giving a low sky view factor, an early heat island is evident on the images. Air pollution levels are highest in the winter dry season, when the dust- and pollution-bearing northerly winds are trapped in the flat coastal plain of Hong Kong island. Natural fresh-air corridors along ridges and valleys, extending from high mountains inland down to the coast and with the potential to moderate the

| <31deg. | 31–33.5 | 34–35.5 | 36–36.5 | 37–38.5 | >38.5 |

FIGURE 12.9 (Color Figure 12.9 follows page 240.) Terrain model looking north across Victoria Harbor to the shoreline at Causeway Bay. Upper and lower left-hand figures show selected buildings extruded to actual height, colored according to image-derived Ts. Lower right hand represents model overlain with IKONOS image. Buildings blocking fresh-air corridors are labeled "A."

build-up of air pollution and the UHI with cooler fresh air at night, are blocked by the transversely oriented tall buildings of Causeway Bay ("A" on Figure 12.9).

Although methods for representing image pixels as GIS database entities are not yet available (Baltsavias and Gruen, 2003), it is potentially possible, using object-oriented data structures, to construct four-dimensional models to simulate changes in space and time across a single building facet, according to solar illumination by time of day and year. The models described here provide animation in Microsoft Video Interlaced format for fly-through simulation. Similar three-dimensional models are generated by Lima noise modeling software used in this study.

12.5 DISCUSSION

Remotely sensed environmental data provide a high-resolution mapping base for UEQ assessment at variable levels of detail according to the characteristics of the study area and of the parameters involved. Methods for combining multiple parameters with others in a single index are investigated, but due to lack of accuracy assessment criteria, only general conclusions can be made. However, the significantly high correlation between UEQ maps obtained from the PCA and the GIS overlay methods suggests consistency between the two methods.

In the study area, the optimum resolution for mapping of UEQ using PC 1 is 64 m, and 91% of the total variance is represented at this resolution when the first three PCs are combined. On the other hand, the method of UEQ mapping by GIS overlay has the advantage of taking into account all parameters and is amenable to mapping at any resolution. However, it ignores the scale at which the parameters operate or interact. Because the PCA method is shown to be sensitive to different resolutions of data integration, its use may be more representative of the scales at which the parameters actually operate, as well as the scales at which they act synergistically to affect UEQ.

The main disadvantage of the PCA and GIS overlay methods as demonstrated in this study is that they fail to consider the relative importance of the parameters. Additionally, the UEQ indices produced are relative scales applicable to each study area. Thus, for Figure 12.4b, which is derived from PCs 1 through 3 at 64-m resolution, the scale of 0 to 82.4 represents the lowest to the highest areas of UEQ in the study area. Because the first three PCs contain 91% of the variability among the six input parameters, this method may be selected if a lower resolution is acceptable because only 61% of the variability is represented by PCs 1 through 3 at 4-m resolution.

Because of the lack of accuracy testing, recommendations are only tentative. However, the combination of the first three PCs at 64-m resolution containing over 90% of the overall data variance is likely to provide a good index of UEQ. The significant correlation of this map with that produced by GIS overlay at the 99% confidence level suggests similarity in the resulting mapped distributions from each method. Additionally, visual analysis of these two maps shows similarity at a general level.

Although use of multicriteria query used in this study was able to demonstrate the practical application of mapping selected parameters at very high resolutions, the usefulness of an index that represents all parameters at resolutions as high as 4 m may be questioned. In addition, accuracy is limited by the resolution of the original data. Thus, notably only 61% of the variability was contained on the first three PCs at 4-m resolution and only 27% on PC 1. Firmer recommendations await the availability of sociological reference data for accuracy testing; to date, no research has included this.

REFERENCES

Baltsavias, E.P. and Gruen, A., Resolution convergence: a comparison of aerial photographs, LIDAR and IKONOS for monitoring cities, in *Remotely Sensed Cities*, Mesev, V., Ed., Taylor & Francis, London, 2003, 474.

Bonaiuto, M., Fornara, F., and Bonnes, M., Indexes of perceived residential environment quality and neighborhood attachment in urban environments, *Landscape Urban Plann.*, 65, 41, 2003.

Fotheringham, A.S. and Wong, D.S.W., The modifiable areal unit problem in multivariate statistical analysis, *Environ. Plann. A*, 23, 1025, 1991.

Fung, T. and Siu, W.L., Environmental quality and its changes using NDVI, *Int. J. Remote Sensing*, 21, 1101, 2000.

Fung, T. and Siu, W.L., A study of green space and its changes in Hong Kong using NDVI, *Geogr. Environ. Syst.*, 5, 111, 2001.

Gallo, K.P. et al., The use of NOAA AVHRR data for assessment of the urban heat island effect, *Appl. Meteorol.*, 32, 899, 1993.

Kaufman, Y.J., Algorithm for remote sensing of tropospheric aerosols from MODIS, product ID MOD04, NASA Goddard Space Flight Center, U.S., 1998, 85.

Lo, C.P. and Faber, B.J., Integration of Landsat Thematic Mapper and census data for quality of life assessment, *Remote Sensing Environ.*, 62, 143, 1997.

Lo, C.P., Quattrochi, D.A., and Luvall, J.C., Application of high-resolution thermal infrared remote sensing and GIS to assess the urban heat island effect, *Int. J. Remote Sensing*, 18, 287, 1997.

Mather, P.M., *Computer Processing of Remotely Sensed Images*, 2nd ed., John Wiley & Sons Ltd, England, 1999, 126.

Nichol, J.E., A GIS-based approach to microclimate monitoring in Singapore's high-rise housing estates, *Photogrammetric Eng. Remote Sensing*, 60, 1225, 1994.

Nichol, J.E., Visualization of urban surface temperatures from satellite images, *Int. J. Remote Sensing*, 19, 1639, 1998.

Nichol, J.E. and Lee, C.M., Urban vegetation monitoring in Hong Kong using high-resolution multispectral images, *Int. J. Remote Sensing*, 26, 903, 2005.

Nichol, J.E. and Wong, M.S., Modeling urban environmental quality in a tropical city, *Landscape Urban Plann.*, 73, 49, 2005a.

Nichol, J.E. and Wong, M.S., Modeling environmental quality in Hong Kong based on multiple parameters, in *Proc. Computers Urban Plann. Urban Manage.*, Batty, S.E., Ed., University College London, London, 2005b, 249.

Nichol, J.E., Remote sensing of urban heat islands by day and night, *Photogrammetric Eng. Remote Sensing*, 71, 613, 2005.

Nichol, J.E. and Wong M.S., Assessment of urban environmental quality in a subtropical city using multispectral satellite images, *Environ. Plann. B*, in press.

Oke, T.R., The distinction between canopy and boundary-layer heat islands, *Atmosphere*, 14, 268, 1976.

Openshaw, S. and Taylor, P.J., A million or so correlation coefficients: three experiments on the modifiable areal unit problem, in *Statistical Applications in the Spatial Sciences*, Wrigley, N., Ed., Pion, London, 1979, 127.

Retalis, A., Cartalis, C., and Athanassiou, E., Assessment of the distribution of aerosols in the area of Athens with the use of Landsat Thematic Mapper data, *Int. J. Remote Sensing*, 20, 939, 1999.

Roth, M., Oke, T.R., and Emery, W.J., Satellite-derived urban heat islands from three coastal cities and the utilization of such data in urban climatology, *Int. J. Remote Sensing*, 10, 1699, 1989.

Sifakis, N.I., Soulakellis, N.A., and Paronis, D.K., Quantitative mapping of air pollution density using Earth observations: a new processing method and application to an urban area, *Int. J. Remote Sensing*, 19, 3289, 1998.

Tzeng, G.H., Tsaur, Y.D., and Opricovic, S., Multicriteria analysis of environmental quality in Taipei: public preferences and improvement strategies, *J. Environ. Manage.*, 65, 109, 2002.

Voogt, J.A. and Oke, T.R., Complete urban surface temperatures, *J. Appl. Meteorol.*, 36, 1119, 1996.

Weng, Q., A remote sensing-GIS evaluation of urban expansion and its impact on surface temperature in Zhujiang Delta, China, *Int. J. Remote Sensing*, 22, 1999, 2001.

Weng, Q., Lu, D., and Schubring, J., Estimation of land-surface temperature–vegetation abundance relationship for urban heat island studies, *Remote Sensing Environ.*, 89, 467, 2004.

Wong, K. and Domroes, M., The visual quality of urban park scenes in Kowloon Park, Hong Kong: likeability and affective appraisal, *Environ. Plann. B*, 32, 617, 2005.

13 Population Estimation and Interpolation Using Remote Sensing

Xiaohang Liu and Martin Herold

CONTENTS

13.1 INTRODUCTION

Knowledge of the size and spatial distribution of the human population in an urban area is essential for understanding and responding to a myriad of social, economic, and environmental problems (Dobson et al., 2000; Sutton et al., 2001). Data on demographic characteristics and socioeconomic conditions are traditionally obtained from census and municipal statistics, questionnaires and interviews, and/or field surveillance. These data sources, however, often lack consistency when applied to a large metropolitan area or across several urban areas or various countries, or they may not be available at all (e.g., in some developing countries). Costs of frequent updates and spatial aggregation and boundary designation problems associated with a census (Openshaw, 1984) have fostered the use of remote sensing to acquire quality population information.

Since the 1950s, various types of remote sensing imagery have been examined in terms of their utility to estimate human population distribution, including aerial photography (Anderson and Anderson, 1973; Adeniyi, 1983; Lo, 1986), Landsat MSS (Iisaka and Hegedus, 1982), TM and ETM (Forster, 1985; Chen, 2002; Li and Weng, 2005), SPOT (Lo, 1995), and DMSP nighttime imagery (Welch, 1980; Dobson et al., 2000; Sutton et al., 1997, 2001). Although the success of these studies varies and remote sensing may not supply all the information needed, they did demonstrate that remote sensing has the potential to assist population estimations from local to global scales and provides a unique perspective from which to extract demographic and socioeconomic attributes of cities.

From a broader research perspective, population studies using remote sensing are a key avenue to improving spatial concepts in social science and fostering approaches to "socializing the pixels." Social sciences outside geography and urban planning are generally more concerned with why things happen than with where they happen; accordingly, not much integration has occurred between remote sensing and social science. For example, it is difficult to quantify the spatial context of social phenomena and measure socially induced spatial phenomena as these evolve over time. Instead, by helping make connections between concepts used by different disciplines and their data sources and techniques, remote sensing can provide additional levels of information. This can help to explore and establish links between land characteristics and changes and a variety of social, economic, and demographic processes (Rindfuss and Stern, 1998).

Recognizing this potential, we will discuss approaches to population estimation and interpolation, emphasize the need for remote sensing-based studies of urban population, introduce different methodological avenues on how to link remotely sensed measurements and population data, and illustrate them using specific examples. In particular, examples will highlight the use of recently available high spatial resolution satellite sensors to study intraurban population characteristics.

13.2 SPATIAL POPULATION ESTIMATION

13.2.1 TRADITIONAL APPROACHES AND PROBLEMS

Practically, large-scale collection of population data can be undertaken in three ways: census, population registers, and remote sensing (Rhind, 1991). In the United States, the decennial census is the primary source of data on population distribution and demographic characteristics. However, when census data is used to conduct geographic research, several problems may arise. First, although the census is conducted at the household level, it is only released as an aggregate of census enumeration units in order to protect confidentiality. The boundaries of the census enumeration units are rather arbitrary because they are primarily designed to ease the enumeration process rather than to represent the geographic distribution of a socioeconomic variable.

The arbitrary nature of census areal partitioning immediately leads to two problems. The modifiable areal unit problem (MAUP) refers to the fact that modifying the spatial basis of data aggregation may significantly affect the results of data analysis; consequently, it is unclear whether the results obtained reflect the individuals living in the region or are simply a function of the particular areal unit used

(Openshaw, 1984). The other problem is visualization of the data. Choropleth mapping of the population density derived from the raw census data creates the impression that population in a census enumeration unit is distributed uniformly. However, the land use within a census areal unit is not always homogenous. Residential as well as nonresidential lands such as open water, derelict land, industrial premises, commercial district, etc. may all be present. This heterogeneity renders choropleth mapping inappropriate to visualize where people actually live.

The arbitrary boundaries of census areal partitioning also create problems for zone transformation or areal interpolation (Flowerdew and Openshaw, 1987; Goodchild and Lam, 1980). This refers to the difficulty of transferring data from one set of zones to a second independent set — for example, to find out the number of children in each school district based on census tract data. Unless the "from" zone and "to" zone are nested hierarchically or the boundaries coincide, an interpolation method is necessary. Similar concerns rise when integrating population data with other spatial data such as hydrological or weather data because the latter is usually collected at the watershed level and watershed boundaries rarely coincide with census boundaries. Another problem associated with census boundaries is that they can be changed, created, or removed over time, making it difficult to study the change of population in an area (Gregory, 2002).

In search of a potential solution to these problems, ancillary data other than population are often employed to assist the estimation of population distributions. In the literature, two types of research have applied remote sensing to population studies. The first category is population interpolation that uses information derived from remote sensing images to refine census-reported population spatially. Dasymetric mapping is a commonly used approach. The other is population estimation, that is, to directly estimate population count and distribution. This type of research differs from population interpolation in that it does not necessarily require census population as the input. Rather, it uses statistical models and other variables (more or less extracted from remotely sensed imagery) — for example, size of residential area, distance to central business district, topographic slope, etc. — to estimate the population count or density of an area.

13.2.2 POPULATION INTERPOLATION WITH REMOTE SENSING

Human population distribution is closely related to other information about the landscape such as land use and transportation facilities. This information can therefore be utilized to better delineate where people actually live in the study area. Dasymetric mapping is a widely used method to interpolate population distribution. The idea for this type of mapping was probably first proposed by Wright (1936), who used land-use information from paper maps to obtain a better description of population distribution. Whereas Wright used a look-up table to implement his dasymetric mapping, Langford et al. (1991) developed a multivariate regression method that uses land-use information derived from remotely sensed images to estimate population distribution:

$$P_i = \sum_j p_{ij} = \sum_j d_j * a_{ij} \tag{13.1}$$

where

P_i is the known total population of source zone i

p_{ij} is the unknown total population of land-use zone j in source zone i

d_j is the unknown average population density of land use j

a_{ij} is the known area of land use j in source zone i

Because there are multiple source zones (e.g., census blocks), linear regression can be applied to estimate d_j and hence p_{ij}. Langford and Unwin (1994) also discussed the idea of dasymetric mapping. Mennis (2003) used remote sensing to improve census-reported population distribution. In his method, the characteristic population density of a land-use category is not obtained from Equation 13.1; rather, it is obtained from the average population density of the census enumeration units with homogenous land use in that category.

Despite the ease of implementation, this method is subject to a few problems. The first is that it associates each land-use class with a single population density. In other words, the differences between land-use classes are recognized, but differences within a land-use class are ignored. Yet, not all residential areas have the same population density, as evidenced by the contrast between detached housing and multiple-unit housing. Even within a detached-housing area, there are the differences among low, medium, and high density. To incorporate such considerations, one can conduct a more detailed classification in residential areas and associate different residential categories with different population densities (see, for example, Donnay and Unwin, 2001). Although this approach improves the estimation, the fundamental problem remains unsolved. There is a continuum of densities and the discrete categorizations are arbitrary.

Furthermore, only the land-use category is incorporated in this estimation process. The location of the land use is not considered, although it is known that geographic location and neighborhood have an impact on population density because of spatial autocorrelation. In remedy, remote sensing-derived information other than land-use category has been examined to inform population interpolation. Examples include Wu and Murray's research (2005), which uses the percentage of impervious surface to allocate residential population; Harvey's research (2002b) on using the reflectance values of the TM images; and Liu's method based on spatial metrics (2003).

13.2.3 POPULATION ESTIMATION BASED ON REMOTE SENSING

The use of remote sensing for population estimation started in the mid-1950s. The initial motivation was to remedy shortcomings such as high cost, low frequency, intense labor, etc. of the decennial census. Research has been conducted using different types of remotely sensed imagery and on different scales. Depending on the degree of detail exhibited and the size of the area covered, the methods may be grouped into three categories (Kraus and Senger, 1974): generalized, semidetailed, or detailed.

Generalized methods are based on the allometric relationship between the size and population of an area. The allometric relationship is presented as a linear regression between the natural logarithmic of population and the size of an urban area (Nordbeck, 1965; Tobler, 1975):

$$\ln P_i = a \ln A_i + b \tag{13.2}$$

where
    A_i is the size of the ith built-up area, which can be inferred from remote sensing imagery
    P_i is the corresponding population to estimate
    a and b are the coefficients to be determined

During the application of the model, existing population data sets from census or field surveys are collected for different regions. A linear regression can then be applied to estimate the coefficients a and b. They are then plugged into the previous equation so that the population of an unknown area can be estimated. Examples of the generalized method can be found in Wellar (1969), Anderson and Anderson (1973), and Holz et al. (1969). The spatial unit or scale used ranges from subcity community (Anderson and Anderson, 1973) to cities within a country (Holz et al., 1969). Sutton and colleagues (2001) examined the utility of nighttime satellite imagery to estimate global population and its distribution at subnational scale. The allometric relationship seems to hold in general, although the coefficients must be tuned for different regions of the world. Vining and Louw (1978) pointed out that the allometric relationship may not be stable or have shifted; consequently, caution must be exercised when estimating population based on the generalized method.

The generalized method appears to be the least accurate for detailed population estimation. In some case studies, the relative error can be as high as 100% or more (Liu and Clarke, 2002). This is primarily due to the exponential form of the allometric model: even though the residual of the linear regression in the log–log space may be small (i.e., the estimation of $\ln P_i$ is fairly accurate, the conversion back to P_i will exponentialize the error and thus will result in significant overestimation or under-estimation).

The estimation of the aggregate population of an area can be fairly accurate because overestimations and underestimations may cancel out; however, for better results at a detailed level, the semidetailed and detailed approaches, which have a linear form, are preferred. Semidetailed methods measure the areas of different land uses (A_{ij}) and associate each with a characteristic population density (d_j) determined from census population data or sample survey. The total population of a spatial unit (P_i) is estimated by:

$$P_i = \sum_j A_{ij} * D_j \tag{13.3}$$

This method is similar to the dasymetric mapping approach described in Equation 13.1, except that population density is obtained through sample data rather than linear regression. Collins and El-Beik (1971) applied the method to estimate intercensal population of the city of Leeds in England and achieved an underestimation of only 2% of the total population. Their study, however, used test sites with a single type of dwelling unit only. Kraus and Senger (1974) also applied the method to four California cities and achieved a rather impressive result. The relative errors of the four cities are 9.17% or less. This method was also used by Mennis (2003) for dasymetric mapping.

The third population estimation method is categorized as *detailed* in the sense that it counts individual dwelling structures and multiplies the figure by an average population per dwelling. The method was first proposed by Green (1957) and was first implemented by Eyre et al. (1970) and Dueker and Horton (1971). The success achieved varies, with overall accuracy ranging between 85 and 98%.

All of these three methods are based on a single land-use-related variable. Other physical or socioeconomic variables can also be incorporated and the result may be significantly better. A prominent example is the LandScan project, which estimates global population at 30-sec × 30-sec resolution using land cover, nighttime DMSP satellite imagery, census data, and other information about demography, topography, and transportation networks (Dobson et al., 2000).

Similar methods can also be applied in small-area population estimation. For example, Liu and Clarke (2002) found that the total population of a small area is correlated with distance to central business district, accessibility to transportation system, slope, and age of the residential areas. Many of these variables can be extracted from remotely sensed imagery. Compared to methods using a single variable, this approach tends to improve overall accuracy. Selection of the variables to be included in the model needs to be guided by theories in urban geography.

A different remote sensing approach is presented by Harvey (2002b) using an iterative method equivalent to expectation maximization to estimate population density at the pixel level from Landsat Thematic Mapper imagery. The method first assigns all residential pixels within a source zone with an equal share of the total population, i.e.,

$$P_{ij} = P_i/n, \quad i = 1,\dots,n \tag{13.4}$$

where P_{ij} is the population initially assigned to the jth pixel in source zone i whose total population is P_i, and n is the number of pixels of residential land use in source zone i.

There are many source zones, each of which has some residential pixels and the digital values of these pixels are not the same; thus, an ordinary least-squares regression can be conducted where the digital value of the pixels and population serve as the independent and dependent variables, respectively. After regression, the population of the jth pixel in source zone i, $P_{ij}(adj)$, is updated as the sum of the least-squared estimate, $\hat{P}_{ij}$, and the average residual in zone i, $\hat{r}$. $P_{ij}(adj)$ then replaces P_{ij} as the dependent variable of a new iteration of linear regression. Because $P_{ij}(adj)$ is closer to $\hat{P}_{ij}$ than P_{ij}, the R^2 of the new iteration should increase. The process is repeated and stops when a predefined threshold is reached.

Other than the physical characteristics extractable from remote sensing imagery, the spectral reflectance values of satellite sensors have also been correlated with population. Iisaka and Hegedus (1982) reported a strong linear correlation between population density of suburban Tokyo and the mean spectral radiance of a 500- × 500-m grid from MSS imagery. Lo (1995) used SPOT imagery to estimate population in Hong Kong and reported a high correlation between population density and SPOT band 3 ($R^2 = 0.91$). Webster (1996) also used SPOT imagery to estimate dwelling densities in the suburbs of Harare, Zimbabwe. His method used TM, SPOT, and various texture measurements in a stepwise regression model. The R^2 for linear and logarithmic models was reported as 0.86 and 0.97, respectively. Harvey (2002a) provided further mathematical and statistical refinements to the regression methodology and evaluated the validity and robustness of the model by applying it to a second image of a nearby culturally and demographically similar area. His method used TM imagery as source data and examined various combinations of the TM bands. Overall, accuracy and robustness were found to improve with increasing model complexity.

13.2.4 CHALLENGES FOR POPULATION REMOTE SENSING

Regardless of whether remote sensing is used to assist population estimation or spatial interpolation, a central issue is the type and analysis technique of remotely sensed imagery to be used to extract the desired population information. Early research focused on aerial photographs and visual interpretation. Because the result obtained from visual interpretation can rarely be reproduced, the impact of the accuracy of image interpretation is hard to assess (Anderson and Anderson, 1973). Starting in the 1980s, satellite imagery gradually replaced aerial photography to become the primary remotely sensed data because of its relatively low cost, large coverage, and timely repetition. Various types of satellite imagery have been examined to study population distribution. However, only recently has the spatial detail of sensors become adequate for fine-scale population applications at the census block and block group level, where a spatial resolution of 0.5 to 5 m is deemed necessary (Jensen and Cowen, 1999). In fact, the advent of satellite sensors with high spatial resolution such as IKONOS and QuickBird has renewed interest in using remotely sensed data for social science research (Donnay et al., 2001). Although improved spatial resolution and data availability bring promise and opportunities, a few challenges remain.

One key issue is the spatial analysis unit. Many of the known methods use polygons of some type and definition as the elementary spatial unit (e.g., census blocks). The use of polygons already causes generalization and, for some applications, information may be too coarse or spatially incompatible (MAUP problem). If more refinement is necessary, modeling can be conducted at the pixel level. Pixel-based modeling has been argued to have several advantages (Harvey, 2002b). For example, existing pixel-based classifiers can be easily applied and the mathematical form of the model is simple and robust. The biggest problem with this approach lies in the difficulty of obtaining reference population data at the pixel level (Harvey, 2002b). Demographic data are usually spatially aggregated. To obtain population data at the pixel level, disaggregation must be conducted.

However, the process is much less straightforward conceptually and computationally than the inverse process of spatially aggregating data. Certainly, characteristics of the urban landscape, existing data sources, and objectives of the study determine the spatial analysis unit.

Regardless of whether classified land type or spectral reflectance information is used, all studies have reported consistent attenuating effects — that is, the remote sensing estimates vary over a narrower range than the actual densities. Low-density populations may be substantially overestimated and high-density populations may be underestimated. This is probably because of the heterogeneity of land cover and population density on the ground. On the other hand, the aggregate population estimation obtained is quite accurate because overestimation and underestimation can cancel out. In general, more research is needed before remote-sensing-based population estimation can be used in an operational context.

It is noteworthy that so far most research, especially that on population density estimation, has used census data as the primary source. The popular approach is to apportion census data according to some criteria. Census population data, however, are only concerned with residential population. In other words, it is based primarily on where people sleep at night rather than where they work or travel during the day. Daytime population distribution can be very different from that described by the census. For example, Las Vegas has a much higher daytime population than that reported by the census because of tourist population. Because some applications (e.g., emergency response) require knowledge of ambient population and others (e.g., urban growth) require residential population, it is desirable for both types of population to be estimated.

To date, little research has attempted to model the ambient population. LandScan (2000), which is based on land-use and nighttime satellite imagery, is one of the few efforts. Theoretically, detailed urban land-use information may be more directly related to daytime population distribution than nighttime lights because the identification of industrial, commercial, and recreational areas and their sizes can provide an idea of where people are during the day. How to relate nonresidential land use with population density has yet to be explored; however, very detailed land-use information in urban areas, together with research on people's travel behavior, may shed light on this field.

13.3 EXAMPLES OF POPULATION ESTIMATION

In the previous sections, we presented several arguments and concepts for using remote sensing data to estimate population distribution within urban areas. This section presents an analysis of high spatial resolution remote sensing data for intraurban population estimation. The goal is to illustrate some of the concepts discussed in the previous section and present innovative approaches in fine-scale population remote sensing. Three methods will be explored: (1) linnear regression model based on land-use/land-cover information to estimate the spatial distribution of population (Donnay and Unwin, 2001), (2) linear regression model based on spatial metrics derived from remote sensing, and (3) still using spatial metrics from remote sensing, but with geographically weighted regression (Fotheringham et al., 2002) instead of conventional linear regression. One objective is to reflect that the advent of high-spatial-resolution satellite imagery (e.g., IKONOS) offers new avenues to obtain urban information on a very detailed level. The data are specific enough to move from census tracts, a common spatial unit in the analysis of Landsat-type data, toward block group and block level analysis in linking remote sensing signals and population data.

The three methods will be illustrated using an example on the southern coastal area of Santa Barbara County, California. This region is located about 170 km northwest of Los Angeles in the foothills of the Santa Ynez Coast Range (Figure 13.1) with a size of about 300 km² and a total population of around 200,000 people. Emphasized are the three urban centers of Santa Barbara, Goleta, and Carpinteria. The area consists of different types of land use, including residential areas with different density and socioeconomic structure, mixed-use areas (e.g., downtown areas), and commercial and industrial districts (Figure 13.1).

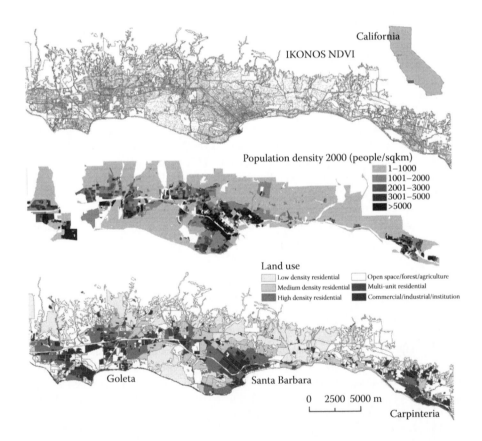

FIGURE 13.1 Santa Barbara south coast area shown with NDVI (from IKONOS image mosaic), the census block group population density for the year 2000, and the major land-use classes derived from the IKONOS data mosaic. (From Herold, M. et al., *Photogrammetric Eng. Remote Sensing*, 69, 991, 2003.)

A mosaic of seven individual multispectral IKONOS images (4-m spatial resolution) acquired in 2001 covering the study area has been used in this case study (Figure 13.1). The data set was first atmospherically and geometrically rectified, then classified into three major land-cover types (buildings, vegetation, and the others) using the recognition object-oriented image analysis system (for more detail, see Herold et al., 2003). Figure 13.1 further shows block-level population data from Census 2000.

13.3.1 LINEAR REGRESSION BASED ON LAND-USE INFORMATION

A comparison between the population density map and the land-use map in Figure 13.1 reveals that clear correlation exists between them. This suggests that land-use information, which can be derived from remote sensing images, can be used to interpolate the spatial distribution of human populations. Langford and Unwin (1994) presented a linear regression model to estimate population based on land-use information. This method was described in Equation 13.1 and is repeated here:

$$P_i = \sum_j P_{ij} = \sum_j d_j * A_{ij} \tag{13.5}$$

In this study, the spatial unit utilized is census block. A_{ij} is the area of land use j within census block i; d_j is the population density of land use j. Because the total population P_i is known from census data and A_{ij} can be easily obtained using GIS overlay function, population density of each land-use class can then be estimated through linear regression. In this analysis, the residential land uses are classified into six types: high-density, single-unit housing (HSU); medium-density, single-unit housing (MSU); low-density, single-unit housing (LSU); multiple-unit housing (MU); mobile homes (mobil); and mixed residential–commercial land use (mix). The land-use data were obtained from a classified IKONOS image (Herold et al., 2003). Using linear regression, the following equation was obtained for the Santa Barbara south coast region:

$$P_i = 31.98 + 0.033 A_{HSU} + 0.017 A_{MSU} + 0.004 A_{LSU} + 0.071 A_{MU}$$
$$+ 0.03 A_{mobile} + 0.051 A_{mixed} \tag{13.6}$$

where P_i is the population of census block i.

The adjusted R^2 of Equation 13.6 is 0.44. The coefficients denote the population density of each land-use type. Note that Equation 13.6 does not guarantee that non-residential areas have population of 0. In fact, it suggests that even when a block has no residential area, 31.98 people are still estimated to live there. To solve this problem, the linear regression can be forced to pass the origin (Langford and Unwin, 1994).

The linear regression method in Equation 13.6 is based on the correlation between land use and population distribution. However, land use is not the only information that can be derived from remote sensing images and may or may not be the best remote sensing surrogate of population distribution. Also in Equation 13.6, only the area or size of a land use is used as the explanatory variable. This essentially

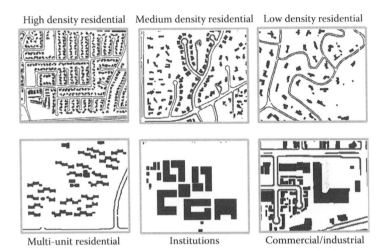

High density residential   Medium density residential   Low density residential

Multi-unit residential   Institutions   Commercial/industrial

FIGURE 13.2 Examples of spatial land configurations (buildings in gray) for major urban land-use categories.

states that the population density of a land-use category is the same across the study area. In reality, this is rarely true because population density is affected by other variables also. For example, given a simple (remote sensing based) land character- ization of "buildings" and "vegetation" (Figure 13.2), the heterogeneity of the class "buildings" can be related to the sizes of structures (small vs. large buildings), their shapes (compact vs. complex and fragmented), and the spatial configurations (regular vs. irregular). Equation 13.6 only included the categorical information of a land use; other aspects such as the geometry of the land use and its spatial relationship with its neighbors are essentially left out.

One method to describe this type of information is spatial metrics. Spatial metrics were first developed in landscape ecology to describe the spatial characteristics of a land. In the case of remote sensing, Herold et al. (2003) demonstrated that spatial metrics combined with image texture can be used to classify different land-use types accurately from IKONOS images (Figure 13.2). This suggests that spatial metrics and land use are correlated. Because land use and the spatial distribution of popu- lation are also correlated, a relationship may be established between spatial metrics and population density. In the following sections, we will first explain what spatial metrics is and then demonstrate how this method can be used by linear regression and geographically weighted regression to estimate population.

13.3.2 SPATIAL METRICS

Techniques to describe the spatial, textural, and contextual information in remotely sensed imagery have been developed and tested in the last three decades. Nearly all of such methods are based on continuous gray-level pixel values (Haralick and Dinstein, 1973; Gong et al., 1992). Spatial metrics is different from these methods in that its calculation is based on a categorical patch-based representation of the landscape within individual land-use regions.

Spatial metrics can be defined as measurements derived from the digital analysis of thematic-categorical maps exhibiting spatial heterogeneity at a specific scale and resolution. They have been developed for categorical, patch-based representations of landscapes. Patches are defined as homogenous regions for a specific landscape property of interest such as land-cover categories "building," "vegetation," or "urban." This landscape perspective assumes abrupt transitions between individual patches that result in distinct edges. Metrics represent spatial heterogeneity at a specific spatial scale, determined by spatial resolution, spatial domain, and thematic definition of the map categories at a given point in time. When applied to multiscale or multitemporal data sets, spatial metrics can be used to analyze and describe change in the degree of spatial heterogeneity (Herold et al., 2005).

Interest in using spatial metric concepts for the analysis of urban environments is growing. So far spatial metrics have been explored for mapping and modeling the urban environment. Most pioneer studies point to the importance of related methods in urban analysis and urge further systematic investigations in this area (Geoghegan et al., 1997; Barnsley and Barr, 1997; Herold et al., 2002). The potential in combining remote sensing, spatial metrics, and urban studies seems obvious. Urban areas are dominated by man-made structures. In contrast to natural environments, man-made structures have been identified as one of the few examples of objects within a landscape that have distinct and crisp boundaries (Couclelis, 1992). This characteristic makes the general approach particularly suitable in spatial metric analysis. Remote sensing can provide the spatially consistent, high-resolution data sets required for analysis of spatial structure and patterns through spatial metrics.

Spatial metrics can be used to quantify the spatial heterogeneity of individual patches, all patches in the same class, and the landscape as a collection of patches. Based on the work of O'Neill et al. (1988), sets of different metrics have been developed, modified, and tested (McGarigal et al., 2002). Many of these quantitative measures are implemented in the public domain statistical package FRAGSTATS (McGarigal et al., 2002; Table 13.1). Some metrics are spatially nonexplicit scalar values, but still capture important spatial properties. Spatially explicit metrics can be computed as patch-based indices (e.g., size, shape, edge length, patch density, fractal dimension) or as pixel-based indices (e.g., contagion) computed for all pixels in a patch (Gustafson, 1998; McGarigal et al., 2002).

Most metrics (Table 13.1) have fairly simple and intuitive values such as percentage of the landscape covered by the class (PLAND), patch density (PD), mean patch size (AREA_MN), and standard deviation (AREA_SD). The largest patch index (LPI) metric describes the percentage of the total area covered by the class concentrated in the largest patch of that class. The contagion index (CONTAG) measures the extent to which landscapes are aggregated or clumped (O'Neill et al., 1988). Landscapes consisting of relatively large, contiguous patches are described by a high contagion index. If a landscape is dominated by a relatively greater number of small or highly fragmented patches, the contagion index is low. The patch COHE-SION measures the physical connectedness of the corresponding land-cover class. The cohesion increases as the patches that comprise a class become more clumped or aggregated, hence more physically connected (Gustafson, 1998).

TABLE 13.1
Examples of Spatial Metrics

Metric	Description/Calculation Scheme	Units	Range
PLAND (percentage of landscape)	PLAND = sum of the areas (m²) of a specific land-cover class divided by total landscape area, multiplied by 100	Percent	0 < PLAND ≤ 100
PD (patch density)	PD = number of patches of a specific land-cover class divided by total landscape area	Numbers per km²	PD ≥ 1, no limit
AREA_MN (mean patch size)	AREA_MN = average size of patches of a land-cover class	Ha	AREA_MN ≥ 0, no limit
AREA_SD (area standard deviation)	AREA_SD = standard deviation in size of patches of a land-cover class	Ha	AREA_SD ≥ 0, no limit
LPI (largest patch index)	LPI = area (m²) of largest patch of the corresponding class divided by total area covered by that class (m²), multiplied by 100 (to convert to a percentage)	Percent	0 < LPI ≤ 100
COHESION	Cohesion is proportional to area-weighted mean perimeter-area ratio divided by area-weighted mean patch shape index (i.e., standardized perimeter-area ratio).	Percent	0 < COHESION < 100
CONTAG (contagion)	CONTAG measures overall probability that a cell of a patch type is adjacent to cells of the same type	Percent	0 < CONTAG ≤ 100

A more detailed exploration of intraurban spatial metric patterns is presented in Figure 13.3. Both presented spatial metric maps show obvious correspondence with the land-use distributions shown in Figure 13.1. High values of patch density reflect high- and medium-density residential land uses with high numbers of individual buildings per area unit. The contagion is lowest for single-unit high-density residential, multiunit residential, and commercial and industrial areas. These land uses represent the most heterogeneous, fragmented type of urban landscape. High contagion

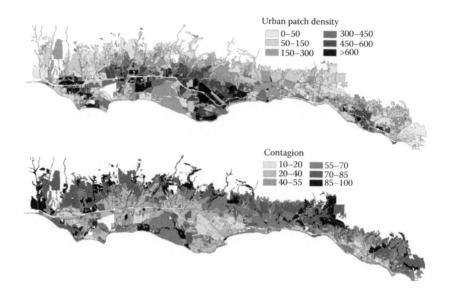

FIGURE 13.3 Spatial urban characteristics of two spatial metrics in the Santa Barbara south coast urban area derived from IKONOS data.

is found for nonurban land uses. Interestingly, the contagion metric follows a concentric pattern with heterogeneous urban environments near the central urban (low contagion) and a gradient of increasing contagion towards the peripheral rural areas. This goes along with a distinct residential gradient of lower contagion for higher residential density (see also Figure 13.1).

13.3.3 Linking Spatial Metrics and Population Density

One problem with regression based on land-use information is that only the categorical information is included in the regression. The location of the land use is omitted, indicating the hidden assumption of spatial independence. To take location information into account, spatial metrics is used instead of the land-use category in the regression. Spatial metrics describes not only the land composition but also its spatial configuration. A comparison between Figure 13.1 and Figure 13.3 suggests a correlation between spatial metrics and population density. This is confirmed by the scatter plots in Figure 13.4 based on four spatial metrics and the census 2000 population density at block level.

It can be seen that positive correlation exists between population density and the amount of built-up area and building patches. This seems intuitive because a higher number of individual buildings and larger area of covered building usually coincide with higher population density for each block. A negative relationship is shown for the contagion and the largest patch index. The more homogenous the urban environment is, the lower the population density is. High-density residential areas represent fragmented urban landscapes with the heterogeneity decreasing for lower density residential areas (Figure 13.2).

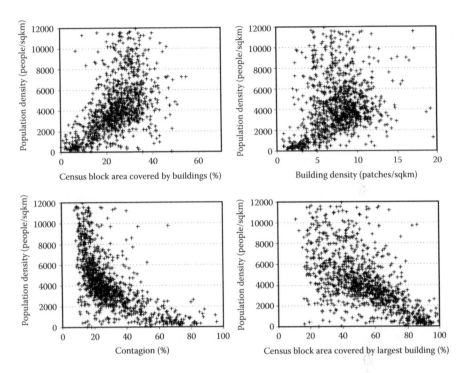

FIGURE 13.4 Relationships between four spatial metrics and population density for $n = 1092$ Census 2000 blocks with residential land use in the Santa Barbara south coast. The spatial metrics describe the land-cover class buildings and have been derived from classified IKONOS data for each census block.

Based on the correlation between spatial metrics and population density as demonstrated by Figure 13.4, the linear regression model similar to Equation 13.6 is applied using the seven spatial metrics in Table 13.1 as explanatory variables. Only three variables were found to be significant: percentage of built-up area, percentage of vegetation in the area, and patch density of built-up area. Other variables such as the largest patch index (LPI) or contagion are thus not included. The logarithmic linear relationship was also explored and found to be more appropriate than the linear regression. The final regression model to estimate population density based on spatial metrics is thus:

$$\ln(d) = 8.819 + 1.772p_1 - 2.612p_2 + 0.0632p_3 , \quad R^2 = 0.55 \quad (13.7)$$

where

d is the population density (people per square kilometer)
p_1 is the percentage of built-up area (PLAND_1)
p_2 is the percentage of vegetation (PLAND_2)
p_3 is the patch density of built-up area (PD_1)

The result in Equation 13.7 is interesting in the sense that the simple measurement of built-up and vegetation percentage together with patch density explains a significant amount of variances in population density. On the other hand, the relatively low R^2 is not surprising because the graphs in Figure 13.4 show a fair amount of scattering, indicating a rather weak linear relationship. The variance in the relationship increases for higher population densities, indicating that spatial metrics would better predict low population densities and are less sensitive to changes in areas for high population densities. This is further emphasized by the nonlinear correlation between contagion and population density. Low contagion values (~20%) do not allow for a clear distinction between areas with 6000 to 12,000 people per square kilometer.

Because the area of a land use can be obtained using GIS, the estimated population density can be easily converted to population count. A comparison between Equation 13.6 and Equation 13.7 suggests two findings: (1) Although Equation 13.7 has a higher R^2 than Equation 13.6, in terms of the accuracy of the estimation of population count, Equation 13.7 may not perform as well as Equation 13.6 because of its logarithmic format. For Equation 13.7, exponential operations must be applied to obtain population density. During the process, small errors will be exponentialized and thereby amplified. (2) In Equation 13.6 and Equation 13.7, the R^2 is not high enough to make predictive estimates. Significant residuals exist after either method is applied. For studies that aim to use remote sensing to assist population interpolation, simple regression-based methods may not be sufficient. Residual modeling may be necessary to improve accuracy.

13.3.4 GEOGRAPHICALLY WEIGHTED REGRESSION

The previous result illustrated existing relationships between spatial metrics and land-use characteristics, and census population densities. However, the correlations are not sufficiently strong to warrant spatial metrics as a direct predictor of urban population densities. A determinant relationship was not necessarily expected because the spatial urban patterns in a larger urban agglomeration like the Santa Barbara south coast are diverse and reflect characteristics of varying evolution processes that shaped the urban landscape in the first place. The socioeconomic and demographic conditions may change over time without showing a thorough impact on the spatial landscape structure. Thus, the metrics reflect an aggregate of different spatial urban characteristics. Assuming spatial autocorrelation (Figure 13.1 and Figure 13.3), the "local" variability in the relationship between urban form (spatial metrics) and demographic characteristics may be significantly higher than shown for the "global" case of a whole urban agglomeration with different urban centers (like the Santa Barbara south coast area, Figure 13.4).

One technique to explore local variability in the link between remote sensing or spatial metric measurements and population densities is geographically weighted regression (GWR). The GWR method has been developed in response to the need for locally specific spatial regression models (Fotheringham et al., 2002). GWR addresses the issue of spatial nonstationarity directly and allows regression relationships to vary over space — for example, the predictors and regression parameters might vary for different urban centers or in urban vs. rural areas.

TABLE 13.2
GWR Analysis Results for Several Bivariate Regression Models of Spatial Metrics vs. Census Population Density and One Multivariate Regression Model Using CONTAG, PLAND, LPI, and COHESION to Estimate Census Population

| | | Landscape | | | Buildings | | | |
	Multivariate	CONTAG	PLAND	PD	AREA_ MN	AREA_ SD	LPI	Cohesion
Units	—	%	%	No./km^2	Ha	Ha	%	%
Global CoD (R-sq)	0.32	0.279	0.213	0.026	0.058	0.059	0.298	0.256
Slope (global)	—	−133	198	240	42095	32258	−116	−788
GWR CoD	0.72	0.627	0.612	0.597	0.592	0.605	0.625	0.632
F-test improve	0.32	6.32	6.80	9.20	8.87	9.24	5.84	6.80

Notes: One metric represents the heterogeneity of the whole landscape (census block); the other metrics describe characteristics of the land-cover class buildings. All regression models are based on 437 samples.

In this case study, GWR was applied to explore the relationship between the spatial metrics (independent predictor variables) and the Census 2000 population densities (dependent variable). Using the data shown in Figure 13.4, the study was conducted for 1092 census blocks with residential land use. The application of GWR is essential to assess the relationship of metrics vs. population density on a more local, intraurban level (e.g., different city districts) rather than as a global model for the whole urban area.

Table 13.2 shows the GWR regression parameters for the multivariate prediction of five spatial metrics from the spatial growth factors. The local sample size was determined with a fixed value of 437 using the automated algorithm provided by the GWR software (Fotheringham et al., 2002). The value global CoD or global coefficient of determination represents the R^2 of the global regression model including all observations. For the metrics presented in Figure 13.4, the global CoD would reflect the R^2 value for linear regression using the data shown in the scatter plots. The values emphasize the rather poor relationship between the individual metrics and the population density. The GWR CoD describes the overall strengths of spatially weighted regression. A comparison between the global CoD and GWR CoD shows the clear improvements gained by using GWR. This result is not surprising because the local regression models better adjust to specific local characteristics and relationships than the global approach. The overall GWR CoDs for these metrics are rather high and range from 59 to 63% of described variance (Table 13.2); hence, the spatial metrics are able to predict most of the spatial urban population density patterns.

The slope (global) value in Table 13.2 represents the slope of the linear regression line for the global case. Values below zero emphasize a negative relationship.

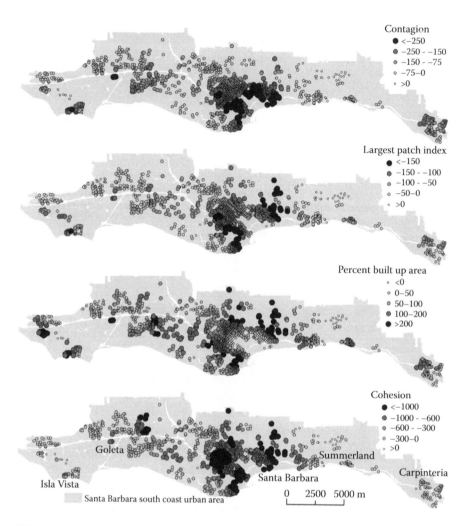

FIGURE 13.5 Spatial distribution of bivariate regression parameter (slope) represented as GWR results for four spatial metrics predicting population density.

The individual values depend on the units of the variables. For example, a slope (global) value of 198 for PLAND metric reflects an average increase of 198 people per square kilometer for each percent increase in area covered by buildings. GWR calculates a regression model for each observation point (a census block), so the regression parameters can be explored in a spatial context.

Figure 13.5 shows that the regression model parameters vary quite significantly in the study area. This again emphasizes how the average slope of the regression line from the global model (Table 13.2) only marginally reflects the local relationships for linking urban form and demographic characteristics. The highest absolute slopes' values are located in areas of dense population (see Figure 13.1) — for

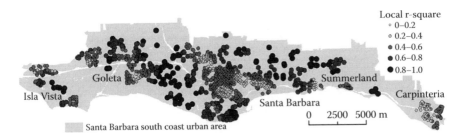

FIGURE 13.6 Spatial distribution of local R^2 value for a multivariate regression model of four spatial metrics predicting census block-level population densities.

example, around the Santa Barbara downtown area. Most interestingly, the highest values for each of the four individual regression results appear for different areas. Thus, each metric represents somewhat different local characteristics. This result points at multivariate regression models combining the predictive value of individual metrics. In fact, the significant multivariate model of the four different metrics shown in Figure 13.5 (percentage built-up area, largest patch index, contagion, and cohesions) reflects a GWR coefficient of determination of 0.72; that is, about 72% of the variance in the census block-level population density for the Santa Barbara south coast area can be explained by the spatial metrics (Table 13.2). In comparison to the global multivariate regression model (R^2 = 0.32), the local regression model presents an improved way of estimating population. Testing the improvement of the local vs. global regression, the F-test of 3.83 (see "F-test improve" in Table 13.2) emphasizes the more robust regression model.

The GWR CoD of 0.72 gives an indication of the overall quality of the relationship; however, the correlation varies in different parts of the urban area. In some areas the relationship is more determined than in others, indicated by spatial distributions of the local CoD (Figure 13.6). The highest local R^2 values (0.8 to 1.0) are found in areas of low and medium population densities in Goleta, Santa Barbara, Isla Vista, and Summerland (see also Figure 13.1). Lower quality regression results appear in densely populated areas. This fact follows previous interpretations related to Figure 13.4 where the metrics show lower sensitivity to population changes in high-density areas.

Overall, the results of applying geographically weighted regression showed a rather high degree of predictability of population density using spatial metrics from IKONOS data. Considering the local characteristics of the urban environment within larger agglomerations, GWR seems to be an appropriate method to gain understanding and improve predictability of demographic information from remote sensing.

13.4 SUMMARY

Remote sensing observations remain one of the few sources potentially to satisfy the need for spatially explicit and updated population information. Based on a variety of sensors and techniques, remote sensing measures have proven to be sensitive to demographic characteristics on different scales: from individual dwellings and census blocks and tracts to global-scale population estimations. The investigated links between remote sensing and population data are versatile but have not resulted in many operational applications. Depending on the scale and the data used, directly linking the remote sensing (spectral) signals and population data has only limited value.

One key is to use remote sensing as an additional source of information for population investigations — for example, as spectral intensity signals, classified land types, or spatial and contextual measures of urban form. Integrated analyses with ancillary spatial data sources that exist for many areas worldwide (e.g., topography and transportation infrastructure) and advanced spatial analysis and modeling methods provide further avenues to assist in estimating population distributions and studying their spatiotemporal dynamics. This is particularly true for the use of new high spatial resolution satellite sensors like IKONOS.

REFERENCES

Adeniyi, P.O. An aerial photographic method for estimating urban population, *Photogrammetric Eng. Remote Sensing*, 49, 545, 1983.

Anderson, D.E. and Anderson, P.N. Population estimates by humans and machines, *Photogrammetric Eng.*, 1, 147, 1973.

Barnsley M.J. and Barr, S.L. A graph-based structural pattern recognition system to infer urban land use from fine spatial resolution land-cover data, *Computers Environ. Urban Syst.*, 21(3/4), 209, 1997

Chen, K. An approach to linking remotely sensed data and areal census data, *Int. J. Remote Sensing*, 23, 37, 2002.

Collins, W.G. and El-Beik, A.H.A. Population census with the aid of aerial photographs: an experiment in the city of Leeds, *Photogrammetric Rec.*, 7, 16, 1971.

Couclelis, H. People manipulate objects (but cultivate fields): beyond the raster–vector debate in GIS, in *Theories and Methods of Spatiotemporal Reasoning in Geographic Space, Lecture Notes in Computer Science*, Frank, A.U., Campari, I., and Formentini, U., Eds., Springer–Verlag, Berlin, 1992, 65–77.

Dobson, J.E., Bright, E.A., Coleman, P.R., Durfee, R.C., and Worley, B.A. LandScan: a global population database for estimating populations at risk, *Photogrammetric Eng. Remote Sensing*, 66, 849, 2000.

Donnay, J.-P. and Unwin, D. Modeling geographical distributions in urban areas, in *Remote Sensing and Urban Analysis*, Donnay, J.-P., Barnsley, M.J., and Longley, P.A., Eds., Taylor & Francis, New York, 2001, 205–224.

Donnay, J.-P., Barnsley, M.J., and Longley, P. Remote sensing and urban analysis, in *Remote Sensing and Urban Analysis*, Donnay, J.-P., Barnsley, M.J., and Longley, P.A., Eds., Taylor & Francis, New York, 2001, 3–18.

Dueker, K. and Horton, F. Toward geographic urban change detection systems with remote sensing inputs; technical papers, *37th Annu. Meeting, Am. Soc. Photogrammetry*, 204–218, 1971.

Eyre, L.A., Blossom, A., and Amiel, M. Census analysis and population studies, *Photogrammetric Eng.*, 36, 460, 1970.

Flowerdew, R. and Openshaw, S. A review of the problems of transferring data from one set of areal units to another incompatible set; report 4, Northern Regional Research Laboratory, Lancaster and Newcastle, 1987.

Forster, B. An examination of some problems and solutions in monitoring urban areas from satellite platforms, *Int. J. Remote Sensing*, 6, 139, 1985.

Fotheringham, A.S., Brunsdon, C., and Charlton, M.E. *Geographically Weighted Regression: The Analysis of Spatially Varying Relationships*, Wiley, Chichester, U.K., 2002.

Geoghegan, J., Wainger, L.A., and Bockstael, N.E. Spatial landscape indices in a hedonic framework: an ecological economics analysis using GIS, *Ecol. Econ.*, 23(3), 251, 1997.

Gong, P., Marceau, D.J., and Howarth, P.J. A comparison of spatial feature extraction algorithms for land-use classification with SPOT HRV data, *Remote Sensing Environ.*, 40, 137, 1992.

Goodchild, M.F. and Lam, N. Areal interpolation: a variant of the traditional spatial problem, *Geo-Processing*, 1, 297, 1980.

Green, N.E. Aerial photographic interpretation and the social structure of the city, *Photogrammetric Eng.*, 23, 89, 1957.

Gregory, I.N. 2002. The accuracy of areal interpolation techniques: Standardising 19th and 20th century census data to allow long-term comparisions. *Computers Environment and Urban Systems* 26: 293–314.

Gustafson, E.J. Quantifying landscape spatial pattern: what is the state of the art? *Ecosystems*, 1, 143, 1998.

Haralick, R.S.K. and Dinstein, I. Texture features for image classification, *IEEE Trans. Syst., Man. Cybernetics*, 3, 610, 1973.

Harvey, J.T. Population estimation models based on individual TM pixels, *Photogrammetric Eng. Remote Sensing*, 68, 1181, 2002a.

Harvey, J.T. Estimating census district populations from satellite imagery: some approaches and limitations, *Int. J. Remote Sensing*, 23, 2071, 2002b.

Herold, M., Clarke, K.C., and Scepan, J. Remote sensing and landscape metrics to describe structures and changes in urban landuse, *Environ. Plann. A*, 34, 1443, 2002.

Herold, M., Liu, X., and Clarke, K. Spatial metrics and local texture for mapping urban land use, *Photogrammetric Eng. Remote Sensing*, 69, 991, 2003.

Herold, M., Couclelis, H., and Clarke, K. The role of spatial metrics in the analysis and modeling of urban land use change, *Computers, Environ. Urban Syst.*, 29, 4, 369, 2005.

Holz, R.K., Huff, D.L., and Mayfield, R.C. Urban spatial structure based on remote sensing imagery, *Proc. 6th Int. Symp. Remote Sensing Environ.*, II, October 15, 819, 1969.

Iisaka, J. and Hegedus, E. Population estimation from Landsat imagery, *Remote Sensing Environ.*, 12, 259, 1982.

Jensen, J.R. and Cowen, D.C. Remote sensing of urban/suburban infrastructure and socio-economic attributes, *Photogrammetric Eng. Remote Sensing*, 65, 611, 1999.

Kraus, S. and Senger, L. Estimating population from photographically determined residential land use types, *Remote Sensing Environ.*, 3, 35, 1974.

Langford M., Maguire, D.J., and Unwin, D.J. The areal interpolation problem: estimating population using remote sensing in a GIS framework, in *Handling Geographical Information: Methodology and Potential Applications*, Masser, I. and Blakemore, M., Eds., Wiley, New York, 1991, 55–77.

Langford, M. and Unwin, D.J., Generating and mapping population density surfaces with a geographical information system, *Cartographic J.*, 31, 21, 1994.

Li, G. and Weng, Q. Using Landsat ETM+ imagery to measure population density in Indianapolis, Indiana, U.S.A., *Photogrammetric Eng. Remote Sensing*, 71(8), 947, 2005.

Liu, X. and Clarke, K.C. Estimation of residential population using high-resolution satellite imagery, *Proc. 3rd Symp. Remote Sensing Urban Areas*, June 2002, Istanbul, Turkey, 153–160, 2002.

Liu, X. Estimation of the spatial distribution of urban population using high spatial resolution satellite imagery, Ph.D. thesis, University of California, Santa Barbara, 2003.

Lo, C.P. Accuracy of population estimation from medium-scale aerial photography, *Photogrammetric Eng. Remote Sensing*, 52, 1859, 1986.

Lo, C.P. Automated population and dwelling unit estimation from high-resolution satellite images: a GIS approach, *Int. J. Remote Sensing*, 16, 17, 1995.

McGarigal, K., Cushman, S.A., Neel, M.C., and Ene, E. *FRAGSTATS: Spatial Pattern Analysis Program for Categorical Maps*, University of Massachusetts, Amherst, Massachusetts, 2002, URL: www.umass.edu/landeco/reesarch/fragstats/fragstats.html, last accessed December 2004.

Mennis, J. Generating surface models of population using dasymetric mapping, *Prof. Geogr.*, 55, 31, 2003.

Nordbeck, S. The law of allometric growth, Michigan Interuniversity Community of Mathematical Geographers, discussion paper 7, Department of Geography, University of Michigan, Ann Arbor, MI, 1965.

O'Neill, R.V., Krummel, J.R., Gardner, R.H., Sugihara, G., Jackson, B., Deangelis, D.L., Milne, B.T., Turner, M.G., Zygmunt, B., Christensen, S.W., Dale, V.H., and Graham, R.L. Indices of landscape pattern, *Landscape Ecol.*, 1, 153, 1988.

Openshaw, S. Ecological fallacies and the analysis of areal census data, *Environ. Plann. A*, 16, 17, 1984.

Rhind, D.W. 1991. Counting the people, in *Geographical Information Systems: Principles and Applications*, Maguire, D.J., Goodchild, M.F., and Rhind, D.W., Eds., Longman, London, 2, 1991, 127–137.

Rindfuss, R.R. and Stern, P.C. Linking remote sensing and social science, in *People and Pixel — Linking Remote Sensing and Social Science*, Liverman et al., Eds., National Academic Press, Washington, D.C., 1998, 1–28.

Sutton, P., Roberts, D., Elvidge, C., and Meij, H. A comparison of nighttime satellite imagery and population density for the continental United States, *Photogrammetric Eng. Remote Sensing*, 63, 1303, 1997.

Sutton, P., Roberts, D., Elvidge, C., and Baugh, K. Census from heaven: an estimate of the global human population using nighttime satellite imagery, *Int. J. Remote Sensing*, 22, 3061, 2001.

Tobler, W. Smooth pycnophylactic interpolation for geographical regions, *J. Am. Statistical Assoc.*, 74, 519, 1979.

Vining, D.R. and S. Louw 1978. A cautionary note on the use of the allometric function to estimate urban populations, *The Professional Geographer*, 30, 365–370.

Webster, C.J. Population and dwelling unit estimates from space, *Third World Plann. Rev.*, 18, 155, 1996.

Welch, R. Monitoring urban population and energy utilization patterns from satellite data, *Remote Sensing Environ.*, 9, 1, 1980.

Wellar, B. The role of space photography in urban and transportation data series, *Proc. 6th Int. Symp. Remote Sensing Environ.*, Vol. II, October 15, 1969, Ann Arbor, Michigan, 831–854.

Wright, J.K. A method of mapping densities of population with Cape Cod as an example, *Geogr. Rev.*, 26, 103, 1936.

Wu, C. and Murray, A. A cokriging method for estimating population density in urban area, *Computer, Environ. Urban Syst.*, 29, 558, 2005.

14 Sociodemographic Characterization of Urban Areas Using Nighttime Imagery, Google Earth, Landsat, and "Social" Ground Truthing

Paul C. Sutton, Matthew J. Taylor, Sharolyn Anderson, and Christopher D. Elvidge

CONTENTS

14.1   INTRODUCTION

Nighttime satellite imagery provided by the Defense Meteorological Satellite Program's Operational Linescan System (DMSP OLS) provides a unique and profound view of urban areas around the globe. Data products derived from the DMSP OLS are unique in their time of observation, spatial resolution, global scope, and availability in time series. Most other satellite observations of the Earth are obtained during daylight hours in which the signal received is reflected solar radiation, emitted thermal radiation, or "bounced" radiation from active sensors. Nighttime satellite imagery captures radiation primarily from lightning, fires, and, most importantly, human sources such as city lights, lantern fishing, and gas flare burns.

All of these different sources of nocturnal emissions of radiation are distinguishable due to the way the data products are produced. Consequently, these products are used to map fire, lightning, lantern fishing, urban extent, population, population density, CO_2 emissions, anthropogenic environmental impact, impervious surface, and economic activity. This chapter provides a summary of ways in which nighttime imagery has been used to study socioeconomic variables and urban environments, suggests potential improvements on these methods if finer resolution sensors are developed, and explores ways in which the DMSP OLS data products can be used to inform our understanding of urban and exurban areas around the world.

14.1.1   BACKGROUND TO DMSP OLS AND DATA PRODUCTS

The DMSP platform was designed as a meteorological satellite for the U.S. Air Force. The system has two bands: a very sensitive panchromatic visible-near-infrared (VNIR) band and a thermal infrared band. The high sensitivity of the sensor was not implemented to see city lights, but rather to see reflected lunar radiation from the clouds of the Earth at night. The variability of lunar intensity as a function of the lunar cycle is one of the reasons why the satellite system's sensors were designed with such a large sensitivity range.

DMSP OLS data were originally produced on mylar film and were not easily analyzed. Nonetheless, the imagery was so interesting that many people worked with the data to characterize relationships between the DMSP OLS imagery and other variables such as population and energy consumption (Welch 1980a, b; Foster 1983). The digital DMSP archive was established in 1992 and has dramatically improved access to and utility of the DMSP OLS data.

The digital archive of the DMSP OLS data is housed at the National Geophysical Data Center (NGDC) in Boulder, Colorado (a subsidiary of NOAA). Elvidge et al. developed algorithms to identify spatiotemporally stable VNIR emission sources utilizing images from hundreds of orbits of the DMSP OLS platform (Elvidge et al.

1996, 1998). The resulting hypertemporal data set is cloud free because the infrared band of the system was used to screen out cloud-affected data. Temporally unstable lights are identified as lightning, fires, or lantern fishing.

A DMSP OLS mosaic of fires would consist of lights on land that were ephemeral. However, it is important to realize that all the fires seen in a DMSP OLS fire data product did not occur at the same time (a global mosaic typically requires orbital data inputs that span several months to a year). Consequently, the fire data set is in this sense more of a "climatology" of fire. Annually composited global data products for city lights are available for every year from 1992 to the present. In addition, there are global fire products, gas flare products, and lantern fishing products.

14.2 SOCIOECONOMIC APPLICATIONS OF NIGHTTIME SATELLITE IMAGERY

Data products derived from nighttime satellite imagery allow for important advances in environmental and socioeconomic application areas that support policy and planning. The following subsections list application areas that we have identified. Many of these areas have been demonstrated with the relatively coarse spatial resolution of DMSP-OLS data (~1 km). We hope to witness the development and implementation of a nighttime satellite that has finer spatial and spectral resolution than the DMSP OLS because we believe the following applications would be greatly augmented with finer resolution imagery.

14.2.1 POPULATION DENSITY

DMSP-OLS nighttime lights are widely used as one of the inputs into regional and global population density grids (Sutton et al. 1997, 2001, 2003; Dobson et al. 2000; Balk et al. 2005). Population density grids are used in modeling population numbers that may be affected by catastrophic events (e.g., earthquakes and tsunamis), diffusion modeling of disease spread, and travel times for patients to reach health care facilities. The coarse spatial resolution and overglow effects present in the OLS lights have emerged as one of the major limiting factors in improving the global mapping of population density.

14.2.2 ECONOMIC ANALYSES

Previous studies have demonstrated a clear correlation between the extent and brightness of nighttime lights and economic activity (Elvidge et al. 1997; Doll and Muller 2000; Ebener 2005). The need for independent economic estimates of parameters such as gross domestic product and per-capita income arise from governments' difficulties in collecting appropriate data in many countries, driven in part by the size of unreported activity and tax evasion. In other cases, governments may manipulate economic reports to attract or retain investments. Subnational per-capita income is widely used in assessing disease risk. Finer resolution data would be a great data source for the independent estimation of economic indices for use in a wide range of human health and socioeconomic applications.

14.2.3 Land-Cover and Land-Use Change

From a mapping perspective, urban areas are arguably the most spatially complex environments. Nighttime imagery can be used as one of the primary data sources in the global mapping of urban land cover and sparse development in rural settings (Sutton 2006). When combined with traditional land-cover products, nighttime imagery enables vastly improved assessments of habitat fragmentation and delineation of the urban–wildland interface (Cova et al. 2004).

14.2.4 Urban Planning

Nighttime imagery also provides information on the patterns of development and forms within human settlements for use in urban planning. Having a consistent high-resolution global data source suitable for use in urban planning could stimulate a substantial amount of regional to international collaboration in planning growth during a time when population is increasingly shifting from rural to urban.

14.2.5 Impervious Surface Areas (ISAs)

The proliferation of ISAs (buildings, roads, parking lots, roofs, etc.) has emerged as a major environmental issue. Impervious surfaces alter sensible and latent heat fluxes, causing urban heat islands. ISA alters the character of watersheds by increasing the frequency and magnitude of surface runoff pulses. Standard flood prediction models use ISA grids as one of the key inputs. Watershed effects of ISA begin to be detectable once 10% of the surface is covered by impervious surfaces (Beach 2002), altering the shape of stream channels, raising water temperatures, and sweeping urban pollutants into aquatic environments.

Consequences of ISA include reduced numbers and diversity of species in fish and aquatic insects and degradation of wetlands. DMSP-OLS nighttime lights have been used with other data sources to estimate that the total area of ISA in the United States approached the size of Ohio in 2000 (Elvidge 2004). When DMSP OLS imagery is combined with higher resolution Landsat style data, these data products improve the accuracy of ISA estimates and enable annual updates.

14.2.6 Anthropogenic Emissions to the Atmosphere

The quantity of nocturnal light is highly correlated with overall energy consumption (Elvidge et al. 1997; Doll et al. 2000). This is important to earth science (biogeochemistry, carbon cycling, and climate change) because of the close relationship between energy consumption and release of pollutants that affect atmospheric composition (e.g., aerosols, NOx, SOx, CO, CO2, and heavy metals). Nighttime imagery provides a data source suitable for modeling the spatial distribution and quantity of emissions to the atmosphere from human settlements when adjusted for the mix of local energy sources. In addition, nighttime imagery can be used to improve monitoring of gas flares — a poorly quantified carbon emission source.

14.2.7 TERRESTRIAL CARBON DYNAMICS

Data products derived from nighttime imagery can be useful in estimating losses or gains in carbon sequestration potential from urban sprawl. Understanding and quantification of the alteration in ecosystem functioning due to urbanization is limited by lack of reliable information on the distribution of urban ecosystems. Although urbanization is a global phenomenon, the ecological consequences are not evenly distributed. They depend dramatically on local climatic conditions and the degree of changes in ecosystem structure compared to pre-urban conditions.

Research using nighttime city lights over the American urban landscape (Imhoff et al. 1997; Milesi 2003) has found that the loss of land to buildings, roads, and infrastructures is typically counterbalanced to some extent by the intensive management of urban vegetation (lawns, trees, etc.), which is pruned, irrigated, and fertilized. This results in urban ecosystems maintaining a significant role in the carbon cycle, often higher than that of the pre-urban vegetation; when not included in regional greenhouse-gas budgets, this may lead to missing important sources or sinks. Development of systems that gather finer resolution data will allow us to extend the accurate quantifications of the ecosystem functioning of urban areas globally.

14.2.8 ECOLOGICAL DISRUPTION FROM LIGHTING

Night lighting may have adverse consequences for species and ecosystems in many different situations (Verheijen 1985; Longcore and Rich 2004, 2005). Documented impacts include enhanced predation of sea turtle hatchlings (Salmon et al. 2000; Salmon 2005) and mortality of seabirds attracted to offshore gas flares (Montevecchi 2005; Wise and Buchanan 2005). In other cases, population declines have been attributed to nocturnal lighting (Perry and Fisher 2005). Attempts have been made to use DMSP nighttime lights to study ecological disruptions from lighting, but coarse spatial resolution, lack of calibration, and overglow effects have proven to be major limitations.

14.2.9 HEALTH IMPACTS OF RAPID URBANIZATION

It is expected that by the end of this century more than half of the world's population will be living in cities. Rapid urban growth is at the root of a large variety of problems (food constraints, water supplies, availability of safe housing, solid waste disposal, and health care services). The term "rapid urbanization" encompasses growth in existing settlements as well as development of new settlements, such as large refugee camps. The World Health Organization (WHO 1993) and many others have noted the potential impacts of rapid urbanization on health and development (disease, malnutrition, substandard education). In this context, higher resolution imagery would provide information on the rates and location of urban expansion, thus enabling improved planning for services and infrastructure development (WHO 2002).

14.2.10 EFFECTS OF LIGHTING ON CRIME

One of the more important rationales for installation of lighting at the local level is to offer protection against crime. Yet, despite the widespread belief that nighttime lighting deters crime, very few studies have been able to evaluate the effectiveness of lighting in varying settings and over time (Weeks 2003). Nighttime imagery could provide data at sufficient spatial resolution so that, when combined with crime data from administrative sources, new scientific evidence could be developed about the relationship between nocturnal lighting and public safety.

14.3 USING DMSP OLS CITY LIGHTS DATA TO MAP "EXURBIA" IN THE UNITED STATES

The city lights data product derived from the DMSP OLS shows expansive areas of low light surrounding almost all major metropolitan areas. In many cases, these areas would be characterized as pure vegetation by 30-m resolution Landsat imagery; however, they often contain large numbers of people who have significant social, economic, and ecological impacts (e.g., traffic congestion and problems associated with the urban–wildlands interface). Conventional wisdom suggests that these "exurbanites" are rich commuters who choose to live in natural settings beyond the city and suburbia; however, given the spatially variable and dynamic real estate situation in the United States today, this may not be the case. Figure 14.1 shows day and nighttime photographs of a typical exurban area southwest of Denver, Colorado.

The exurban areas southwest of Denver have substantial populations as evidenced by the fact that there are three high schools (Evergreen, Conifer, and Platte Canyon) in the southwest exurbs alone. In addition, the second busiest Safeway grocery store in the state of Colorado happens to be in the exurban community of Conifer. The residences in exurbia tend to have private water wells, no street lights, and septic tanks. Needless to say, exurban development is significant, substantial, and growing. Nighttime imagery is uniquely suited for making systematic assessments of exurban development because even 30-m Landsat data cannot identify this kind of sparse development. It takes fine resolution imagery (e.g., ~1-m IKONOS) to see development in

FIGURE 14.1 Day and night photos of Pine Park Estates southwest of Denver, Colorado.

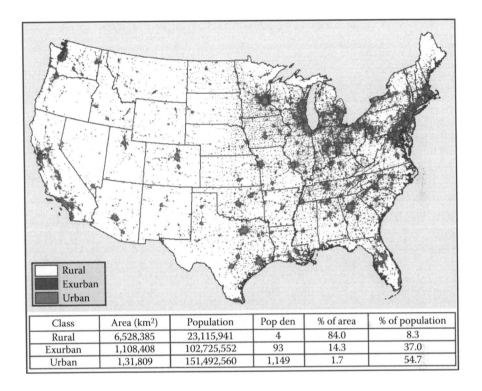

Class	Area (km²)	Population	Pop den	% of area	% of population
Rural	6,528,385	23,115,941	4	84.0	8.3
Exurban	1,108,408	102,725,552	93	14.3	37.0
Urban	1,31,809	151,492,560	1,149	1.7	54.7

FIGURE 14.2 Urban, exurban, and rural areas of the United States based on city lights data.

exurban areas. Broad-scale, systematic assessment of exurbia would be extremely difficult to conduct with the data volumes associated with 1-m imagery.

By using the ~1-km resolution imagery from the DMSP OLS city lights data product, we broadly classified an image of the conterminous United States into three classes based on light intensity thresholds: urban (bright lights), exurban (dim lights), and rural (no lights). This image was overlaid on the LandScan global population data set to determine how many people lived in low-light (exurban) areas that were within 100 km of major urban areas. By doing this, we could answer questions such as (Figure 14.2 and Table 14.1): 1) How many people live in exurbia? 2) What is the population density and areal extent of exurban areas? 3) How variable are these from one urban area to another?

Historically, populations have commonly been characterized as urban or rural. Changing urban form in the United States since the 1950s has suggested that perhaps populations should be characterized as urban, suburban, and rural. Changes in planning policy driven by concerns of urban sprawl have increased the density of recent suburban developments in many areas. For the conterminous United States, 55% of the population live in urban areas (1.7% of the land), 37% in exurban areas (14% of the land), and 8% in rural areas (84% of the land). The social, economic, environmental, and institutional challenges created by these exurban areas are significant and suggest this new classification scheme.

TABLE 14.1
Summary of Urban and Exurban Populations and Areas for Major Metro Areas

Metro Area	Urban Area (km²)	Urban Population	Exurban Area (km²)	Exurban Population	Total Population	Total Area (km²)	Exurban Pop. Density	Urban Pop. Density	Urban/Exurban Pop. Density Ratio
New York	6366	10,443,497	17,555	2,055,822	12,499,319	23,921	117	1641	14
Los Angeles	5060	9,056,845	13,733	1,883,055	10,939,900	18,793	137	1790	13
Chicago	5313	4,970,777	22,321	1,131,023	6,101,800	27,634	51	936	18
Miami	2987	3,708,042	4,619	452,270	4,160,312	7,606	98	1241	13
Baltimore–Washington	3638	3,444,641	26,552	1,985,356	5,429,997	30,190	75	947	13
Philadelphia	3498	3,055,041	19,618	1,836,717	4,891,758	23,116	94	873	9
Dallas–Ft. Worth	3320	2,800,483	11,433	1,003,921	3,804,404	14,753	88	844	10
San Francisco	1639	2,739,945	10,997	1,606,460	4,346,405	12,636	146	1672	11
Detroit	3143	2,579,779	25,042	1,235,907	3,815,686	28,185	49	821	17
Houston	3081	2,472,011	10,959	893,508	3,365,519	14,040	82	802	10
Phoenix	1750	1,838,568	5,523	464,487	2,303,055	7,273	84	1051	12
Boston	1845	1,718,846	14,599	1,496,005	3,214,861	16,444	102	932	9
San Diego	907	1,646,181	2,359	541,241	2,187,422	3,266	229	1815	8
Atlanta	2119	1,482,878	16,741	1,873,347	3,356,225	18,860	112	700	6
Minneapolis–St. Paul	2470	1,378,028	25,790	628,514	2,006,542	28,260	24	558	23
Denver	1397	1,238,270	7,225	449,862	1,688,132	8,622	62	886	14
Cleveland	1527	1,180,484	12,438	943,042	2,123,526	13,965	76	773	10
St. Louis	1930	1,173,411	14,128	723,077	1,896,488	16,058	51	608	12
Seattle	1375	1,094,569	8,984	997,220	2,091,789	10,359	111	796	7
El Paso	572	1,074,367	2,269	171,841	1,246,208	2,841	76	1878	25
San Antonio	905	858,794	6,720	583,086	1,441,880	7,625	87	949	11
Las Vegas	744	855,447	2,669	111,235	966,682	3,413	42	1150	28
Orlando	1256	832,389	10,401	1,126,446	1,958,835	11,657	108	663	6

Sacramento	723	754,018	6,213	621,201	1,375,219	6,936	100	1043	10
Kansas City	1290	748,418	10,239	573,373	1,321,791	11,529	56	580	10
St. Petersburg	850	739,454	3,922	494,657	1,234,111	4,772	126	870	7
Milwaukee	1130	737,248	16,435	484,727	1,221,975	17,565	29	652	22
Portland	874	718,111	5,850	624,159	1,342,270	6,724	107	822	8
New Orleans	624	706,159	9,679	709,583	1,415,742	10,303	73	1132	15
Indianapolis	1317	697,775	18,494	673,913	1,371,688	19,811	36	530	15
Columbus (OH)	859	669,368	13,458	648,093	1,317,461	14,317	48	779	16
Cincinnati	938	668,061	9,168	632,654	1,300,715	10,106	69	712	10
Memphis	869	549,916	8,166	444,341	994,257	9,035	54	633	12
Tampa	637	516,387	4,121	434,104	950,491	4,758	105	811	8
Norfolk	602	508,661	3,505	465,231	973,892	4,107	133	845	6
Salt Lake City	588	503,984	6,277	478,635	9,82,619	6,865	76	857	11
Buffalo	630	487,221	10,044	535,930	1,023,151	10,674	53	773	14
Louisville (KY)	729	471,376	11,267	619,255	1,090,631	11,996	55	647	12
Nashville	1005	433,648	14,195	647,703	1,081,351	15,200	46	431	9
Providence	502	415,876	5,559	583,961	999,837	6,061	105	828	8
Oklahoma City	736	412,966	8,048	424,786	837,752	8,784	53	561	11
Omaha	573	384,682	6,584	225,102	609,784	7,157	34	671	20
New Haven	501	376,065	9,464	1,095,397	1,471,462	9,965	116	751	6
Jacksonville	577	373,287	6,751	642,031	1,015,318	7,328	95	647	7
Charlotte	620	357,651	15,533	1,457,915	1,815,566	16,153	94	577	6
Birmingham	675	352,683	10,055	668,578	1,021,261	10,730	66	522	8
Dayton	587	345,293	10,186	476,234	821,527	10,773	47	588	13
Youngstown	502	227,605	15,423	1,159,728	1,387,333	15,925	75	453	6

14.4 CHARACTERIZING EXURBIA IN LATIN AMERICA

Urban form and development in Latin America are clearly not the same as in the United States. (Bahr and Mertins 1992; Arreola and Curtis 1993). In an attempt to identify and characterize exurban areas in Latin America, we used the same low-light parameters that we used in the United States. The objective here was to see whether low-light areas in Latin America reflect similar kinds of development and/or sociodemographic characteristics that are associated with exurban areas in the United States. To do this we took two approaches: (1) we compared nighttime images of several urban areas to corresponding Google Earth images; and, (2) for one city, Ciudad Victoria in Mexico, we compared Landsat and DMSP OLS imagery with ground-based, geo-located observations and interviews with local residents.

14.4.1 COMPARING NIGHTTIME IMAGERY WITH GOOGLE EARTH IMAGERY

We used Google Earth in conjunction with nighttime imagery to look for and characterize exurban areas in Latin America. We used Google Earth because it strikes us as a powerful tool that can provide relatively up-to-date imagery — often at finer spatial resolution than Landsat — of any place in the world. Google Earth can be used for streamlining or perhaps eliminating ground verification of other satellite image classifications such as finding exurban areas with low light in DMSP OLS data. We chose the following cities to explore the utility of Google Earth imagery to evaluate our characterizations based on the nighttime imagery: Morelia, Puebla, and Mexico City in Mexico, and Guatemala City.

Morelia, Mexico (year 2000 population: ~1 million). Morelia is the capital of the state of Michoacan, Mexico. Its population, based on official figures from INEGI and from local residents, runs anywhere from 800,000 to 1.5 million. This city typifies the fast growth of many Mexican cities. During a visit to Morelia in October 2005, we traveled to various low-light areas and saw, just as in other Latin American places we have studied, that these areas vary dramatically (Figure 14.3). For example, in the hills southeast of Morelia, low-light areas are characterized by large American-style homes on large lots perched on foothills with a view of the city. Clearly, the

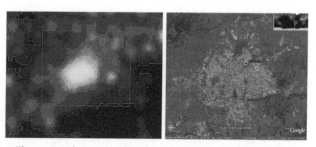

These are nighttime and Google Earth images of Morelia, Mexico

FIGURE 14.3 DMSP OLS image and Google Earth image of Morelia, Mexico.

The image on the left is a nighttime image of Puebla, Mexico. The image on the right was obtained from Google Earth for the red boxed area of the Puebla. It is clear from this comparison that many areas climbing up the side of the Volcano are agricultural. It is also clear that there is some residential development (Vicente Guerrero) in these areas.

FIGURE 14.4 DMSP OLS image and Google Earth image of Puebla, Mexico.

nighttime imagery can be thresholded to capture truly urban areas; however, exurban areas are much more difficult to characterize with nighttime imagery alone. At the other extreme, in the areas west of Morelia, low-light areas are characterized by a mixture of light industry, small-scale agriculture, and low-income housing.

Puebla, Mexico (year 2000 population: ~1,346,000). The nighttime image of Puebla shows a clear boundary between urban and nonurban areas (Figure 14.4). Upon inspection of this boundary using Google Earth, one discovers that the low-light area is characterized by agricultural fields and a small community (Vicente Guerrero). In this case, the nighttime imagery picks up small towns as well as large urban areas; however, characterizing the low-light areas as exurban in the North American sense is again problematic.

Guatemala City, Guatemala (year 2000 population: ~2 million). The largest city in Central America, Guatemala City contains extreme socioeconomic diversity. Our knowledge of this city allows us to state that the nighttime lights can be used to map the urban extent of Guatemala City fairly well (Figure 14.5). Here, low-light

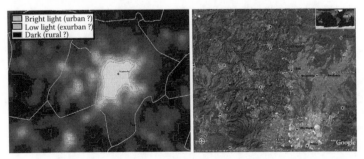

These are nighttime and Google Earth images of Guatemala City

FIGURE 14.5 DMSP OLS image and Google Earth image of Guatemala City, Guatemala.

areas do in fact (in some cases) capture exurban areas typical of the United States For example, high-income housing areas in the mountains east and west of Guatemala City fall into the low-light areas of the nighttime image. The low-light areas also capture sparsely populated rural communities outside the city.

Mexico City, Mexico (year 2000 population: ~18 million). This composite of images of Mexico City illustrates the utility and limitations of DMSP OLS nighttime satellite imagery for characterizing socioeconomic parameters such as exurban land-scapes (Figure 14.6). Insets A through E are derived from Google Earth. Inset A shows that the imagery does capture the "edge of urbanization"; however, the low-light areas beyond inset A do not appear to have anything resembling exurban development and probably represent "bleeding" of the lights from the large urban area that is Mexico City. Inset B is an area of lower light intensity within the urban core of Mexico City. It does capture an interior mountainous area that is experiencing urban encroachment. Also, the low-light areas in this inset do seem to be exurban in the sense that roads and buildings are scattered throughout this interior mountain area.

Inset C shows an *ejido* near the remnants of Lake Texcoco. Despite the fact that the interior areas of Lake Texcoco are not developed, there are measurable low levels of light over the whole lake, again suggesting a bleeding of light from the greater urban area. Inset D shows a small village outside the greater Mexico City area surrounded by agricultural areas. Again, all these agricultural areas show low levels of light, which betray limitations of the coarse spatial resolution of the DMSP OLS nighttime imagery for characterizing certain patterns of development.

Inset E struck us as a particularly interesting pattern of development. It appears to be urban development outside the slopes of an extinct volcano; the interior of the

Nighttime image of Mexico City (center) with insets of Google Earth images
of various locations in and around the city

FIGURE 14.6 (Color Figure 14.6 follows page 240.) DMSP OLS image and Google Earth image of Mexico City, Mexico.

volcano is an agricultural area. The coarse spatial resolution of the DMSP OLS imagery would characterize all this complexity as merely urban. These limitations encouraged us to explore a Latin American city (Ciudad Victoria of Tamaulipas, Mexico) at an even greater depth.

14.4.2 "Social Ground-Truthing" in Ciudad Victoria, Mexico

In an attempt to obtain a greater appreciation of the ability of remotely sensed imagery to characterize social, economic, and demographic characteristics of intraurban areas, we traveled to Ciudad Victoria in Tamaulipas, Mexico. Via the use of Landsat imagery from 1980, 1990, and 2000 and nighttime imagery from 1990 and 2000, we traveled around the city with a GPS and interviewed people on the ground. Figure 14.7 shows the 1990 and 2000 DMSP OLS imagery in 50% transparency over a false color composite of the year 2000 Landsat image of the city of Ciudad Victoria.

Our approach to gaining greater understanding of the growth, morphology, and nature of Ciudad Victoria was really quite simple. We drove around the city with a hand-held GPS to locate ourselves and took photographs of various locations. In addition, we contacted local people and asked them about what was going on in a particular place. The following six sites capture the complexity and diversity of the reality on the ground in Ciudad Victoria; each figure shows a true-color Landsat image, a DMSP OLS nighttime image, and a photograph of the area noted with a red circle.

This first site is a brand-new *colonia* on the northwestern edge of the city (Figure 14.8). This plot of land measuring 7 × 18 m cost its owners $1000 (U.S.). Because it is a new colonia, it has no water, sewage, or electricity. It is believed that with time the city and/or national government will provide these services. This is not a squatter settlement because residents own legal title to the land. Land is cleared of trees before subdivision and subsequent construction. This is an area of relatively

1992 50% transparency   2000 50% transparency

FIGURE 14.7 Transparent DMSP OLS image of Ciudad Victoria for 1992 and 2000 over 2000 Landsat image.

Site #1: New Colonia

FIGURE 14.8 Ciudad Victoria site 1: new colonia.

high levels of light in the DMSP OLS nighttime satellite image despite the rudimentary level of development that has taken place here.

Our second site represents an area of second homes used by affluent residents of Ciudad Victoria (Figure 14.9). They use these houses as weekend getaways or as sites for recreation and picnics while they improve their plots and begin house construction. This is an area of low housing density that we call Latin American "protoexurbia." Also note the lack of trees here. This area was agricultural prior to land subdivision. This area was dark in the 1990 DMSP OLS image and became a low-light area in the 2000 image. This photograph was taken in 2004.

The third site is clearly in the heart of suburban Ciudad Victoria (Figure 14.10). This residential area is over 10 years old and comes with all municipal services like

Site #2: Proto-suburbia or exurbia

FIGURE 14.9 Ciudad Victoria site 2: protosuburbia.

Site #3: Suburbia

FIGURE 14.10 Ciudad Victoria site 3: suburbia.

electricity, waste water, telephone, and neighborhood shops. This area was in a brightly lit part of the 1990 and 2000 DMSP OLS imagery.

The fourth site is interesting in that it is a foreign-owned *maquiladora* (Figure 14.11). One of the buildings shows up in the Landsat image as a 400 x 200 m bright area. This area is a sparsely developed area of maquiladoras and day-care facilities for the employees of the maquiladoras. Again, there are very few trees and low levels of light in the DMSP OLS image (although it is very close to bright levels of light).

Our fifth site is deep in the heart of town along the Rio San Marcos (Figure 14.12). Not surprisingly, there is a great deal of lush vegetation here beside the river. (Ciudad Victoria is located in a semiarid environment.) The river runs from the southwest to the northeast through the city. This is an established residential area with shops and

Site #4: Maquila-Urbia

FIGURE 14.11 Ciudad Victoria site 4: maquila-urbia.

Site #5: Heart of the city
Is this intra-urban exurbia?

FIGURE 14.12 Ciudad Victoria site 5: middle of the city.

all municipal services available (water, sewage, electricity, etc.). In the DMSP OLS imagery, this is a very bright area in the heart of the city. We found it somewhat ironic that this was an area of mature and abundant vegetation.

Our sixth and final site is probably most similar to exurban areas in the United States in that it represents relatively low levels of development (according to Mexican standards) among a great deal of native vegetation (Figure 14.13). This area is an older colonia (20 years old). Electrical services were introduced 12 years ago. Roads remain unpaved and are rather bumpy due to a recent installation of sewage in the fall of 2004. It is an area of steep hillsides and established vegetation. People here have title to their land but did not when they colonized this area.

Site #6: Older Colonia

FIGURE 14.13 Ciudad Victoria site 6: older colonia.

14.5 DISCUSSION

Urban areas are one of the most difficult areas to characterize with remotely sensed imagery alone. They are complex, dynamic, and growing rapidly (Pezzoli 1998). Nonetheless, as practitioners of the art and science of remote sensing, we believe that pictures are worth thousands of words for many reasons. Our explorations of exurbia in the United States and Latin America were humbling with respect to how different low levels of light in the DMSP OLS nighttime satellite imagery were in the United States and Latin America. In addition, the functionality of Google Earth as a tool for visualizing urban areas is profound yet limited now, but will increase as it gets "fleshed out" with respect to spatial and temporal resolution. We are convinced that urban areas around the world are sufficiently complex that remotely sensed imagery alone, regardless of the spatial, spectral, and temporal resolution, will never adequately characterize them from a socioeconomic–demographic perspective. However, we believe that remotely sensed imagery, used in conjunction with primary survey data, will be increasingly used to map, measure, and characterize many social, economic, and demographic characteristics of urban and intraurban areas in the future.

We envision a day in which economic, demographic, political, and social surveys are routinely conducted with GPS or GIS (i.e., every survey respondent is georeferenced). The data sets that result from surveys of this nature will provide for a vibrant area of research with respect to smart interpolation of the information they are attempting to obtain. Characterization of the regional variation of the spatial autocorrelation of variables such as income, fertility, car ownership, and numerous other variables will be a challenging yet tractable research agenda in the future. Successful characterization of variables such as these will allow for dramatically improved ability to interpolate such information spatially using primary survey data in conjunction with remotely sensed imagery. The ability to accomplish such tasks skillfully is a double-edged sword with respect to the fact that the information can be used for good and bad purposes. We optimistically envision this kind of information and estimation to be used to provide health services, respond to natural disasters, focus economic development efforts, and improve civil society. In light of governmental response to Hurricane Katrina in 2005, we understand that these efforts may be for naught. Nonetheless, they are a worthy pursuit for academics and, we hope, will be utilized for the public good.

14.6 CONCLUSIONS

Remote sensing of urban areas is challenging. Delineating urban extent is becoming a relatively trivial task; however, characterizing the intraurban variation of important socioeconomic and demographic variation remains difficult using remotely sensed imagery alone. Primary survey data are expensive and difficult to obtain. Questions associated with "people and pixels" raised by several authors warrant continued research (Liverman et al. 1998). This chapter provided a preliminary investigation suggesting that the use of primary survey data in conjunction with remotely sensed imagery may be a fruitful area of research in the future. We believe that Google

Earth or a similar application will eventually provide high spatial resolution imagery of the whole world. This will change the nature of urban remote sensing.

High-resolution daytime images of urban areas can provide a great deal of information with respect to where and how we live. However, they cannot provide other information with respect to who we are; how much money we have; how many children we have; whom we support politically; etc. Nighttime satellite imagery provided by the DMSP OLS has been useful historically; however, is this utility waning in light of applications such as Google Earth (although the DMSP OLS imagery is easily used for global assessments in ways that Google Earth imagery is not)? Moderate resolution nighttime imagery (30 to 100 m) may still prove to be useful for enabling spatial interpolation of primary survey data regarding variables such as energy consumption, income, transportation infrastructure, crime, and population density.

In addition, we feel that moderate-resolution nighttime satellite imagery would be useful for mapping and estimating many socioeconomic variables in times when primary survey data are unavailable for financial or political reasons. In any case, urban areas represent a growing fraction of the human population. The absolute size and spatially varying behavior of the human population is increasingly important for myriad social, economic, environmental, and political reasons. Ascertaining ways in which we can cost effectively improve our ability to map and characterize the human population is clearly a worthwhile effort.

ACKNOWLEDGMENT

This research was partly supported by The University of Denver's Office of Internationalization. We are grateful for their continued financial support.

REFERENCES

Arreola, D. and Curtis, J.R. (1993). *The Mexican Border Cities: Landscape Anatomy and Place Personality.* Tucson, The University of Arizona Press.

Bahr, J. and Mertins, G. (1992). The Latin American city colloquium. *Geographicum* 22: 65–75.

Balk, D., Pozzi, F., et al. (2005). The distribution of people and the dimension of place: methodologies to improve the global estimation of urban extent. Urban Remote Sensing Conference, Tempe, AZ.

Beach, D. (2002). Coastal sprawl: the effects of urban design on aquatic ecosystems of the United States. Pew Oceans Commission.

Cova, T.J., Sutton, P.C., and Thebald, D.M. (2004). Exurban change detection in fire-prone areas with nighttime satellite imagery. *Photogrammetric Eng. Remote Sensing* 70(11): 1249–1257.

Dobson, J.E., Bright, E.A., et al. (2000). LandScan: a global population database for estimating populations at risk. *Photogrammetric Eng. Remote Sensing* 66(7; July): 849–857.

Doll, C.N.H. and Muller, J.-P. (2000). A comparison of different techniques applied to the UK to map socioeconomic parameters: implications for modeling the human dimensions of global change. *IAPRS* XXXIII (Part B4): 222–229.

Doll, C.N.H., Muller, J.-P., et al. (2000). Nighttime imagery as a tool for global mapping of socioeconomic parameters and greenhouse gas emisssions. *Ambio* 29(3): 159–164.

Ebener, S., Murray, C., Tandon, A., Elvidge, C.D. (2005). From wealth to health: modeling the distribution of income per capita at the subnational level using night-time imagery. *Int. J. Health Geogr.* 4(5).

Elvidge, C., Baugh, K., et al. (1997). Relationship between satellite observed visible-near infrared emissions, population, economic activity, and electric power consumption. *Int. J. Remote Sensing* 18: 1373–1379.

Elvidge, C.D., Baugh, K.E., et al. (1998). Radiance calibration of DMSP-OLS low-light imaging data of human settlements. *Remote Sensing Environ.* 68: 77–88.

Elvidge, C.M.C., Dietz, J., Tuttle, B., Sutton, P., Nemani, R., and Vogelmann, J. (2004). U.S. constructed area approaches the size of Ohio. *EOS* 85(24): 233–234.

Foster, J.L. (1983). Observations of the Earth using nighttime visible imagery. *Int. J. Remote Sensing* 4: 785–791.

Imhoff, M.L., Lawrence, W.T., et al. (1997). Using nighttime DMSP/OLS images of city lights to estimate the impact of urban land use on soil resources in the United States. *Remote Sensing Environ.* 59(1): 105–117.

Liverman, D., Moran, E.F., et al., Eds. (1998). *People and Pixels: Linking Remote Sensing and Social Science*. Washington, D.C., National Research Council.

Longcore, T. and Rich, C. (2004). Ecological light pollution. *Frontiers Ecol. Environ.* 2(4): 191–198.

Longcore, T. and Rich, C. (2005). Synthesis. In *Ecological Consequences of Artificial Night Lighting*. C. Rich and T. Longcore, Eds. Covelo, CA, Island Press.

Milesi, D., Elvidge, C.D., Nemani, R.R., and Running, S.W. (2003). Assessing the impact of urban land development on net primary productivity in the southeastern United States. *Remote Sensing Environ.* 86: 273–432.

Montevecchi, W. (2005). Influences of artificial lights on marine birds. In *Ecological Consequences of Artificial Night Lighting*. C. Rich and T. Longcore, Eds. Covelo, CA, Island Press.

Perry, G. and Fisher, R.N. (2005). Night lights and reptiles: observed and potential effects. In *Ecological Consequences of Artificial Night Lighting*. C. Rich and T. Longcore, Eds. Covelo, CA, Island Press.

Pezzoli, K. (1998). *Human Settlements and Planning for Ecological Sustainability: the Case of Mexico City*. Cambridge, MA, MIT Press.

Salmon, M., Witherington, B., et al. (2000). Artificial lighting and the recovery of sea turtles. In *Sea Turtles of the Indo-Pacific: Research, Management and Conservation*. N. Pilcher and G. Ismail, Eds. London, Asian Academic Press.

Salmon, M. (2005). Protecting sea turtles from artificial night lighting at Florida's oceanic beaches. In *Ecological Consequences of Artificial Night Lighting*. C.L. Rich and T. Longcore, Eds. Covelo, CA, Island Press.

Sutton, P., Roberts, D., et al. (1997). A comparison of nighttime satellite imagery and population density for the continental United States. *Photogrammetric Eng. Remote Sensing* 63(11; November): 1303–1313.

Sutton, P., Roberts, D., et al. (2001). Census from heaven: an estimate of the global population using nighttime satellite imagery. *Int. J. Remote Sensing* 22(16): 3061–3076.

Sutton, P.C., Elvidge, C.D., et al. (2003). Building and evaluating models to estimate ambient population density. *Photogrammetric Eng. Remote Sensing* 69(5): 545–553.

Sutton, P.C. and Elvidge, C.D. (2006). Mapping "exurbia" in the conterminous United States using nighttime satellite imagery. *Geocarto Int.* In press.

Verheijen, F.J. (1985). Photopollution: artificial light optic spatial control systems fail to cope with: incidents, causations, remedies. *Exp. Biol.* 44: 1–18.

Weeks, J. (2003). Does nighttime lighting deter crime? In *Remotely Sensed Cities*. V. Mesev, Ed., London, Taylor & Francis.

Welch, R. (1980a). Monitoring urban population and energy utilization patterns from satellite data. *Remote Sensing Environ.* 9(1): 1–9.

Welch, R.Z.S. (1980b). Urbanized area energy utilization patterns from DMSP data. *Photogrammetric Eng. Remote Sensing* 46(2): 201–207.

WHO (1993). *Urban Health Crisis: Strategies for Health for All in the Face of Rapid Urbanization*. Geneva, World Health Organization.

WHO (2002). Innovative care for chronic conditions, building blocks for action. Global report, noncommunicable diseases and mental health. Geneva, World Health Organization.

Wise, S. and Buchanan, B. (2005). Influence of artificial illumination on the nocturnal behavior and physiology of salamanders. In *Ecological Consequences of Artificial Night Lighting*. C. Rich and T. Longcore, Eds., Covelo, CA, Island Press.

15 Integration of Remote Sensing and Census Data for Assessing Urban Quality of Life: Model Development and Validation

Guiying Li and Qihao Weng

CONTENTS

15.1   INTRODUCTION

The world has experienced dramatic urban growth in recent decades. According to statistics from the Population Reference Bureau (2005), about 47% of the world's population lived in urban areas in 2000. Spatial variation, social stratification, and segregation persist throughout all postindustrial cities. Therefore, study of quality of urban life has drawn increasing interest from a variety of disciplines such as planning, geography, sociology, economics, psychology, political science, behavioral medicine, marketing, and management (Andrew, 1999; Foo, 2001); such information is an important tool for policy evaluation, rating of places, and urban planning and management. Various concepts concerning urban quality of life (QOL) are encountered in the literature, such as urban environmental quality, livability, living quality, quality of place, residential perception and satisfaction, sustainability, etc. Because different disciplines address different aspects of urban quality of life based on different notions and theories, neither a comprehensive conceptual framework concerning urban QOL and human well-being nor any agreed-upon indicator system to evaluate physical, spatial, and social aspects of urban quality exists (Kamp et al., 2003). Practically, the quality of life is a collective attribute that adheres to groups of people. Bonaiuto et al. (2003) studied the perceived residential environment quality and neighborhood attachment in the city of Rome from the environmental psychological view and proposed two distinctive instruments.

These instruments consist of 11 scales measuring the perceived environmental qualities of urban neighborhoods and one scale measuring neighborhood attachment. They are grouped into four categories: spatial (architectural-planning space, organization, and accessibility of space, green space), human (people and social relations), functional (welfare, recreational, commercial, and transport services), and contextual (pace of life, environmental health, upkeep). Pacione (2003) addressed the issues of urban environmental quality and human well-being from a sociogeographical perspective and presented a five-dimensional model for quality of life research in which the major theoretical and methodological issues confronting quality of life research were examined. He also pointed out two distinct types of social indicators to be used to measure social and individual well-being: objective indicators describing environments such as health care provision, crime, education, facilities, and housing, and subjective indicators describing people's perception and evaluation of their living conditions. Usually, subjective measures are based on questionnaires investigating people's satisfaction about their lives; they are affected by many factors, such as individual characteristics, experience, and cultural background. Objective measures are built on "hard" variables such as population, material welfare, the social system, and physical environment. Smith (1973) suggested that objective indicators such as sufficient income, decent housing, high-quality education and health services, and pleasant ambient environment are very important indicators of QOL.

The traditional approach of QOL was based on socioeconomic variables derived from census or survey and rarely considered the biophysical environment. Remote sensing data provide a great resource for extracting environmental variables for QOL. Green (1957) pioneered incorporation of urban biophysical variables derived from remote sensing data with socioeconomic variables extracted from census data to assess QOL. He employed aerial photography to extract physical data, including housing density, number of single-family homes, land uses adjacent to or within residential areas, and distance of residential areas to central business districts. These data were then combined with socioeconomic data such as education, crime rate, and rental rates to rank each residential area of Birmingham, Alabama, in terms of "residential desirability." Later, Green and Monier (1957) used the same method in other U.S. cities.

Mumbower and Donoghue (1967) and Metivier and McCoy (1971) have also studied poverty in cities as an aspect of QOL based on housing density and other indicators derived from aerial photography. The advances of remote sensing and the geographic information system (GIS) allow QOL research to be conducted more efficiently based on integration digital remote sensing imagery with census socioeconomic data. For example, Weber and Hirsch (1992) developed urban life indices by combining remotely sensed SPOT data with conventional census data, such as population and housing data, for Strasbourg, France. They found some strong correlations between census and remotely sensed data, mostly with housing-related data. Three urban QOL indices were developed based on mixed data. They were interpreted as housing index, attractiveness index, and repulsion index. However, each of these describes only one aspect of quality of life and could not give a whole picture of QOL for a specific unit.

Lo and Faber (1997) created QOL maps for Athens–Clarke County, Georgia, by integrating environmental factors including land use and land cover (LULC), surface temperature, and vegetation index (NDVI) derived from Landsat Thematic Mapper (TM) and census variables such as population density, per-capita income, median home value, and percentage of college graduates. They used principal component analysis (PCA) and GIS overlay methods, respectively. However, the first principal component, which the authors interpreted as "the greenness of environment," only explained 54.2% of total variance; therefore, to measure quality of life based only on the first principal component was not complete because it did not incorporate the second principal component of "personal traits." On the other hand, the GIS overlay method, which sums up ranked data layers (variables) in GIS, used redundant data due to the high correlation between selected variables.

15.2 BACKGROUND

Remotely sensed data and census data are two essential data sources for urban analyses. Remote sensing data effectively record the physical properties of the environment, provide large quantities of timely and accurate spatial information, and are widely used in mapping and monitoring changes in LULC (Welch, 1982; Forster, 1985; Pathan et al., 1993; Weng, 2002; Harris and Ventura, 1995) and in modeling urban environmental quality (Fung, 2000; Nichol and Wong, 2005). Census data offer a wide range of demographic and socioeconomic information and are used in racial and ethnic diversity research (Frey, 2001) and urban planning and management.

The introduction of GIS techniques, especially recent advances, provided an effective environment for manipulating large amounts of digital data of spatial phenomena from different sources (Burrough, 1986). The integration of remote sensing and GIS has received widespread and extensive recognition, and research has increasingly been conducted based on integrating remote sensing with GIS (Donnay et al., 2001). According to Wilkinson (1996), remote sensing and GIS are complementary to each other in three main ways:

(1) Remote sensing is used as a tool to gather data for use in GIS.
(2) GIS data are used as ancillary information to improve the products derived from remote sensing.
(3) Remote sensing and GIS are used together for environmental modeling and analysis.

Census data collected within arbitrary spatial extents with corresponding spatial information can be viewed as GIS data. Integration of remote sensing and census data has three applications in urban study. First, remote sensing images are used in extracting and updating transportation networks (Lacoste et al., 2002; Doucette et al., 2004; Harvey et al., 2004; Kim et al., 2004; Song and Civco, 2004), mapping land use and land cover (Haack et al., 1987; Ehlers et al., 1990; Treitz et al., 1992), and detecting urban expansion (Yeh and Li, 1997; Weng, 2002; Cheng and Masser, 2003). Second, census data are used to improve land-use and land-cover classifications in urban areas (Harris and Ventura, 1995; Mesev, 1998).

The third and most important application of integration of census and remote sensing is to model urban growth patterns (Pathan, et al., 1993; Cheng and Masser, 2003) and estimate population and residential density (Langford et al., 1991; Lo, 1995; Sutton, 1997; Yuan et al., 1997; Harris and Longley, 2000; Harvey, 2000, 2002; Martin et al., 2000; Qiu et al., 2003; Li and Weng, 2005). In addition, they were also used in poverty pocket detection (Hall et al., 2001), low-income housing site identification (Thomson, 2000), and QOL studies (Weber and Hirsch, 1992; Lo and Faber, 1997).

This study focuses on development of a methodology based on integrating Landsat Enhanced Thematic Mapper Plus (ETM+) images with Census 2000 in a GIS framework. The quality of life was assessed in Marion, Monroe, and Vigo Counties in Indiana. A QOL model from Marion County was created and then validated by applying it to Monroe and Vigo Counties.

15.3 STUDY AREA

Marion County, Monroe County, and Vigo County, Indiana, (Figure 15.1) were chosen to implement this study. Marion County is located in the center of Indiana. It is the core of the Indianapolis metropolitan area and highly concentrates major employers and manufacturing, as well as professional, technical, and educational services of the state. With its moderate climate, rich history, excellent education, social services, arts, leisure, and recreation, Indianapolis was named one of America's best places to live and work (*Employment Review's* August 1996 issue). According to the 2000 Census results, there were 860,454 people and a total 387,183 housing units.

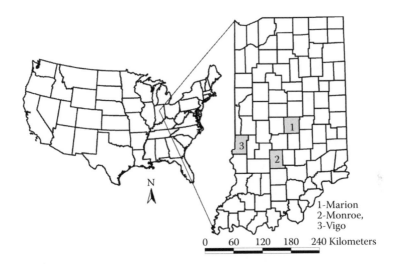

FIGURE 15.1 Location of the study area.

In 1999, the unemployment rate was 3.7%, and 8.7% of families were under poverty level. Diversities in incomes, education levels, and environments within the city cause significant differences in quality of life.

Compared with Marion County, Monroe County and Vigo County have less population. Monroe County covers 1020.8 km² in land area; population density is 118.5 per square kilometer; median household income is $33,311 and per-capita income is $18,534. About 7.1% of families are under poverty level. The county seat is Bloomington, which is the home of the main campus of Indiana University. With the influence of the university and the beauty of surrounding natural resources such as Hoosier Forest and Monroe Lake, Bloomington is the sixth best place to settle down in the United States, according to *Money* magazine. Vigo County covers 1044.1 km² with population density of 98.8 per square kilometer; median household income and per-capita income are $33,184 and $17,620, respectively. The county seat is Terre Haute, which is home to nationally recognized colleges Rose–Hulman Institute of Technology, St. Mary of the Woods College, Indiana State University, and Ivy Tech College. Vigo County is also considered a regional shopping area that draws people from surrounding Indiana and Illinois counties. This study focused on the urbanized areas within these three counties, especially Indianapolis, Bloomington, and Terre Haute.

15.4 DATA SET AND METHODOLOGY

15.4.1 DATA SET

The primary data sources are Census 2000 and Landsat ETM+. Census 2000 data from the U.S. Bureau of the Census used in this study include (1) tabular data stored in summary file 3, which contains the information about population, housing, income, education, etc.; and (2) spatial data, called topologically integrated geographic encoding and referencing (TIGER), which contain data representing the position and boundaries of legal and statistical entities. These two types of data are linked by census geographical entity code. The U.S. Census has a hierarchical structure composed of ten basic levels: United States, region, division, state, county, county subdivision, place, census tract, block group, and block. The block group level was selected in this study as the work level.

This study focused on urban areas; thus, urbanized area boundaries obtained from census data were used to determine the study area. Block groups completely or mostly contained within urbanized boundaries were selected to conduct QOL analysis.

Landsat 7 ETM+ images dated June 22, 2000, were used in this research. These images were acquired through the USGS Earth Resource Observation Systems Data Center and then coregistered to 1:24,000 topographic maps with RMS errors less than 0.5 pixel.

Because of the different coordinate systems between census vector data and ETM+ images, the geographic coordinates of census vectors were converted to the UTM as the same coordinate system with the ETM+ image.

15.4.2 METHODOLOGY

Figure 15.2 gives the processing procedures for development of a QOL index and prediction model. The detailed descriptions of processing are given next.

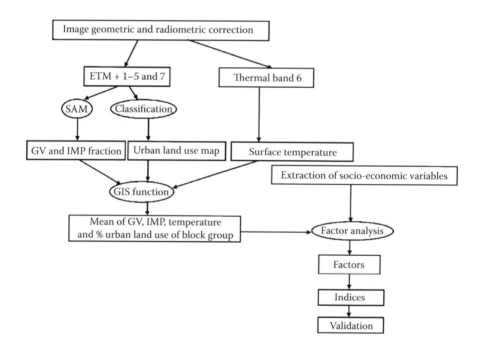

FIGURE 15.2 Flowchart of processing.

15.4.2.1 Extraction of Socioeconomic Variables from Census Data

Selection of socioeconomic variables was based on common agreed-upon variables used in previous studies (Weber and Hirsch, 1992; Lo and Faber, 1997). These variables included population density, housing density, median family income, median household income, per-capita income, median house value, median number of house rooms, percentage of college graduates, unemployment rate, and percentage of families under poverty level. Initially, 26 variables were extracted from Census 2000 summary file 3, and a series of processes was performed to obtain the variables selected. TIGER shape files of block group were downloaded from the Internet. The socioeconomic variables were integrated with TIGER shape files by geographic entity code as attributes of shape file.

15.4.2.2 Extraction of Environmental Variables from ETM+

15.4.2.2.1 Greenness, Impervious Surface, and Urban Land Use

Urban vegetation is a desirable urban environmental parameter that contributes to urban QOL (McPherson, 1992, 1994; Weber and Hirsch, 1992; Lo and Faber, 1997; Fung, 2000; Nichol and Wong, 2005) for its providing aesthetics, mitigating urban heat island effect, and reducing air pollution (Dwyer et al., 1992; Akbari et al., 2003). Greenness relates to vegetation, which can be recorded by remote sensors. It can be measured using different forms of vegetation index — for example, normalized difference vegetation index (NDVI). However, NDVI values are affected by many other external factors such as view angle, soil background, seasons, and differences in row direction and spacing in agricultural fields; therefore, it does not measure the amount of vegetation (Weng et al., 2004). Thus, green vegetation abundance (to be introduced later) was employed to measure the amount of vegetation as greenness indicator instead of NDVI in this study.

Impervious surfaces are mainly constructed surfaces, including rooftops, sidewalks, roads, and parking lots covered by impenetrable materials such as asphalt, concrete, brick, and stone. These materials seal surfaces, repel water, prevent precipitation, and melt water from infiltrating soils. Impervious surface is considered a key environmental indicator for its impacts on water quantity and quality, energy balances and local climates, habitat degradation, streams, and landscape aesthetics.

Green vegetation abundance and impervious surface can be extracted from remote sensing images by spectral mixture analysis (SMA). SMA is a technique based on the assumption that the reflectance spectrum by a sensor is the linear combination of the spectra of the materials within the sensor's field of view, frequently called endmembers. SMA has been extensively applied in estimation of vegetation cover (Smith et al., 1990; Asner and Lobell, 2000; McGwire et al., 2000; Small, 2001), in vegetation or land-cover classification and change detection (Adams et al., 1995; Roberts et al., 1998; Cochrane and Souza, 1998; Aguiar et al., 1999; Lu et al., 2003), and in urban studies (Rashed et al., 2001; Phinn et al., 2002; Wu and Murray, 2003; Lu and Weng, 2004).

In this study, three endmembers, including green vegetation, impervious surface, and water, were initially identified from the ETM+ image based on high-resolution aerial photographs. The water endmember was identified from the areas of clear and deep water, while green vegetation was selected from the areas of dense grass and cover crops. Different types of impervious surfaces were selected from building roofs and highway intersections. The radiances of these initial endmembers were compared with those of the endmembers selected from the scattergram of TM 3 and TM 4 and the scattergram of TM 4 and TM 5. Endmembers whose curves were similar but had extreme values in the triangle of the scattergram were finally selected. A constrained least-squares solution was used to decompose the six ETM+ bands (1 through 5, and 7) into three fraction images (vegetation, impervious surface, and water) in ENVI software. Of the three endmembers, green vegetation and impervious surface fractions were used as environmental parameters.

Urban use such as transportation and commercial/industrial is considered an undesirable parameter (Weber and Hirsch, 1992; Lo and Faber, 1997) to environmental quality. It can be obtained by classifying an ETM+ image into different land-use or land-cover types. In this study, ETM+ images were classified into six categories: urban (transportation, commercial, and industrial), residential, water, grass, forest, and agricultural land, with total accuracy over 85%.

15.4.2.2.2 Temperature

Urban heat island is a common phenomenon in the cities; temperature of urban area is higher than that of rural areas. Temperature is an important factor affecting human comfort. High surface temperature is regarded as undesirable by most people; therefore, it can be used as an indicator of environmental quality (Lo and Faber, 1997; Nichol and Wong, 2005). A thermal band (band 6) of ETM+ records thermal infrared emission from the land surface and provides the source to extract surface temperature. The procedure to develop the surface temperature involves three steps: (1) converting the digital number of Landsat ETM+ band 6 into spectral radiance; (2) converting spectral radiance to at-satellite brightness temperature, which is also called blackbody temperature; and (3) converting blackbody temperature to land surface temperature. A detailed description of the procedures for extracting temperature images from Landsat ETM+ imagery can be found in Weng et al. (2004).

Because census data and ETM+ data have different formats and spatial resolution, they need to be integrated. With help of the GIS function in ERDAS Imagine, remote sensing data were aggregated at block group level; the mean values of green vegetation, impervious surface and temperature, and percentage of urban use were calculated for each block group. All these data were exported into SPSS software for further analyses.

15.4.2.3 Factor Analysis and Development of QOL Indices

Factor analysis attempts to identify underlying variables called factors that explain the pattern of correlations within a set of observed variables. It is often used in data reduction to identify a small number of factors that explain most of the variance observed in a much larger number of observed variables. The first factor explains most of the variance in the data, and each successive factor explains less of the variance (Tabachnick and Fidell, 1996). The number of factors to be selected depends on the percentage of variance explained by each factor.

There are different factor extraction methods. The principal component is one that is used here. Factors whose eigenvalue is greater than 1 were selected. Each factor can be explained as one aspect of QOL. Therefore, factor scores can be used as a single index indicating the aspect with which the factor associates. The synthetic QOL index is a composite of different aspects and is computed by the following equation:

$$QOL = \sum_{1}^{n} F_i W_i$$

where
  n = number of factors selected
  F_i = factor i score
  W_i = weight of factor, represented by the percentage of variance that factor i
      explains

The directions of contribution of each factor to QOL should be considered when adding weighted factor scores. Finally, the QOL map is created showing geographic patterns.

15.4.2.4 Creation of QOL Estimation Model from Marion County

Ideally, QOL scores developed based on factor analyses should be related to real quality of life and further develop the prediction model. However, no such data are available so far. Therefore, in this study, QOL scores were eventually related to original indicators by regression model. The predictors used to develop the QOL model were variables that have the highest loading on respective factors.

15.4.2.5 Validation of Model

The QOL index model from Marion County was applied to Monroe and Vigo Counties to predict QOL scores. The predicted scores were then compared with the QOL index created from factor analysis by correlation analysis.

15.5 RESULTS

15.5.1 Extraction of Environmental and Socioeconomic Variables from ETM+ and Census

As mentioned in the methodology section, ten socioeconomic variables were extracted from census data. The distributions of per-capita income, median family income, and median household income by block groups in Marion County have similar trends; that is, the highest incomes are found in the northern part of the county and most of the lowest incomes are found in the county's center. There are also several block groups with high per-capita income located in the center. In contrast to income findings, higher population and housing density are located in the downtown area (except for the central business district), and the lower density is located in suburban areas. Percentage of college graduates and above has the same trend as income, as does the median house value. For Monroe County, block groups with higher per-capita income are found in the southeastern parts of Bloomington city and those of lower income are found in the downtown area of Bloomington. Conversely, high population density is found in the downtown area, especially on the Indiana University campus. For Vigo County, block groups of higher income are found to the east and south of Terre Haute, while the lower income groups are located in the center of Terre Haute. Block groups with high population density are found in the city; low density is in suburban areas.

Four environmental variables were extracted from ETM+. Three fraction images, including green vegetation, impervious surface, and water, were produced from ETM+ bands 1 through 5 and band 7 using SMA. Green vegetation and impervious surface fraction images are used as two environmental indicators. The highest values of green vegetation are located in forest, grass, and crop areas, while the lowest values are found in the urban and water areas. In contrast, the highest values of impervious surface are found in the urban area and the lowest are in the forest, grass, and water. Temperature images derived from ETM+ band 6 indicate that urban heat islands definitely exist in the cities of Indianapolis, Bloomington, and Terre Haute; the high surface temperatures are in urban areas, especially downtown, and low temperature is in the vegetated areas and water bodies. Land-use and land-cover maps for three counties also were produced using an unsupervised classification method.

These four remote sensing variables then were aggregated at block group level. The means of green vegetation, impervious surface, surface temperature, and percentage of urban use were calculated for each block group.

15.5.2 CORRELATION BETWEEN VARIABLES

Pearson's correlation was used to analyze the correlations between all variables. For example, in Marion County (Table 15.1), there are very strong correlations between most of the variables. The highest correlations are found between the variables of same sector — for example, the correlation between population density and housing density, between environmental variables, and between economic parameters. Also, there are strong correlations between environmental variables and economic variables; for example, green vegetation has a significantly positive relationship with all income variables ($r = 0.328$ to 0.458), median house value ($r = 0.328$), and median number of rooms ($r = 0.483$); it has negative relationships with density variables ($r = -0.195$ and -0.237) and percentage of poverty ($r = -0.405$). Percentage of college graduates has very high correlations with income variables and house characteristics, indicating that highly educated people make more money and may live with high quality. The relationships of impervious surface, percentage of urban use, and temperature with other variables are in contrast to those of vegetation.

TABLE 15.1
Correlation Matrix between Variables for Marion County

	PD	HD	GV	IMP	UR	TEMP	MHI	MFI	PCI	PCG	UNEMP	POV	MR
HD	0.912[a]												
GV	-0.195[a]	-0.237[a]											
IMP	0.047	0.070	-0.878[a]										
UR	0.499[a]	0.495[a]	-0.843[a]	0.779[a]									
TEMP	0.514[a]	0.515[a]	-0.850[a]	0.772[a]	0.979[a]								
MHI	-0.259[a]	-0.293[a]	0.458[a]	-0.522[a]	-0.562[a]	-0.553[a]							
MFI	-0.243[a]	-0.235[a]	0.408[a]	-0.505[a]	-0.534[a]	-0.518[a]	0.924[a]						
PCI	-0.258[a]	-0.178[a]	0.328[a]	-0.472[a]	-0.502[a]	-0.475[a]	0.811[a]	0.857[a]					
PCG	-0.261[a]	-0.161[a]	0.242[a]	-0.376[a]	-0.451[a]	-0.407[a]	0.674[a]	0.735[a]	0.825[a]				
UNEMP	0.226[a]	0.177[a]	-0.261[a]	0.254[a]	0.330[a]	0.305[a]	-0.446[a]	-0.474[a]	-0.439[a]	-0.458[a]			
POV	0.315[a]	0.329[a]	-0.405[a]	0.366[a]	0.477[a]	0.464[a]	-0.616[a]	-0.617[a]	-0.522[a]	-0.419[a]	0.564[a]		
MR	-0.054	-0.153[a]	0.483[a]	-0.524[a]	-0.413[a]	-0.415[a]	0.684[a]	0.590[a]	0.446[a]	0.308[a]	-0.300[a]	-0.351[a]	
MHV	-0.191[a]	-0.140[a]	0.328[a]	-0.442[a]	-0.476[a]	-0.443[a]	0.721[a]	0.737[a]	0.790[a]	0.699[a]	-0.343[a]	-0.361[a]	0.466[a]

[a]Correlation at 99% confidence level (two tailed).

Notes:

PD = population density
HD = housing density
GV = green vegetation
IMP = impervious surface
UR = percentage of urban use
TEMP = temperature
MHI = median household income
MFI = median family income
PCI = per-capita income
PCG = percentage of college above graduates
UNEMP = unemployment rate
POV = percentage of families under poverty level
MR = median number of rooms
MHV = median house value

Because the high correlations exist between these variables, it is necessary to reduce the data dimension and redundancy. Factor analysis provides such a function.

15.5.3 FACTOR ANALYSIS INTERPRETATION

As a general guide in interpreting factor analysis results, first the suitability of data for factor analysis is checked based on Kaiser–Meyer–Olkin (KMO) and Bartlett's test of sphericity significance. Only when KMO is greater than 0.5 and significance level of Bartlett's test is less than 0.1 are the data acceptable for factor analysis. The second step is to validate the variables based on communalities, which can be interpreted as the reliability of the indicator. Small values indicate that variables do not fit well with the factor solution and should be dropped from the analysis. Initially, all 14 variables were input in processing. For Marion County, KMO (0.863) and Bartlett's test (significance level 0.000) indicated the data were suitable for factor analysis. However, three variables (i.e., median number of rooms, unemployment rate, and percentage of families under poverty level) had low communality values, so these variables were eliminated from analysis. Then the second factor analysis was conducted for the remaining 11 variables; the results are given in Table 15.2. Based on the rule that the minimum eigenvalue should not be less than 1, three factors were extracted from factor analysis (Table 15.3). The first factor (factor 1) explains about 37.52% of total variance; the second (factor 2) accounts for 28.16%, and the third (factor 3) explains 20.32%. Thus, the first three factors explain 89.01% of all variance.

TABLE 15.2
Communalities from Factor Analyses for Marion County

Variables	Communality	
Population density	0.943	0.952
Housing density	0.933	0.947
Green vegetation	0.935	0.940
Impervious surface	0.924	0.929
Percentage of urban land use	0.927	0.946
Surface temperature	0.940	0.956
Median household income	0.849	0.826
Median family income	0.880	0.872
Per-capita income	0.852	0.896
Education level	0.731	0.774
Median value of house	0.694	0.752
Rate of poverty	0.500	
Unemployment rate	0.363	–
Median number of rooms	0.450	
KMO	0.863	0.842
Bartlett's test of sphericity significance	0	0

TABLE 15.3
Factor-Loading Matrix for Marion County

Variables	Factor 1	Factor 2	Factor 3
Population density	−0.168	0.124	0.953
Housing density	−0.101	0.167	0.953
Green vegetation	0.132	−0.956	−0.096
Impervious surface	−0.304	0.909	−0.100
Percentage of urban land use	−0.325	0.830	0.390
Surface temperature	−0.291	0.838	0.412
Median household income	0.836	−0.308	−0.181
Median family income	0.886	−0.259	−0.141
Per-capita income	0.920	−0.185	−0.120
Education level	0.867	−0.129	−0.080
Median value of house	0.842	−0.201	−0.058
Eigenvalue	4.128	3.428	2.235
% Variance explained	37.524	31.162	20.321
Cumulative %	37.524	68.686	89.007

The key in factor analysis is to interpret factor loadings, which are measurements of relationships between variables and factors. Comrey and Lee (1992) suggested a range of values to interpret the strength of the relation between variables and factors. Loadings with absolute values of 0.71 and higher are considered excellent; 0.63 is very good, 0.55 good, 0.45 fair, and 0.32 poor. Table 15.3 presents the factor loadings on each variable from factor analysis for Marion County: Factor 1 has strong positive loadings (greater than 0.8) on five variables, including median household income, median family income, per-capita income, median house value, and percentage of graduates from college and above. Therefore, factor 1 is associated with economic conditions. The higher the score of factor 1 is, the better the quality is in an economic aspect. Factor 2 has high negative factor loadings on green vegetation (−0.956) and positive loadings on impervious surface (0.909), surface temperature (0.830), and percentage of urban use (0.838). Therefore, factor 2 relates to environmental conditions; the higher the score is, the worse the quality of the environment is. Factor 3 shows high positive loadings on population density and housing density; thus, it relates to crowdedness. The higher the score of factor 3 is, the smaller is the space in which a person lives and the poorer is the quality of the living space.

15.5.4 QOL Index Development

Factor scores can be used as indices representing the QOL of different dimensions; therefore, the spatial variations of factor scores indicate the geographic pattern of life quality. In ArcView, the distribution of each factor was mapped (Figure 15.3, Figure 15.4, and Figure 15.5) for Marion County. Factor 1, called "economic sector,"

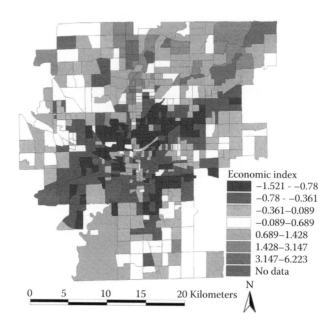

FIGURE 15.3 Index of economic quality in Marion County.

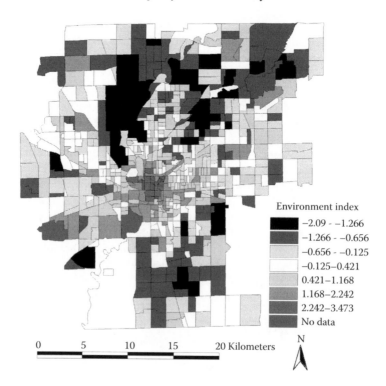

FIGURE 15.4 Index of environmental quality in Marion County.

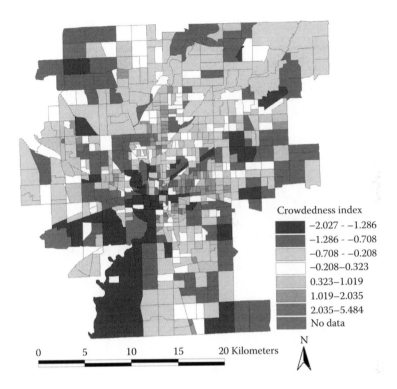

FIGURE 15.5 Index of crowding in Marion County.

has distribution similar to that of income — that is, block groups with higher scores are located in the northern part of the county, and very few are found in the inner city. Factor scores range from –1.5 to 6.2. Factor 3, which represents crowdedness, has distribution similar to that for population density and housing density. The crowed block groups are found near the city's center. In contrast, factor 2, called "environmental sector," has inverse distribution compared to that of green vegetation; the high scores are found in the city's center, and park, forest, and low-density residential have low scores.

Developing synthetic QOL involves the combination of three factors that represent different aspects of quality of life. Factor 1 has a positive contribution to QOL, while factors 2 and 3 have negative contributions. The final QOL index is obtained by adding the weighted scores of three factors with their direction of contribution. The following equation is used for Marion County:

$$QOL = (37.524 * Factor\ 1 - 31.162 * Factor\ 2 - 20.321 * Factor\ 3)/100$$

The index ranges from a low of –1.05 to a high of 2.74. About 11% of block groups have scores greater than 0.7, and most of them are located in the northern part of the county. Block group 1809733003002 has the highest QOL score (2.74), characterized by low population density, large portion of green vegetation (58%),

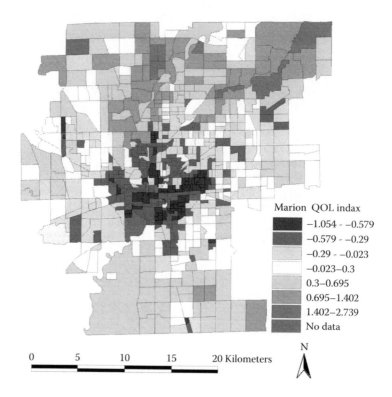

FIGURE 15.6 (**Color Figure 15.6 follows page 240.**) Synthetic QOL index in Marion County.

low impervious surface, and high family income. The important characteristics of block groups 180973304012 and 180973203012, which have the second and fourth highest scores, are their highest median house values and income. Although their vegetation cover is not as high as others', the existence of a water body makes them more attractive to high-income people. About 30.5% of block groups have scores of less than –0.3; most of them are located near the center of the city, where high impervious surface, low vegetation cover, high surface temperature, high density, and low income are found. Block group 180973401081 has the lowest QOL score because it has the highest population density in the city. Figure 15.6 illustrates the geographical pattern of QOL in Marion County.

The same procedures were applied to Monroe and Vigo Counties, and QOL index maps were created (Figure 15.7 and Figure 15.8). For Monroe County, QOL scores range from –0.95 to 0.96. The lower scores are found in the Bloomington downtown area, especially the Indiana University sites, because of concentrations of students who have low incomes and high population density. Higher scores are located in the southeastern section of Bloomington. For Vigo County, QOL scores range from –1.18 to 1.53; the lower score is found in downtown Terre Haute and the higher score is in the southern and eastern parts of Terre Haute.

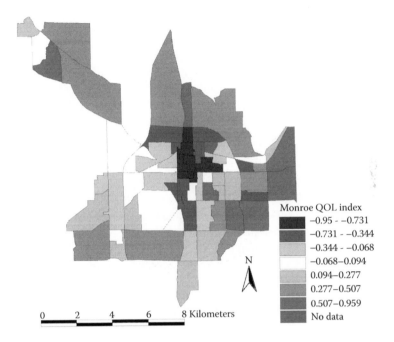

FIGURE 15.7 QOL index in Monroe County based on factor analysis.

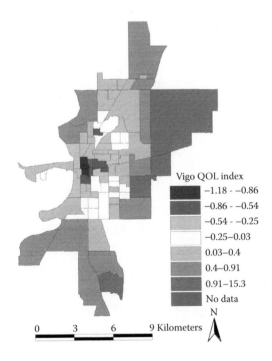

FIGURE 15.8 QOL index in Vigo County based on factor analysis.

15.5.5 Regression Analysis and Model Development

Once the single index and synthetic QOL index were created based on factors, regression analyses then were applied to relate them to original environmental and socioeconomic variables. As Table 15.3 shows, each factor has been closely related to some variables; therefore, regressing factor score of QOL against the variables that have the largest contribution to it can produce regression models. These could possibly be applied to other regions to estimate QOL. In this study, we tried to develop an estimation model from Marion County and then applied it to Monroe and Vigo Counties. For the economic index of Marion County, the per-capita income has highest correlation (0.920) with factor 1; thus, per-capita income is used as a predictor. For the environmental index, green vegetation has the strongest correlation (−0.956) and vegetation is used as a predictor. For crowdedness, housing density (correlation coefficient 0.953) is a good predictor. All three variables were used in developing estimation models for combined QOL. Table 15.4 presents four models based on these variables. For the synthetic QOL model, R^2 reaches 0.948, which indicates the model fits the data very well.

15.5.6 Model Validation

The model developed (Table 15.4) from Marion County was applied to estimate QOL in Monroe and Vigo Counties. For Monroe County, estimated scores range from −1.38 to 0.94 with mean of −0.062; for Vigo County, estimated QOL score ranged from −1.06 to 1.22 with mean of −0.231. The estimated QOL spatial distributions are shown in Figure 15.9 and Figure 15.10. Comparing estimated scores with QOL index from factor analysis, we found that distribution patterns were similar. This can be confirmed by analyzing the correlations between QOL scores from factor analysis and predicted scores. Figure 15.11 shows that estimated QOL scores are highly correlated to QOL index developed from factor analysis for both counties ($r = 0.968$ and 0.964 for Monroe and Vigo Counties, respectively). This indicates that the QOL model has potential to be applied to other regions to explain the spatial pattern of QOL.

In SPSS, the factor analysis function saves standardized factor scores as variables in the corresponding data set; therefore, QOL scores from factor analysis for different regions could not be compared to each other directly to test whether they have significant

TABLE 15.4
QOL Estimation Models from Marion County

Model	Predictor	R^2	Constant	Coefficient
Economic	Per-capita income	0.847	−1.775	0.00008759
Environmental	Green vegetation	0.913	3.782	−0.111
Crowdedness	Housing density	0.908	−1.337	0.002
	Per-capita income	0.948	−1.319	3.098×10^{-5}
QOL	Green vegetation			0.026
	Housing density			−0.00027

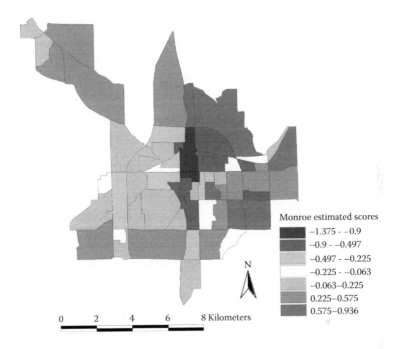

Monroe estimated scores

- −1.375 - −0.9
- −0.9 - −0.497
- −0.497 - −0.225
- −0.225 - −0.063
- −0.063−0.225
- 0.225−0.575
- 0.575−0.936

0   2   4   6   8 Kilometers

FIGURE 15.9 Estimated QOL scores for Monroe County.

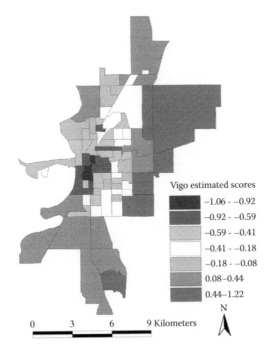

Vigo estimated scores

- −1.06 - −0.92
- −0.92 - −0.59
- −0.59 - −0.41
- −0.41 - −0.18
- −0.18 - −0.08
- 0.08−0.44
- 0.44−1.22

0   3   6   9 Kilometers

FIGURE 15.10 Estimated QOL scores for Vigo County.

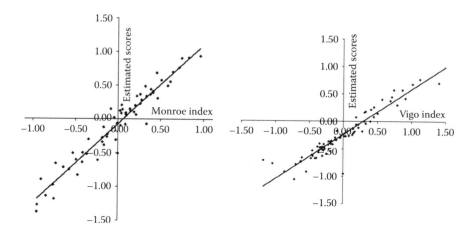

FIGURE 15.11 Correlation between QOL index developed from factor analysis and estimated QOL scores. Correlation coefficients are 0.968 for Monroe County and 0.964 for Vigo County.

differences. However, the estimated QOL had this potential, so T-tests were conducted based on estimated QOL. Table 15.5 shows the results of T-tests. There are significant differences between Monroe and Vigo and between Marion and Vigo; the former is better; there is no significant difference between Marion and Monroe. Table 15.6 shows mean values of selected block groups for three predictors. For Marion and Monroe Counties, we see no significant difference between means of green vegetation coverage, but housing density and per-capita income are different: Marion is more crowded and has higher per-capita income. However, they have cancelled each other due to their opposite impacts on QOL, thus resulting in no significant difference between overall QOLs of these two counties. For Vigo County, per-capita income is significantly lower than that for the other two counties; even though it has lower housing density, weight of housing density in QOL is too low compared to the impact of per-capita income. This results in a significant difference, with Vigo the poorest.

TABLE15.5
T-Test for QOL among Three Counties

	Marion	Monroe	Monroe	Vigo	Marion	Vigo
Mean	−0.0015	−0.0622	−0.0622	−0.2305	−0.0015	−0.2305
Variance	0.2783	0.2964	0.2964	0.1900	0.2783	0.1900
Observations	615	65	100	117	615	82
Degree of freedom		77		121		115
T-statistic		−0.86		−2.03		−4.35
T-critical two tail		1.99		1.98		1.98

Note: Confidence level is 95%.

TABLE 15.6
Mean of Three Predictors of QOL

Predictors	Marion	Monroe	Vigo
Housing density (per km²)	765	650	603
Green vegetation coverage (%)	33.9	33.1	29.4
Per-capita income ($)	19,974	18,074	15,458

15.6 DISCUSSION AND CONCLUSIONS

15.6.1 DISCUSSION

This research demonstrates that GIS provides a powerful platform for integrating different data models from different data sources, such as remote sensing and census socioeconomic data, and for creating comprehensive databases combining socioeconomic and environmental components. Furthermore, it provides a realistic and practical way to assess QOL in cities. This would help managers and policy makers in formulating strategies for urban development plans. However, the issues that arise in the integration of disparate data, variable selection, the weight associated with selected variables, and the model's expansion should be of concern.

Remote sensing and census data are collected for "different purpose[s], at different scale[s], and with different underlying assumptions about the nature of the geographic features" (Huang and Yasuoka, 2000). Remote sensing data are digital records of spectral information about ground features with raster format and often exhibit continuous spatial variation; census data usually relate to administrative units (such as block, block group, tract, county, and state) and tend to be more discrete and have sharp discontinuities between adjacent areas. Usually, socioeconomic data are integrated into vector GISs as attributes of spatial units for various mapping and spatial analysis. Integration between remote sensing and GIS/socioeconomic data involves conversion between data models.

One method is to aggregate remote sensing data to vector polygons, which assumes value is constant throughout each polygon. That would lead to some loss of spatial information existing in remote sensing data. Generally speaking, the finer the spatial unit is, the better the opportunity for extracting remote sensing information. However, socioeconomic data such as income and school attendance at the individual level are not released to the public by the Census Bureau in consideration of confidentiality. Block group level is the finest level to access these socioeconomic data. In addition, the census has different scales (levels); integrating remote sensing data with different scales of census data will produce so-called modifiable area unit problems (Liang, 2005). Therefore, to find a suitable aggregation unit is very important in order to balance between loss of spatial information of remote sensing data and the operational scale of phenomena. Another method of integration is one through vector to raster by rasterization or surface interpolation to produce raster layers for each socioeconomic variable. Therefore, more research is needed on disaggregating census data to individual pixela to match remote sensing data for data integration in the future.

QOL is a huge topic in social study and consists of different aspects of human beings' lives; thus, variable or indicator selections are the key to assessing QOL. However, opinions on the indicators and ingredients of QOL still differ considerably. Generally, a combination of objective and subjective indicators is considered to be preferable; however, acquisition of subjective indicators involves face-to-face interviews or comprehensive questionnaires, which are time consuming and costly, especially for large areas at more detailed levels, such as block groups. Collaboration between different agencies and organizations is necessary to conduct such projects. Also, health and accessibility to cultural and social facilities are important sectors in measuring QOL. Therefore, incorporation of such data in the future should be considered.

The interpretation of variables raises another issue in QOL due to their dual nature. For example, transportation shapes the urban form and plays an important role in urbanization. It provides access to opportunities; supports urban economic activities; facilitates social interactions and development of industrial and commercial stimulants to the economy; and provides more employment opportunities for residents. From this point of view, it improves QOL. On the other hand, transportation infrastructure is built at the expense of decreasing agriculture and vegetation, and it changes urban morphology. In turn, it degrades quality of life through urban environment changes, congestion, pollution, accidents, and community severance. In this study, urban land use and impervious surface are considered as negative factors in assessing QOL in two main considerations. First, the focus of this study is to assess spatial distribution of QOL from livability and environmental and socioeconomic relationships. Second, most people in the city would like to live away from heavy traffic, noise, and pollution, where high impervious or urban land uses dominate because transportation is not a problem any more due to the availability of personal vehicles. They care more about beauty of the physical environment, such as more vegetation and water bodies.

The weight associated with each variable is another challenging task. Different groups of people have different criteria for the same indicator. For example, for low-income residents, income and housing probably are most important, while high-income residents pursue high quality of recreation and leisure. In this study, the percentages of variance factors explained were used as weights, resulting in high QOL scores in suburban areas due to high vegetation and low density there.

The model developed from Marion County was applied to Vigo and Monroe Counties and did predict QOL scores, which have strong correlation with quality of life produced from factor analysis. The comparison among three counties indicated Monroe and Marion have better overall QOL than Vigo. As we know, geographical locations, landscape, and economy of these three counties are significantly different; other important indicators, such as health service, cultural, recreation, and entertainment facilities, were not taken into account in this study. Generally speaking, the model developed should be validated by applying it to similar physical and socioeconomic settings; also, different nations or regions may have their own criteria on a specific indicator due to different cultures and limited resources. Therefore, validation of the model needs to be conducted in the future.

15.6.2 CONCLUSIONS

This research explored the relationships between environmental and socioeconomic characteristics of Marion, Monroe, and Vigo Counties in Indiana, presented a practical way to measure quality of life based on integration of remote sensing and census data, and provided a potential model to predict QOL in other regions. It was found that green vegetation has a strong positive correlation with income, house value, and education level and a negative relationship with temperature, impervious surface, and population or housing density.

Factor analysis provided an effective way to reduce data dimension and redundancy. Three factors were produced from 11 variables for Marion County, among which the first factor was related to the economic aspect of QOL, the second was associated with environment, and the third represented crowdedness. The application of factor techniques has advantages over simply using GIS overlays or first-principal components. Application of a regression model from Marion County to Monroe and Vigo Counties indicated that the model has potential use in predicting QOL for analyzing the spatial variations of QOL within a specific region. Comparison of overall quality of life indicated significant differences among overall QOL of three counties, with Vigo the worst.

REFERENCES

Adams, J.B., Sabol, D.E., Kapos, V., Filho, R.A., Roberts, D.A., Smith, M.O., and Gillespie, A.R., 1995. Classification of multispectral images based on fractions of endmembers: application to land-cover change in the Brazilian Amazon. *Remote Sensing Environ.*, 52, 137–154.

Aguiar, A.P.D., Shimabukuro, Y.E., and Mascarenhas, N.D.A., 1999. Use of synthetic bands derived from mixing models in the multispectral classification of remote sensing images. *Int. J. Remote Sensing*, 20, 647–657.

Akbari, H., Shea Rose, L., and Haider, T., 2003. Analyzing the land cover of urban environment using high-resolution orthophotos. *Landscape Urban Plann.*, 63, 1–14.

Andrew, K., 1999. Quality of life in cities. *Cities*, 16, 221–222.

Asner, G.P. and Lobell, D.B., 2000. A biogeophysical approach for automated SWIR unmixing of soils and vegetation. *Remote Sensing Environ.*, 74, 99–112.

Bonaiuto, M., Fornara, F., and Bonnes, M., 2003. Indexes of perceived residential environment quality and neighborhood attachment in urban environment: a confirmation study on the city of Rome. *Landscape Urban Plann.*, 65, 41–52.

Burrough, P.A., 1986. *Principles of Geographical Information Systems for Land Resource Assessment.* Clarendon Press, New York.

Cheng, J. and Masser, I., 2003. Urban growth pattern modeling: a case study of Wuhan City, PR China. *Landscape Urban Plann.*, 62, 199–217.

Cochrane, M.A. and Souza, C.M., Jr., 1998. Linear mixture model classification of burned forests in the eastern Amazon. *Int. J. Remote Sensing*, 19, 3433–3440.

Comrey, A.L. and Lee, H.B., 1992. *A First Course in Factor Analysis*, 2nd ed. Erlbuum, Hillsdale, NJ.

Donnay, J.P., Barnsley, J.M., and Longley, A.P., 2001. Remote sensing and urban analysis. In J.-P. Donnay, M.J. Barnsley, and P.A. Longley (Eds.), *Remote Sensing and Urban Analysis.* Taylor & Francis Inc., New York.

Doucette, P., Agoouris, P., and Stefanidis, A., 2004. Automated road extraction from high-resolution multispecteal imagery. *Photogrammetric Eng. Remote Sensing*, 70, 1405–1416.

Dwyer, J.F., Macpherson, G.E., Schroeder, H.W., and Rowntree, R.A., 1992. Assessing the benefits and costs of the urban forest. *J. Arboric.*, 18, 227–234.

Ehlers, M., Jadkowski, M.A., Howard, R.R., and Brostuen, D.E., 1990. Application of SPOT data for regional growth analysis and local planning. *Photogrammetric Eng. Remote Sensing*, 56, 175–180.

Foo, T.S., 2001. Quality of life in cities. *Cities*, 18, 1–2.

Forster, B., 1985. An examination of some problems and solutions in monitoring urban areas from satellite platforms. *Int. J. Remote Sensing*, 6, 139–151.

Frey, H.W., 2001. Melting pot suburbs: a census 2000 study of suburban diversity. The Brookings Institution, Census 2000 Series, June.

Fung, T., 2000. Environmental quality and its changes, an analysis using NDVI. *Int. J. Remote Sensing*, 21, 1011–1024.

Green, N.E., 1957. Aerial photographic interpretation and the social structure of the city. *Photogrammetric Eng.*, 23, 89–96.

Green, N.E. and Monier, R.B., 1957. Aerial photographic interpretation and the human geography of the city. *Prof. Geogr.*, 9, 2–5.

Haack, B., Bryant, N., and Adams, S., 1987. Assessment of Landsat MSS and TM data for urban and near-urban land-cover digital classification. *Remote Sensing Environ.*, 21, 201–213.

Hall, G.B., Malcolm, N.W., and Piwowar, J.M., 2001. Integration of remote sensing and GIS to detect pockets of urban poverty: the case of Rosario, Argentina. *Trans. GIS*, 5, 235–253.

Harris, P.M. and Ventura, S.J., 1995. The integration of geographic data with remote sensed imagery to improve classification in an urban area. *Photogrammetric Eng. Remote Sensing*, 61, 993–998.

Harris, R.J. and Longley, P.A., 2000. New data and approaches for urban analysis: modeling residential densities. *Trans. GIS*, 4, 217–234.

Harvey, J., 2000. Small area population estimation using satellite imagery. *Trans. GIS*, 4, 611–633.

Harvey, J., 2002. Estimation census district population from satellite imagery: some approaches and limitations. *Int. J. Remote Sensing*, 23, 2071–2095.

Harvey, W., McGlone, J.C., McKeown, D.M., and Irvine, J.M., 2004. User-centric evaluation of semiautomated road network extraction. *Photogrammetric Eng. Remote Sensing*, 70, 1353–1364.

Huang, T. and Yasuoka, Y., 2000. Integration and application of socioeconomic and environmental data within GIS for development study in Thailand. http://www.gisdevelopment.net/aars/acrs/2000/ts7/gdi004pf.htm.

Kamp, I.V., Leidelmeijer, K., Marsman, G., and Hollander A.U., 2003. Urban environmental quality and human well-being towards a conceptual framework and demarcation of concepts; a literature study. *Landscape Urban Plann.*, 65, 5–18.

Kim, T., Park, S., Kim, M., Jeong, S., and Kim, K., 2004. Tracking road centerlines from high-resolution remote sensing images by least-squares correlation matching. *Photogrammetric Eng. Remote Sensing*, 70, 1417–1422.

Lacoste, C., Descombes, X., and Zerubia., J., 2002. A comparative study of point processes for line network extraction in remote sensing. Research Report, 4516, INRIA, France.

Langford, M., Maguire, D.J., and Unwin, D.J., 1991. The areal interpolation problem: estimating population using remote sensing in a GIS framework. In: L. Masser and M. Blakemore (Eds.), *Handling Geographical Information: Methodology and Potential Applications*. Longman Scientific & Technical, co-published in the United States with John Wiley & Sons, Inc., New York.

Li, G. and Weng, Q., 2005. Using Landsat ETM+ imagery to measure population density in Indianapolis, Indiana, USA. *Photogrammetric Eng. Remote Sensing*, 71, 947–958.

Liang, B., 2005. A multi-scale analysis of census based land surface temperature variation and determinants in Indianapolis, Indiana, master's thesis, Indiana State University.

Lo, C.P., 1995. Automated population and dwelling unit estimation from high-resolution satellite images: a GIS approach. *Int. J. Remote Sensing*, 16, 17–34.

Lo, C.P. and Faber, B.J., 1997. Integration of Landsat Thematic Mapper and census data for quality of life assessment. *Remote Sensing Environ.*, 62, 143–157.

Lu, D., Moran, E., and Batistella, M., 2003. Linear mixture model applied to Amazonian vegetation classification. *Remote Sensing Environ.*, 87, 456–469.

Lu, D. and Weng, Q., 2004. Spectral mixture analysis of the urban landscapes in Indianapolis City with Landsat ETM+ imagery. *Photogrammetric Eng. Remote Sensing*, 70, 1053–1062.

Martin, D., Tate, N.J., and Langford M., 2000. Refining population surface models: experiments with Northern Ireland census data. *Trans. GIS*, 4, 343–360.

McGwire, K., Minor, T., and Fenstermaker, L., 2000. Hyperspectral mixture modeling for quantifying sparse vegetation cover in arid environments. *Remote Sensing Environ.*, 72, 360–374.

McPherson, E.G., 1992. Accounting for benefits and cost of urban greenspace. *Landscape Urban Plann.*, 22, 41–51.

McPherson, E.G., 1994. Cooling urban heat islands with sustainable landscape. In: Platt, R.H., Rowntree, R.A., and Muick, P.C. (Eds.), *The Ecological City: Preserving and Restoring Urban Biodiversity*, University of Massachusetts Press, Amherst, 151–171.

Mesev, V., 1998. The use of census data in urban image classification. *Photogrammetric Eng. Remote Sensing*, 64, 431–438.

Metivier, E.D. and McCoy, R.M., 1971. Mapping urban poverty housing from aerial photographs. *Proc. 7th Int. Symp. Remote Sensing Environ.*, Ann Arbor, MI, pp.1563–1569.

Mumbower, L.E. and Donoghue, J., 1967. Urban poverty study. *Photogrammetric Eng.*, 33, 610–618.

Nichol, J. and Wong, M.S., 2005. Modeling urban environmental quality in a tropical city. *Landscape Urban Plann.*, 73, 49–58.

Pacione, M., 2003. Urban environmental quality and human wellbeing — a social geographical perspective. *Landscape Urban Plann.*, 65, 19–30.

Pathan, S.K., Sastry, S.V.C., Dhinwa, P.S., Mukund Rao, and Majumdar, K.L., 1993. Urban growth trend analysis using GIS techniques — a case study of the Bombay metropolitan region. *Int. J. Remote Sensing*, 14, 3169–3179.

Phinn, S., Stanford, M., Scarth, P., Murray, A.T., and Shyy, P.T., 2002. Monitoring the composition of urban environments based on the vegetation–impervious surface–soil (VIS) model by subpixel analysis techniques. *Int. J. Remote Sensing*, 23, 4131–4153.

Population Reference Bureau, 2005. http://www.prb.org/Content/NavigationMenu/PRB/Educators/Human_Population/Urbanization2/Patterns_of_World_Urbanization1.htm.

Qiu, F., Woller, K.L., and Briggs, R., 2003. Modeling urban population growth from remotely sensed imagery and TIGER GIS road data. *Photogrammetric Eng. Remote Sensing*, 69, 1031–1042.

Rashed, T., Weeks, J.R., Gadalla, M.S., and Hill, A.G., 2001. Revealing the anatomy of cities through spectral mixture analysis of multispectral satellite imagery: a case study of the Greater Cairo region, Egypt. *Geocarto Int.*, 16, 5–15.

Roberts, D.A., Batista, G.T., Pereira, J.L.G., Waller, E.K., and Nelson, B.W., 1998. Change identification using multitemporal spectral mixture analysis: applications in eastern Amazonia. In: R.S. Lunetta and C.D. Elvidge (Eds.), *Remote Sensing Change Detection: Environmental Monitoring Methods and Applications*, Ann Arbor Press, Ann Arbor, MI, 137–161.

Small, C., 2001. Estimation of urban vegetation abundance by spectral mixture analysis. *Int. J. Remote Sensing*, 22, 1305–1334.

Smith, D.M., 1973. *The Geography of Social Well-Being in the United States*. McGraw–Hill, New York.

Smith, M.O., Ustin, S.L., Adams, J.B., and Gillespie, A.R., 1990. Vegetation in deserts: I. A regional measure of abundance from multispectral images. *Remote Sensing Environ.*, 31, 1–26.

Song, M. and Civco, D., 2004. Road extraction using SVM and image segmentation. *Photogrammetric Eng. Remote Sensing*, 70, 1365–1372.

Sutton, P., 1997. Modeling population density with nighttime satellite imagery and GIS. *Computer, Environ. Urban Syst.*, 21, 227–244.

Tabachnick, B. and Fidell, L., 1996. *Using Multivariate Statistics*, 3rd ed. Harper Collins College Publishers, New York.

Thomson, C.N., 2000. Remote sensing/GIS integration to identify potential low-income housing sites. *Cities*, 17, 97–109.

Treitz, P.M., Howard, P.J., and Gong, P., 1992. Application of satellite and GIS technologies for land-cover and land-use mapping at the rural-urban fringe: a case study. *Photogrammetric Eng. Remote Sensing*, 58, 439–448.

Weber, C. and Hirsch, J., 1992. Some urban measurements from SPOT data: urban life quality indices. *Int. J. Remote Sensing*, 13, 3251–3261.

Welch, R., 1982. Spatial resolution requirements for urban studies. *Int. J. Remote Sensing*, 3, 139–146.

Weng, Q. 2002. Land use change analysis in the Zhujiang Delta of China using satellite remote sensing, GIS, and stochastic modeling. *J. Environ. Manage.*, 64, 273–284.

Weng, Q., Lu, D., and Schubring, J., 2004. Estimation of land surface temperature–vegetation abundance relationship for urban heat island studies. *Remote Sensing Environ.*, 89, 467–483.

Wilkinson, G.G., 1996. A review of current issues in the integration of GIS and remote sensing data. *Int. J. Geogr. Inf. Sci.*, 10, 85–101.

Wu, C. and Murray, A.T., 2003. Estimating impervious surface distribution by spectral mixture analysis. *Remote Sensing Environ.*, 84, 493–505.

Yeh, A.G.O. and Li, X., 1997. An integrated remote sensing and GIS approach in the monitoring and evaluation of rapid urban growth for sustainable development in the Pearl River Delta, China. *Int. Plann. Stud.*, 2, 193–210.

Yuan, Y., Smith, R.M., and Limp, W.F., 1997. Remodeling census population with spatial information from Landsat imagery. *Computer, Environ. Urban Syst.*, 21, 245–258.

Part V

Progress, Problems, and Prospects

16 Mapping Human Settlements Using the Middle Infrared (3–5 μm): Advantages, Prospects, and Limitations

Geoffrey M. Henebry

CONTENTS

16.1   INTRODUCTION

Characterization of urban environments using remote sensing has relied primarily on three spectral regions: visible through near infrared, thermal infrared, and microwave. This chapter explores the potential use of an unfamiliar spectral region for characterization of human settlements, particularly urban and suburban environments.

Despite the presence of myriad atmospheric windows between 3 and 5 μm (Meier et al. 1997, 2000), relatively little terrestrial remote sensing work has exploited this middle infrared (MIR) spectral region, other than applications for hotspot detection, such as fire and volcano monitoring (Cracknell 1997; Roy et al. 1999). This spectral region does, however, have significant utility for atmospheric remote sensing and meteorological monitoring (Cracknell 1997). The Geostationary Operational Environmental Satellite (GOES) spacecraft have carried an MIR imaging capability since the launch of GOES-8(I) in April 1994. Channel 2 of GOES provides spectral coverage from 3.78 to 4.03 μm with a nominal 4-km spatial

resolution (CIRA 2005). For many years, the AVHRR (advanced very high resolution radiometer) has been the principal sensor with MIR imaging capability in
a polar orbit. Despite its moderate spatial resolution — 1.1 km at nadir — and
chronic problems with image noise (Dudhia 1989; Warren 1989; Simpson and
Yhann 1994), some promising applications for terrestrial remote sensing have
emerged.

Tucker and coworkers (1984) discovered that AVHRR channel 3 (3.55 to 3.93 μm)
was effective in detecting deforestation in the Brazilian Amazon. Kerber and Schutt
(1986) explored the potential of AVHRR channel 3 data for land-cover mapping.
They noted distinct urban and rural contrasts in analysis of a single summer scene
of the Chesapeake Bay region. In contrast to the thermal channels, the MIR channel
exhibited greater within-scene variability, thus increasing discriminatory power
for classification. They concluded that most of the variation was due to differential
emittance, not reflectance, based on analysis of the differential thermal response
of AVHRR's middle and thermal IR channels. MIR has been shown to be effective
for detecting biomass burning (Matson and Holben 1987; Pereira 1998; Roy et al.
1999), for monitoring forest structure (Malingreau et al. 1989; Kaufman and Remer
1994; LaPorte et al. 1995; Foody et al. 1996; DeFries et al. 1997; Boyd et al.
1999, 2000a, b; Phipps and Boyd 2000), and for retrieving forest biogeophysical
data through inversion of semiempirical models (Boyd et al. 1999; Weimann and
Boyd 1999).

Kaufman and Remer (1994) demonstrated that MIR could be useful for
studying what they termed "dense, dark vegetation" and promoted the use of a
vegetation index, VI3, similar to NDVI with red reflectance substituted by mid-IR
reflectance. They argued that red and MIR reflectances are correlated due to
similar "surface-darkening processes" related to absorption by liquid water and
that MIR is better able to penetrate anthropogenic haze and smoke (particle radii =
0.1 to 0.4 nm), although worse for dust (particle radii = 1000 to 2000 nm). They
further suggested an emissitivity correction due to latitudinal and altitudinal
variations in emittance.

The daytime signal mixing of solar reflectance and terrestrial emission has been
suggested as a principal reason for the scant record of MIR applications (Cracknell 1997).
Yet, Boyd and Curran (1998) argued that exploiting this portion of the electromagnetic spectrum may significantly improve inventories of aboveground carbon stocks,
especially in forests. Researchers have sought to retrieve the reflectance component
by subtracting the emission component (Kaufman and Remer 1994; Roger and
Vermote 1998; Boyd et al. 2000a, b; Phipps and Boyd 2000; Petitcolin and Vermote
2002). Recently, Boyd and Petitcolin (2004) provided a useful review of the progress
made toward exploiting the MIR for terrestrial remote sensing, particularly of land-
cover and land-use change studies in forested ecosystems. The utility of the MIR
for terrestrial remote sensing, however, is not limited to the information in the solar
signal.

Strong spectral contrasts exist in the 3- to 5-μm region between the reflectance
of active vegetation on the one hand and reflectance of senesced vegetation, soils,
rocks, minerals, and anthropogenic surfaces on the other (Salisbury and D'Aria 1994;
Figure 16.1; Table 16.1).

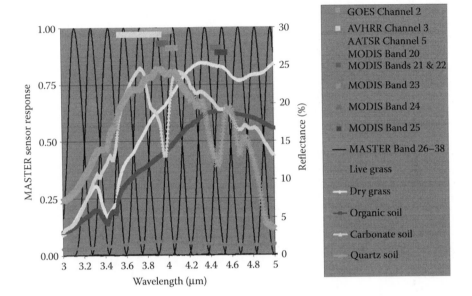

FIGURE 16.1 Spectral coverage of current MIR sensors and representative MIR spectra of natural materials. (From the ASTER Spectral Library, 1998, http://speclib.jpl.nasa.gov. California Institute of Technology. Accessed: 04 Nov 2005.)

Spectral contrasts in the MIR may offer novel ways to estimate vegetation fraction and impervious surface fraction. These contrasts offer several improvements over vegetation index approaches. First, the MIR loses sensitivity at much higher levels of leaf area index (LAI) than vegetation indices, such as NDVI (but see Gitelson, 2004, and Viña et al., 2004). Second, plant canopies exhibit negligible anisotropy in the MIR. Third, the strong spectral contrast between live plant tissue, on the one hand, and soils, rocks, and building materials, on the other, suggests the feasibility of linear mixing models. Furthermore, detection of various types of surface perturbations may be possible using MIR spectral information, including: (1) removal or reduction of green biomass in forests, shrublands, grasslands, croplands, and suburban and urban settings; (2) soil disturbance: subsurface soils exposed or brought to surface during excavation; (3) disturbance or accumulation of dried plant materials at the surface; and (4) change in surfaces of the built environment: concrete, asphalt, roofing materials, paints.

16.2 TERRESTRIAL BIOGEOPHYSICS OF 3- TO 5-μM RADIATION

What follows is a concise survey of the key aspects of MIR biogeophysics relevant to terrestrial remote sensing. Laboratory studies of MIR spectral reflectance reveal consistent patterns among natural materials commonly encountered (Table 16.1).

TABLE 16.1
Categories of Broadband Spectral Contrasts in 3.55 to 3.93 μm (AVHRR Channel 3 and MODIS Band 20)

		Soils	General Construction Materials[a]	Dried Grass[a]	Rocks	Plant Litter	Roofing Materials[a]	Paving Materials[a]	Plants	Deciduous[a]	Graminoids[a]	Conifers[a]
Soils	0.45	1	1.02	1.14	1.29	1.40	1.57	1.71	2.28	3.42	3.48	4.07
General[a]	0.44	0.98	1	1.12	1.26	1.36	1.53	1.67	2.23	3.34	3.40	3.98
Dried grass[a]	0.39	0.87	0.90	1	1.13	1.22	1.37	1.49	2.00	2.99	3.05	3.56
Rocks	0.34	0.77	0.79	0.89	1	1.08	1.21	1.32	1.77	2.65	2.70	3.15
Plant litter	0.32	0.72	0.73	0.82	0.93	1	1.12	1.22	1.63	2.45	2.50	2.92
Roofing[a]	0.28	0.64	0.65	0.73	0.82	0.89	1	1.09	1.46	2.19	2.23	2.60
Paving[a]	0.26	0.59	0.60	0.67	0.76	0.82	0.92	1	1.34	2.01	2.04	2.39
Plants	0.20	0.44	0.45	0.50	0.57	0.61	0.69	0.75	1	1.50	1.53	1.79
Deciduous[a]	0.13	0.29	0.30	0.33	0.38	0.41	0.46	0.50	0.67	1	1.02	1.19
Graminoids[a]	0.13	0.29	0.29	0.33	0.37	0.40	0.45	0.49	0.65	0.98	1	1.17
Conifers[a]	0.11	0.25	0.25	0.28	0.32	0.34	0.38	0.42	0.56	0.84	0.86	1

[a] From ASTER Spectral Library.

Source: Ratios of square-root-arcsine-transformed average percent reflectances from Salisbury, J.W. and D.M. D'Aria, 1994, *Remote Sensing Environ.*, 47:345–361, or ASTER Spectral Library.

Green leaves exhibit very low reflectance — 2 to 5% — comparable to that of liquid water, and plant canopies are likely to exhibit even lower reflectance due to multiple scattering (Salisbury and D'Aria 1994; Figure 16.1). MIR is better able to penetrate anthropogenic haze and smoke (particle radii = 0.1 to 0.4 nm) than red wavelengths, although it is worse for dust (particle radii = 1000 to 2000 nm) (Kaufman and Remer 1994). Soils are MIR bright, reflecting up to one third of the incident radiation (Salisbury and D'Aria 1994; ASTER Spectral Library 1998; Figure 16.2). Rocks are generally less reflective than soils but exhibit more definite spectral signatures (Salisbury and D'Aria 1994; ASTER Spectral Library 1998). Cellulose, lignin, hemicelluloses, and other structural plant molecules exhibit enhanced reflectivity between 3.6 and 4.3 µm (Elvidge 1988; Salisbury and D'Aria 1994), but less than soils (Salisbury and D'Aria 1994; Snyder et al. 1997; Figure 16.1). View angle dependence of emissivity is greater in the MIR than thermal IR (Labed and Stoll 1991), but there is little change in the spectral features (Snyder et al. 1997). View angle dependence becomes an issue when the scene is composed of objects of sharply contrasting reflectance and emission characteristics (Smith et al. 1997). Changes in spectral emissivity across 3 to 5 µm are greater than the sensitivity to temperature changes (Wan and Dozier 1996).

Finally, MIR spectral contrast between vegetation and soils is a seasonal phenomenon due to canopy development. At the beginning of the growing season in

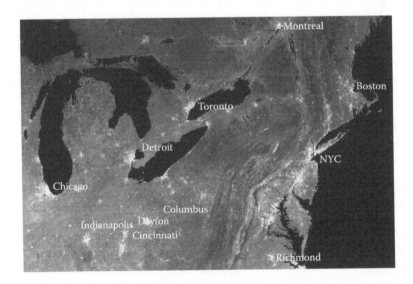

FIGURE 16.2 AVHRR channel 3 average brightness temperature following canopy closure in 1997. Bright patches correlate with high concentrations of anthropogenic building materials. Bright lines correlate with highway rights-of-way. Increasingly darker gray tones correlate with increasing canopy density. Water bodies have been masked to black. Note the transition to forests in northern Wisconsin, the expanse of Adirondack State Park in northern New York, and the pervasive pattern of settlements from Virginia to Boston and on across the upper Midwest.

temperate humid to semiarid climates, there is a low green vegetation fraction amid much exposed soil and/or senesced plant litter. These MIR-bright surfaces are gradually obscured as vegetation canopies develop through the growing season. Canopy closure may occur slowly or rapidly, but as the green vegetation fraction approaches 100%, a significant change is observed in the MIR reflectance + emittance signal, analogous to the spectral phase transition observed as an increasingly off-nadir view loses sight of the soil background (Snyder and Wan 1998). Near-peak cover of the spectral MIR contrast across the landscape can be striking and reminiscent of the "nighttime lights" maps, except with a finer spatial articulation, especially along transportation corridors (Figure 16.2).

For the purposes of spectral contrast, the mixing of solar reflectance and terrestrial emittance can be considered a benefit rather than a problem because it stabilizes the contrast signal for several reasons. First, there is apparently very low (<1%) variation in MIR solar irradiance despite significant periodic variation in other regions of the solar spectrum (Lean 1991). Second, the differential partitioning of the surface energy budget between latent and sensible heat fluxes on vegetated vs. bare soil surfaces leads to strong thermal contrasts that reinforce the MIR spectral contrast. (Lambin and Ehrlich, 1996, effectively exploited a thermal contrast for monitoring and classification using a vegetation index–surface temperature (VI-ST) change space.) Third, the variation in spectral emissivity due to material differences is greater in the MIR than temperature dependence (Wan and Dozier 1996), suggesting that even subpixel surface alterations may be detectable, if they result in significantly different proportions of component materials. Fourth, the thickness of the MIR thermal skin is only a few millimeters (Wan and Dozier 1996).

Across materials, the strongest broadband MIR contrast is between green vegetation and soils, with dried plant materials, rocks, and anthropogenic surfaces constituting intermediate contrasts, as seen in Table 16.1. Reasoning from this table, we expect to see strong MIR spectral contrasts between urban and suburban areas and their rural vegetated context at the regional scale and, at local scale, between parks, recreational areas, cemeteries, and green spaces and their urban matrix of impervious surfaces.

16.3 KEY SPECTRAL FEATURES IN THE MIR

The ASTER Spectral Library offers high-resolution spectra for many materials, including rocks, minerals, soils, building materials, vegetation, water in several forms, and even materials of extraterrestrial origin (ASTER Spectral Library 1998). Of particular interest here are the MIR spectra of building materials common in human settlements, soils, and vegetation. There are five key spectral features relevant to the remote sensing of human settlements (Figure 16.1). The MIR reflectance of green vegetation is very low and spectrally featureless, thus providing an important element of spectral contrast within MIR imagery (Table 16.1; Figure 16.2). Dried plant materials exhibit two MIR spectral features: distinct cellulose absorption features near 3.4 µm and significantly higher reflectance near 4.4 µm than below 4.0 µm.

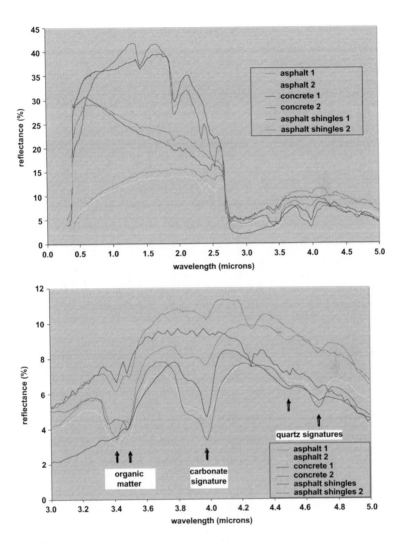

FIGURE 16.3 Common urban surface materials that exhibit very different spectral responses in the VNIR have more comparable responses in the MIR (upper). Zooming into the MIR reveals multiple distinct absorption features (lower). (Data from ASTER Spectral Library, 1998, http://speclib.jpl.nasa.gov. California Institute of Technology. Accessed: 04 Nov 2005.)

Carbonate soils and surfaces exhibit a strong, distinct absorption feature near 4.0 μm. Soils and surfaces containing significant amounts of quartz display an absorption feature near 4.5 μm. Materials that are mixtures can exhibit MIR reflectances that display one or more of these key features (Figure 16.1 and Figure 16.3). This spectral information can be retained even across broad MIR bands, as evidenced by reflectance range bracketing the average MIR reflectances for building materials in Figure 16.4 (see Appendix 16.1).

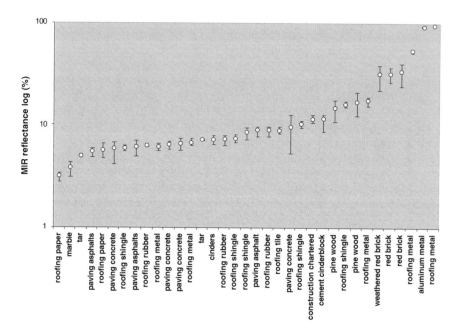

FIGURE 16.4 Average MIR reflectance of the 35 building materials in the ASTER Spectral Library calculated for MODIS bands 20 to 25. Bars indicate maximum and minimum values across MODIS bands. Note the log scale for percent reflectance.

16.4 SOURCES OF MIR IMAGERY

In addition to channel 3 on AVHRR and channel 2 on GOES, MIR imagery is available from the MODIS sensors on the Terra and Aqua platforms. Although MODIS offers five MIR channels at a nominal spatial resolution of 1 km, there are only four distinct MIR bands: 21 and 22 share the 3.929- to 3.989-μm window, but band 21 is designated for hotspot detection. All the MIR bands on MODIS Terra have problems with electronic cross-talk and bands 21 and 24 also have striping (MCST 2005a). The MIR bands on MODIS Aqua are in better shape, although the calibration for band 21 is incomplete (MCST 2005b). The channel 5 of the AATSR on board the European Space Agency's ENVISAT has a 1-km MIR band centered at 3.75 μm to enable continuity with AVHRR data (ESA 2005).

Two airborne MODIS simulators, MASTER (http://masterweb.jpl.nasa.gov/; Hook et al. 2001) and MAS (http://mas.arc.nasa.gov; King et al. 1996), offer higher spectral resolution in the MIR at finer spatial resolution than MODIS: 13 bands at 50 m vs. 4 + 1 bands at 1000 m. Data from these sensors are available only for select areas in conjunction with NASA field campaigns and calibration/validation exercises associated with MODIS and ASTER; however, there are multiple acquisitions of urban areas. These sensors currently offer the greatest opportunity to explore the potential of MIR imagery for urban remote sensing. Although the finer spatial resolution will enhance the contrast between surfaces, view angle dependencies, including terrain effects, may become more important as well (Smith et al. 1997).

APPENDIX 16.1
Thirty-Five Samples of Building Materials Ordered by Increasing Average Reflectance across MODIS MIR Bands (Compare Figure 16.4), with Descriptions from the ASTER Spectral Library

Rank	Reflectance (%)	Sample Description
1	3.2	Roofing paper, extremely weathered black paper, thickly tarred and deeply cracked
2	3.8	Marble, white, weathered construction marble from beach sidewalk in Naxos, Greece
3	5.0	Tar, black matte, weathered
4	5.5	Paving asphalts, weathered runway, black with little aggregate showing
5	5.7	Roofing paper, unweathered black, thickly tarred
6	5.9	Paving concrete, quartz and limestone aggregate showing, soiled, runway side strips
7	6.0	Roofing shingle, weathered, reddish, granule covered
8	6.1	Paving asphalts, soiled and weathered, some limestone and quartz aggregate showing
9	6.2	Roofing rubber, black and weathered
10	6.2	Roofing metal, weathered, galvanized steel from roof and vent covers
11	6.4	Paving concrete, quartz aggregate showing, smooth and clean
12	6.5	Paving concrete, weathered runway
13	7.0	Roofing metal, slightly weathered bare copper flashing
14	7.2	Tar, black and glossy
15	7.3	Cinders, black, ashen from railroad bed
16	7.3	Roofing rubber, 5-year-old weathered, white, rubberized coating
17	7.4	Roofing shingle, unweathered, reddish, granule-covered asphalt shingle
18	8.5	Roofing shingle, white granule-covered asphalt shingle, white sand over construction asphalt
19	8.9	Paving asphalts, soiled and weathered, some limestone and quartz aggregate showing
20	8.9	Roofing rubber, white, weathered, fiberglass reinforced
21	9.1	Roofing tile, weathered terra cotta ceramic composite, orange
22	9.5	Paving concrete, asphaltic, mostly limestone and some quartz showing, rough and black, runway side strips
23	10.2	Roofing shingle, 15-year-old roofing shingle, weathered, soiled, gray, granule covered
24	11.6	Construction concrete, 30-year-old runway, granite aggregate showing, light gray and weathered
25	11.6	Cement cinderblock, light brown
26	14.7	Pine wood, sanded and buffed
27	16.0	Roofing shingle, weathered gray slate
28	17.1	Pine wood, bare and weathered
29	17.6	Roofing metal, extremely weathered bare copper flashing
30	31.8	Weathered red brick, smooth
31	32.0	Red brick, smooth faced
32	33.6	Red brick, smooth, bare
33	53.7	Roofing metal, weathered galvanized steel from roof and vent covers
34	92.2	Aluminum metal, bare, weathered, scored
35	93.8	Roofing metal, 10-year-old weathered, bare copper flashing

16.5 EXPLOITING KEY SPECTRAL FEATURES
WITH MIR INDICES

An important question to address is whether the information content in the MIR is different from that in the more readily available thermal infrared (TIR) bands. Figure 16.5 clearly demonstrates that the answer is yes: there is different spectral information in the MIR. The MASTER scene depicts a complex urban environment in early summer. The upper left panel provides the familiar false color composite in which dense green vegetation appears in saturated reds. The upper right panel shows a false color composite of three thermal bands. The uniformity of the longwave

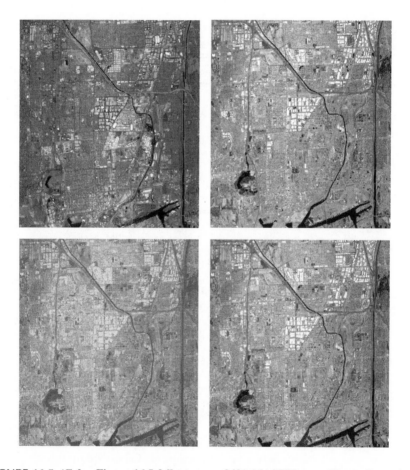

FIGURE 16.5 (Color Figure 16.5 follows page 240.) MASTER upwelling radiance from Los Angeles, California, on 06 June 2000: (upper left) reflected shortwave VNIR (R = 0.910, G = 0.658, B = 0.542 μm); (upper right) emitted longwave TIR (R = 10.6, G = 11.2, B = 12.1 μm); (lower left) reflected + emitted MIR 3 to 5 μm. (R = avg, G = stdev, B = skewness); (lower right) MIR + TIR (R = avg MIR, G = 11.2, B = 12.1 μm). The lower panels demonstrate that, within this scene, there is more information content in the MIR than in the TIR.

TABLE 16.2
Candidate MIR Indices for Use with MODIS MIR Bands

MODIS MIR Index	Simple Normalized Difference
Organic matter index (OMI)	(B25 – B20)/(B25 + B20)
Carbonate surface index (CSI)	(B23 – B22)/(B23 + B22)
Quartz surface index (QSI)	(B23 – B24)/(B23 + B24)

emittance field makes this false color composite appear nearly grayscale. The lower left panel provides a statistical summary of the information in the MIR bands. Red encodes the average MIR radiance and green and blue map, respectively, the standard deviation, and skewness of the MIR radiance across the MASTER MIR bands. This quick dimension reduction reveals significant spatial heterogeneity in MIR radiance. The lower right panel combines the average MIR radiance in red with radiances from two TIR bands in the green and blue. Were the MIR spectral information to be redundant, then the lower right panel would appear nearly identical to the upper right panel.

Given the relatively low flux of MIR photons, view angle dependencies, atmospheric effects, and differences among sensors, there is a need to develop MIR indices that target key spectral features to enhance object discrimination and classification. Table 16.2 describes three spectral indices based on the MODIS MIR bands that target surfaces with high levels of apparent high organic matter content in the form of dried vegetation and/or soils (OMI), apparent carbonate (CSI), and apparent quartz (QSI). These MODIS MIR indices are in the familiar form of the normalized difference: the difference in a measured quantity (typically radiance or reflectance) between two spectral bands divided by their sum.

How do these indices behave? Figure 16.6 displays the coverage of building materials in the metric spaces formed by pairs of MODIS MIR indices. Several features are noteworthy. First, the significant negative correlation between OMI and QSI (Pearson's $r = -0.84$, $p < 0.05$) is not an index artifact — each index uses a different pair of MODIS MIR bands. Furthermore, more than 10% of the samples fall significantly off the trend: all are weathered pavements. Second, the joint distributions of CSI × QSI and OMI × CSI lie almost entirely in the positive half of the CSI space. Third, many samples lie along the axes, indicating almost no contribution from one MIR index. Fourth, the SW quadrant is nearly empty, suggesting that the MIR indices are doing a good job of characterizing the building materials.

These initial results are promising, but they are based on high-resolution spectra acquired in a laboratory setting. To evaluate the robustness and sensitivity of the indices, it is necessary to summarize "populations" of spectra — whether gathered in the lab or in the field — using dimension reduction approaches such as principal components analysis and/or projection pursuit. Moreover, the limited number of MODIS bands (four) offers relatively little flexibility for index construction.

When more than one spectral contrast is associated with a key feature, such as MIR absorption by carbonate bonds, and sufficiently fine spectral resolution is

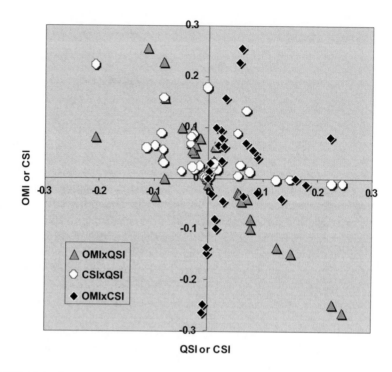

FIGURE 16.6 Three joint distributions of MODIS MIR index pairs for the 35 building materials in the ASTER Spectral Library.

available, it is possible to enhance the signal-to-noise ratio of individual normalized difference indices by summing them into a composite normalized difference (CND) index. The spectral resolution of the airborne sensors MAS/MASTER enables construction of CND indices that use three MIR bands to highlight a spectral feature (Table 16.3). It is not reliable to calculate indices on the basis of specific bands of these complex airborne sensors because band number assignments and band passes

---

TABLE 16.3

Candidate MIR Indices for Use with MAS/MASTER Bands Based on a July 1999 Calibration

MAS/MASTER MIR Index	Composite Normalized Difference[a]
Organic matter index (OMI)	$(R_{4.21} - R_{3.59})/(R_{4.21} + R_{3.59}) + (R_{4.52} - R_{3.59})/(R_{4.52} + R_{3.59})$
Carbonate surface index (CSI)	$(R_{3.75} - R_{3.9})/(R_{3.75} + R_{3.9}) + (R_{4.06} - R_{3.9})/(R_{4.06} + R_{3.9})$
Quartz surface index (QSI)	$(R_{3.9} - R_{4.99})/(R_{3.9} + R_{4.99}) + (R_{4.06} - R_{4.99})/(R_{4.06} + R_{4.99})$

[a] R_{λ} = upwelling radiance measured with band centered at wavelength λ in micrometers.

FIGURE 16.7 (Color Figure 16.7 follows page 240.) False color composites from a MASTER scene of Albuquerque, New Mexico, acquired June 3, 1999: (left) conventional RGB composite displaying near infrared/red/green bands; (right) the experimental composite normalized difference MIR indices displayed as red = organic matter index (OMI), green = quartz surface index (QSI), and blue = carbonate surface index (CSI). Magenta tones indicate high levels of dried plant materials at the surface. Green tones indicate concrete surfaces. Bright blue tones reveal exposed subsoils (caliche).

of the sensors are reconfigurable. Instead the composite normalized difference indices for MAS/MASTER data are calculated from the upwelling radiance values (R_λ = upwelling radiance at sensor band centered at wavelength λ in micrometers) and thus combine contributions from solar reflectance and terrestrial emittance. These MIR indices can yield interpretable images: Figure 16.7 offers a MASTER scene of an edge of Albuquerque, New Mexico, in early summer 1999. On the left is a conventional false color composite of the VNIR bands. On the right is a false color composite of three MIR CND indices based on the MASTER bands. A quick glance at this figure reveals an apparent anomaly: the same highly vegetated golf courses and parks that appear bright red on the left appear as intense magenta on the right. Why is this?

Given that green vegetation has very low MIR reflectance, it is not possible that surfaces with dense green biomass could yield high OMI values; however, there is more than simply green vegetation in these areas. Especially for closely cropped turfgrasses, a significant amount of senesced material (litter), soil background (O-horizon in particular), and nonphotosynthetic tissue is visible from a near-nadir view, and these materials are MIR bright. It is also notable that the OMI indicates significant quantities of biomass on the slope surfaces in the lower portion of the image. From the image alone it is not clear whether this is an artifact induced by topographic shading or a correct index response due to aspect-induced habitat differences or a combination of both. Further investigation would be needed to resolve this question.

There is little in the conventional image on the left to enable discrimination between high-albedo surfaces. On the right, however, the CSI and QSI appear to distinguish between some of these surfaces. Isolated bright green surfaces indicated high QSI values and likely correspond to roofs with intact aggregate and slight weathering. The bright bluish areas on the right correspond to high CSI values and likely indicate concrete pavement or areas of sufficient soil disturbance to expose the caliche layer.

16.6 CONCLUSIONS

The MIR (3 to 5 μm) spectral region has been little exploited in terrestrial remote sensing, but it may offer certain advantages for mapping human settlements, characterizing the urban to rural gradient, and monitoring built environments that complement the rich information available in more familiar spectral regions. Advantages include resistance to obscuring smoke and haze, spectral contrasts between MIR-dark vegetation and MIR-bright surfaces in the built environment that arise from reflectance and emittance, sensitivity to concrete due to carbonate absorption feature, and hotspot detection for urban fires and industrial effluents. Limitations stem from (1) the paucity of MIR photons in solar and terrestrial radiation fluxes, (2) challenging physical optics that lead to low signal-to-noise ratios, (3) obscuring atmospheric effects due to dust and some cloud types, and (4) soils possibly too bright in semiarid to arid environments to enable separation of human settlements from surfaces with sparse green vegetation.

What are the prospects for an expansion of terrestrial remote sensing using the MIR? A new AVHRR system was deployed with the launch of NOAA K in May 1998. AVHRR/3 has two spectral windows observed through channel 3. At night, acquisitions occur through channel 3B, which is in the heritage region of 3.55 to 3.93 μm, but during the day the system shifts to channel 3A in the region of 1.58 to 1.64 μm (Goodrum et al. 2000). Thus, there are no more daytime collections in the MIR from the POES platforms. The Terra and Aqua MODIS MIR bands offer the highest spectral resolution and spatial currently available from synoptic sensors. Two MIR channels (3.7 and 4.05 μm) will be available on the visible infrared imager/radiometer suite (VIIRS) on the National Polar-Orbiting Operational Environmental Satellite System (NPOESS).

It is clear that for the foreseeable future the development of an urban remote sensing capability in the MIR region will depend on the continued availability of the MAS and MASTER airborne sensors. At the same time, airborne sensors cannot be expected to collect image time series over large extents; thus, in the near time, the MODIS pair will remain the primary way in which the temporal development of MIR spectral contrast can be studied at regional to continental scales.

Finally, scientific investigation of MIR terrestrial phenomena requires the development of field-portable spectrometers to enable calibration and validation of the information retrieved using airborne and space-borne sensors. This is not a small step. Such products have been developed but the technical challenges are significant, the community experience is low, and the market is weak. Development of spectral mixing models for the MIR awaits the capability to build a library of spectral endmembers acquired in the field. In the interim, the MIR spectral indices offer a means to link data acquired from airborne and space-borne sensors with data gathered from laboratory or field spectrometers.

ACKNOWLEDGMENTS

This research was supported, in part, by the NASA LCLUC program and by SBIR DOE Contract # DE-FG02-00ER86109 to OptoMechanical Enterprises, Inc. Special thanks to Dr. Art Poulos of OME, Inc. Spectral data reproduced from the ASTER Spectral Library through the courtesy of the Jet Propulsion Laboratory, California Institute of Technology, Pasadena, California. Copyright 1999, California Institute of Technology. All rights reserved.

REFERENCES

ASTER Spectral Library Version 1.1 1998. http://speclib.jpl.nasa.gov. California Institute of Technology. Accessed: 04 Nov 2005.

Boyd, D.S., and P.J. Curran. 1998. Using remote sensing to reduce uncertainties in the global carbon budget: the potential of radiation acquired in the middle infrared wavelengths. *Remote Sensing Rev.* 16:293–327.

Boyd, D.S., G.M. Foody, and P.J. Curran. 1999a. The relationship between the biomass of Cameroonian tropical forests and radiation reflected in middle infrared wavelengths (3.0 to 5.0 μm). *Int. J. Remote Sensing* 20:1017–1023.

Boyd, D.S., T.E. Wicks, and P.J. Curran. 1999b. Estimation of LAI using middle infrared reflectance (3 to 5 μm). In *Earth Observation: From Data to Information, Proceedings of the 25th Annual Conference and Exhibition of the Remote Sensing Society*, P. Pan and M. Barnsley, Eds. pp. 591–598. The Remote Sensing Society: Nottingham, U.K.

Boyd, D.S., P.C. Phipps, G.M. Foody, and P.J. Curran. 2000a. Remote sensing biophysical indicators of ENSO-related drought stress in Sabah rainforests. In *Adding Value to Remotely Sensed Data, Proceedings of the 26th Annual Conference and Exhibition of the Remote Sensing Society*, P. Fisher, J. Wellens, and K. Moore, Eds. On CD-ROM. The Remote Sensing Society: Nottingham, U.K.

Boyd, D.S., T.E. Wicks, and P.J. Curran. 2000b. Use of middle infrared radiation to estimate the leaf area index of a boreal forest. *Tree Physiol.* 20:755–760.

Boyd, D.S. and F. Petitcolin. 2004. Remote sensing of the terrestrial environment using middle infrared radiation (3.0 to 5.0 μm). *Int. J. Remote Sensing* 25(17):3343–3368.

CIRA (Cooperative Institute for Research in the Atmosphere). 2005. GOES 3.9-μm channel tutorial http://rammb.cira.colostate.edu/training/goes_39um/default.asp. Accessed: 04 Nov 2005.

Cracknell, A.P. 1997. *The Advanced Very High Resolution Radiometer (AVHRR)*. Taylor & Francis: Bristol, PA.

DeFries R., M. Hansen, M. Steininger, R. Dubayah, R. Sohlberg, and J. Townshend. 1997. Subpixel forest cover in central Africa from multisensor, multitemporal data. *Remote Sensing Environ.* 60:228–246.

Dudhia, A. 1989. Noise characteristics of the AVHRR infrared channels. *Int. J. Remote Sensing* 10:637–644.

Elvidge, C.D. 1988. Thermal infrared reflectance of dry plant materials: 2.5 to 20 μm. *Remote Sensing Environ.* 26:265–285.

ESA (European Space Agency). 2005. The AATSR Products User Guide. http://envisat. esa.int/dataproducts/aatsr/. Accessed 09 Nov 2005.

Foody, G.M., D.S. Boyd, and P.J. Curran. 1996. Relations between tropical forest biophysical properties and data acquired in AVHRR channels 1–5. *Int. J. Remote Sensing* 17:1341–1355.

Gitelson, A.A. 2004. Wide dynamic range vegetation index for remote quantification of biophysical characteristics of vegetation. *J. Plant Physiol.* 161:165–173.

Goodrum, G., K.B. Kidwell, and W. Winston. 2000. NOAA KLM Users' Guide with NOAA-N, N- Supplement. http://www2.ncdc.noaa.gov/docs/klm/. Accessed 09 Nov 2005.

Hook, S.J., J.J. Myers, K.J. Thome, M. Fitzgerald, and A.B. Kahle. 2001. The MODIS/ASTER airborne simulator (MASTER) — a new instrument for earth science studies. *Remote Sensing Environ.* 76:93–102.

Kaufman, Y.J. and L.A. Remer. 1994. Detection of forest using mid-IR reflectance: an application for aerosol studies. *IEEE Trans. Geosci. Remote Sensing* 32:672–683.

Kerber, A.G. and J.B. Schutt. 1986. Utility of AVHRR channels 3 and 4 in land-cover mapping. *Photogrammetric Eng. Remote Sensing* 52:1877–1883.

King, M.D., W.P. Menzel, P.S. Grant, J.S. Myers, G.T. Arnold, S.E. Platnick, L.E. Gumley, S.C. Tsay, C.C. Moeller, M. Fitzgerald, K.S. Brown, and F.G. Osterwisch. 1996. Airborne scanning spectrometer for remote sensing of cloud, aerosol, water vapor and surface properties. *J. Atmos. Oceanic Technol.* 13:777–794.

Labed, J. and M.P. Stoll. 1991. Angular variation of land surface spectral emissivity in the thermal infrared: laboratory investigations on bare soils. *Int. J. Remote Sensing* 12:2299–2310.

Laporte, N., C. Justice, and J. Kendall. 1995. Mapping the dense humid forest of Cameroon and Zaire using AVHRR satellite data. *Int. J. Remote Sensing* 16:1127–1145.

Lean, J. 1991. Variations in the Sun's radiative output. *Rev. Geophys.* 29:505–535.

Malingreau, J.P., C. Tucker, and N. Laporte. 1989. AVHRR for monitoring global tropical deforestation. *Int. J. Remote Sensing* 10:855–867.

Matson, M. and B.N. Holben. 1987. Satellite detection of tropical burning in Brazil. *Int. J. Remote Sensing* 8:509–516.

MCST (MODIS Characterization Support Team). 2005a. Terra MODIS instrument performance history. http://www.mcst.ssai.biz/mcstweb/performance/ terra instrument. html. Accessed 09 Nov 2005.

MCST (MODIS Characterization Support Team). 2005b. Aqua MODIS instrument performance history. http://www.mcst.ssai.biz/mcstweb/performance/ aqua_instrument. html. Accessed 09 Nov 2005.

Meier, A., G.C. Toon, C.P. Rinsland, and A. Goldman. 1997. *Spectroscopic Atlas of Atmospheric Microwindows in the Middle Infrared.* IRF Preprint No. 123. Kiruna: Swedish Institute of Space Physics.

Meier, A., G.C. Toon, C.P. Rinsland, and A. Goldman. 2000. *Supplement to Spectroscopic Atlas of Atmospheric Microwindows in the Middle Infrared.* Supplement to IRF Preprint No. 123. Kiruna: Swedish Institute of Space Physics.

Pereira, J.M.C. 1998. A comparative evaluation of NOAA/AVHRR vegetation indices for burned surface detection and mapping. *IEEE Trans. Geosci. Remote Sensing* 37:217–226.

Phipps, P.C. and D.S. Boyd. 2000. The use of spatiotemporal variability in middle infrared reflectance as an indicator of moisture status in a coniferous forest ecosystem. In *Adding Value to Remotely Sensed Data, Proceedings of the 26th Annual Conference and Exhibition of the Remote Sensing Society,* P. Fisher, J. Wellens, and K. Moore, Eds. On CD-ROM. The Remote Sensing Society: Nottingham, U.K.

Roger, J.C. and E.F. Vermote. 1998. A method to retrieve the reflectivity signature at 3.75 μm from AVHRR data. *Remote Sensing Environ.* 64:103–114.

Roy, D.P., L. Giglio, J.D. Kendall, and C.O. Justice. 1999. Multi-temporal active-fire based scar detection algorithm. *Int. J. Remote Sensing* 20:1031–1038.

Salisbury, J.W. and D.M. D'Aria. 1994. Emissivity of terrestrial materials in the 3- to 5-μm atmospheric window. *Remote Sensing Environ.* 47:345–361.

Simpson, J.J. and S.R. Yhann. 1994. Reduction of noise in AVHRR channel 3 data with minimum distortion. *IEEE Trans. Geosci. Remote Sensing* 32:315–328.

Smith, J.A., N.S. Chauhan, T. Schmugge, and J.R. Ballard. 1997. Remote sensing of land surface temperature: the directional viewing effect. *IEEE Trans. Geosci. Remote Sensing* 35(4):972–974.

Snyder, W.C., Z. Wan, Y. Zhang, and Y.-Z. Feng. 1997. Thermal infrared (3 to 14 μm) bidirectional reflectance measurements of sands and soils. *Remote Sensing Environ.* 60:101–109.

Snyder, W.C. and Z. Wan. 1998. BRDF models to predict spectral reflectance and emissivity in the thermal infrared. *IEEE Trans. Geosci. Remote Sensing* 36:214–225.

Tucker, C.J., B.N. Holben, and T.E. Goff. 1984. Intensive forest clearing in Rondonia, Brazil, as detected by satellite remote sensing. *Remote Sensing Environ.* 15:255–261.

Viña, A., G.M. Henebry, and A.A. Gitelson. 2004. Satellite monitoring of vegetation dynamics: sensitivity enhancement by the wide dynamic range vegetation index. *Geophys. Res. Lett.* 31:L04503. doi:10.1029/2003GL019034.

Wan, Z. and J. Dozier. 1996. A generalized split-window algorithm for retrieving land-surface temperature from space. *IEEE Trans. Geosci. Remote Sensing* 34:892–905.

Warren, D. 1989. AVHRR channel-3 noise and methods for its removal. *Int. J. Remote Sensing* 10:645–651.

Weimann, A. and D.S. Boyd. 1999. Assessment of a vegetation canopy model of the interaction between middle infrared reflectance (3 to 5 μm) and leaf area. In *Earth Observation: From Data to Information, Proceedings of the 25th Annual Conference and Exhibition of the Remote Sensing Society*, P. Pan and M. Barnsley, Eds. pp. 357–364. The Remote Sensing Society: Nottingham, U.K.

17 New Developments and Trends for Urban Remote Sensing

Manfred Ehlers

CONTENTS

17.1 INTRODUCTION

Due to the high spatial resolution requirements for urban information systems, aerial photography has been used as standard imaging input. However, the advent of new satellites with a resolution of better than 1 m (e.g., IKONOS, Quickbird) and digital airborne scanners with excellent geometric fidelity and high spatial resolutions in the centimeter range (e.g., HRSC, ADS, DMC) challenges the analog airphoto techniques. These airborne and space-borne high-resolution sensors offer an advanced potential for generating and updating GIS databases, especially for urban areas. Moreover, digital airborne stereo sensors are capable of producing digital surface models (DSMs) by automated techniques. Coupled with differential GPS and inertial navigation systems (INS), these sensors generate georeferenced ultrahigh resolution multispectral image data together with their accompanying DSMs.

    Another source of three-dimensional information for city modeling is laser scanning (LIDAR) sensors that produce accurate height information of high and very high resolution. This chapter presents an overview of the new sensor developments and examples of integrated processing. Highest resolution for all imaging sensors is obtained in their panchromatic mode, whereas multispectral information is acquired at lower resolutions. The ratio between panchromatic and multispectral resolution is usually in the order of 1:2 to 1:8. Especially for vegetation analyses and urban sprawl quantification, multispectral information is a mandatory requirement. To obtain multispectral image data with high spatial resolution, multispectral and panchromatic images must be merged (pansharpened). Image transforms such as the intensity hue saturation (IHS) or principal component (PC) transforms are widely used to fuse panchromatic images of high spatial resolution with multispectral images of lower resolution. These techniques create multispectral images of higher spatial resolution but usually at the cost that these transforms do not preserve the original color or spectral characteristics of the input image data. This chapter presents a new method for image fusion based on the standard IHS transform combined with filtering in the Fourier domain. This method preserves the spectral characteristics of the lower resolution images. Using this fusion technique, multispectral images can be pansharpened without changing their spectral characteristics.

17.2 SATELLITE SENSORS

The advent of commercial very high resolution satellite programs has opened new application fields for space-based remote sensing. Satellite data offer, for the first time, the potential for large-scale applications such as urban planning and environmental monitoring at the highest level of detail (Ehlers et al., 2003a; Möller, 2003). Spatial resolutions of 0.60 to 1.00 m (panchromatic) and 2.5 to 4 m (multispectral) have begun to challenge aerial photography. Companies such as Digital Globe or Space Imaging promise extremely fast processing. Data should be delivered within days (or even hours if downloading via the Internet is possible). Tiltable cameras offer short revisit periods of 2 to 3 days and across-track as well as along-track stereo capabilities. Ikonos II was launched in September 1999, making it the first commercial very high resolution satellite in orbit (see Table 17.1). Figure 17.1 shows the level of detail provided by the Ikonos sensors and Figure 17.2 shows a pansharpened multispectral Quickbird image at 70-cm resolution. The potential of these sensors for urban application is demonstrated by individual buildings that are clearly discernible in the images.

17.3 DIGITAL AIRBORNE SENSORS

After a long period of development, we now see the emergence of operational digital camera systems that are replacing aerial frame cameras. Advanced technologies such as GPS coupled navigation systems and advanced digital sensor technologies have overcome the strongest impediment of aircraft scanners: the lack of geometric stability. Public and private research has concentrated on development of digital line

TABLE 17.1
Satellite Programs of Very High Resolution

Company	Space Imaging		Digital Globe		Orbimage		Imagesat
System	Ikonos II launch 9/99		QuickBird 2 launch 11/01		OrbView-3 launch 6/03		EROS A1 launch 12/00
URL	www.spaceimaging.com		www.digitalglobe.com		www.orbimage.com		www.imagesat.com
Modus	Pan 11 bit	Multispectral 11 bit	Pan 11 bit	Multispectral 11 bit	Pan 11 bit	Multispectral 11 bit	Pan 11 bit
Geometric resolution	1 m	4 m	0.61 m	2.44 m	1 m	4 m	1.8 m (hypersampling 1 m)
Spectral resolution (nm)	525–929	445–516 (b) 506–595 (g) 632–698 (r) 767–853 (nir)	450–900	450–520 (b) 520–600 (g) 630–690 (r) 760–900 (nir)	450–900	450–520 (b) 520–600 (g) 630–690 (r) 760–900 (nir)	500–900
Scale for applications			1:5,000–1:25,000				
Swath width	11 km		16.5 km		8 km		13.5 km
Image scene size	11 × 11 km²		16.5 × 16.5 km² Strip: 16.5 × 165 km²		8 × 8 km²		13.5 × 13.5 km² Vector scene: 13.5 × 40 km²
Orbit altitude	681 km		450 km		470 km		480 km
Inclination	98.1		97.2		97		97.3
	Sun synchronous		Sun synchronous		Sun synchronous		Sun synchronous

(continued)

TABLE 17.1 (CONTINUED)
Satellite Programs of Very High Resolution

Company	National Space Organization Taiwan		Indian Space Research Organization		CNES	
System	Formosat-2 launch 5/2004		Cartosat-1 launch 5/2005		SPOT-5 Launch 5/2002	
URL	www.nspo.org.tw/2005e		www.isro.org		www.spotimage.fr	
Modus	Pan	Multispectral	Pan fore 10 bit	Pan aft 10 bit	Pan 8 bit	Multispectral 8 bit
Geometric resolution	2 m	8 m	2.54 m	2.54 m	2.5 m (enhanced supermode) 5 m (normal mode)	10 m (swir 20 m)
Spectral resolution (nm)	450–900	450–520 (b) 520–600 (g) 630–690 (r) 760–900 (nir)	500–850	500–850	480–710	500–590 (g) 610–680 (r) 780–890 (nir) 1580–1750 swir
Scale for applications	1:12,000–1:30,000		1:12,500–1:25,000		1:20,000–1:50,000	
Swath width	24 km		29.4 km (fore) 26.2 km (aft)		60 km	
Image scene size	24 × 24 km²		25 × 25 km² or floating scenes		60 × 60 km²	
Orbit altitude	891 km		618 km		822 km	
Inclination	99.14		97.87		98.7	
	Sun synchronous		Sun synchronous		Sun synchronous	

Source: Based on Ehlers, M., in Ehlers, M., H.J. Kaufmann, and U. Michel (Eds.) *Remote Sensing for Environmental Monitoring, GIS Applications, and Geology III, Proc. SPIE,* 5239, Bellingham, WA: 1–13 2004.

In January 2006 Orbimage Acquired Space Imaging to form the company Geo Eye.

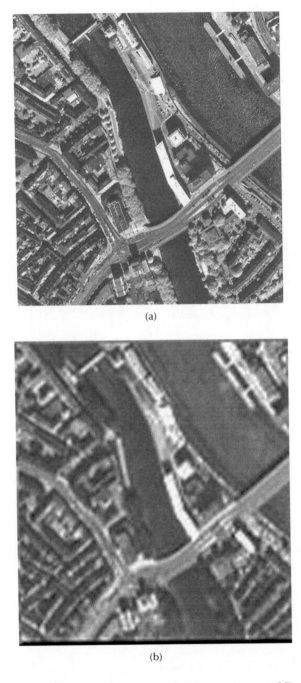

(a)

(b)

FIGURE 17.1 (Color Figure 17.1 follows page 240.) Ikonos image of Bremen, Germany. The panchromatic resolution of 1 m offers suitability for many urban applications (a). The multispectral bands allow true-color (b) and false-color (c) infrared information at a resolution of 4 m. (Images courtesy of OHB Systems, Bremen, Germany.)

(c)

FIGURE 17.1 (Continued).

array or matrix scanners that actually serve as successors to the "classical" air cameras. Companies such as Leica Geosystems, Zeiss Imaging (Z/I), or Vexcel offer commercial systems; research centers such as the German Space Center (DLR) fly their own prototypes. Such systems must establish their market somewhere between the satellite image user seeking higher resolution and the airphoto user seeking digital input and GIS compatibility. Consequently, airborne scanner systems have to offer stereo capability and multispectral recording (Figure 17.3).

17.3.1 TECHNICAL PARAMETERS OF ULTRAHIGH RESOLUTION AIRBORNE SYSTEMS

Two different technologies are employed to accomplish an airborne digital recording system. Z/I and Vexcel make use of two-dimensional arrays and a set of coupled nadir looking lenses to emulate a standard frame camera's central perspective (Dörstel, 2003; Leberl and Gruber, 2003). Leica Geosystems and the DLR employ triplet scanner technology with one-dimensional line arrays arranged in fore, nadir, and aft looking modes (Fricker et al., 2000; Hoffmann and Lehmann, 2000). The advantage of a two-dimensional matrix camera is that all standard photogrammetric techniques can be used in a digital environment. The advantage of a stereo triplet solution is that photogrammetric preprocessing (i.e., digital surface model [DSM] and orthoimage generation) is performed before the user receives the data, thus alleviating the need to run sophisticated software at the user's organization. The image

(a)

(b)

FIGURE 17.2 (Color Figure 17.2 follows page 240.) Quickbird image of the pyramids of Giza, Egypt. The panchromatic resolution is 0.70 cm due to the camera tilt for the acquisition (maximum resolution for nadir view is 0.61 cm). The information content challenges aerial photography (a). The Quickbird multispectral CIR bands were resampled to the panchromatic pixel size. Even the resolution of 2.80 m (maximum resolution for nadir view is 2.44 m) can prove sufficient for many urban remote sensing applications (b). (Images courtesy of Globe View, Longmont, Colorado, and distributed by Leica Geosystems Geospatial Imaging, Norcross, Georgia.)

data are provided in the required coordinate system and can be easily integrated into an existing GIS database.

Figure 17.4 shows one test site that was recorded by three different ultrahigh resolution sensors at pixel sizes of 10, 25, and 50 cm, respectively. Figure 17.5 demonstrates the geometric differences between the central perspective for the area CCDs (left) and the line CCDs' cameras (right). Which of the two approaches is more suitable will largely depend on user demands and the price-performance ratios of the respective systems. Table 17.2 presents five selected ultrahigh resolution

FIGURE 17.3 Three-dimensional perspective view of a subset of the nature protection area Heuckenlock near Hamburg, Germany, captured by the German Space Agency's High Resolution Stereo Camera HRSC-AX. The image has a ground resolution of 25 cm. The perspective view is a computer-generated drape of the RGB image data over the 50-cm resolution digital surface model (DSM) also generated by the HRSC-AX.

FIGURE 17.4 Comparison of ultrahigh resolution airborne sensors (from left to right): DMC (10-cm pixel size), ADS 40 (25-cm pixel size), and Falcon (50-cm pixel size). The image subsets are 50 × 50 m (top) and 20 × 20 m (bottom). (After Schiewe, J. and M. Ehlers, *Photogrammetrie-Fernerkundung-Geoinf.* (PFG), 6: 463–474, 2004.)

FIGURE 17.5 Comparison of the different recording geometries: central perspective DMC (left) and the line scanner ADS (right) with central perspective distortion only in cross-track direction. Digital cadastral data (houses) are overlaid for better clarification.

airborne digital camera systems. The advantages of digital cameras are widely understood: no film, no photo processing, no scanning, better radiometric quality through direct sensing, "nonaging" storage and direct integration into GIS and image processing systems. The disadvantages of digital scanners, most notable of which are geometric distortions and monoscopic imaging mode, no longer exist due to the stereo capabilities of the new sensors and the use of integrated INS and differential GPS technology during image acquisition.

17.3.2 Multisensor Systems

The progress of GPS and inertial navigation systems for direct orientation is also responsible for development of operational laser scanning or LIDAR systems (Lemmens, 2004). The simultaneous use of LIDAR and imaging technology creates multisensor systems that produce accurate digital surface models and image data at the same time. Table 17.3 presents a selection of multisensor systems that have increasingly been used in Europe.

The acquisition of elevation data can be achieved by different techniques: For area CCD sensors, such as the DMC or the Ultracam, standard digital photogrammetric techniques (image matching or stereocorrelation) can be employed to create digital elevation models from overlapping frame images (see, for example, Spiller, 2000). The line CCD sensors, as ADS or HRSC, make use of a triplet along-track stereo geometry for three-dimensional information extraction, which requires specific software for preprocessing (see, for example, Fricker et al., 2000; Ehlers et al., 2003b).

Combined laser scanners and imaging sensors create a very accurate and dense digital surface model that can be easily coregistered with the image data. All techniques allow creation of orthophotos that can be readily interfaced with GIS and digital map data. The differences in elevation determination for stereocorrelation and LIDAR

TABLE 17.2
Selected Digital Airborne Sensors

Sensor	HRSC-AX	DSS	ADS 40	UltraCam-D	DMC
Company	DLR	Applanix (Emerge)	Leica Geosystems	Vexcel Corp.	Z/I Imaging
URL	www.dlr.de	www.emergedss.com	www.gis.leica-geosystems.com/	www.vexcel.com	www.ziimaging.com
Sensor type	Line CCD	Area CCD	Line CCD	Area CCD	Area CCD
Introduction	2000	2004	2000	2003	2002
Focal length	151 mm	55 mm (color and CIR) 35 mm (only color)	62.7 mm	100 mm (28 mm multispectral)	120 mm (25 mm multispectral)
Field-of-view	29°	37° × 55.4°	64°	55° × 37°	74° × 44°
No. CCD lines/matrix camera	9	1	7	9	8
No. CCDs across track	12,172	4,077	2 × 12,000 (pan) 12,000 (ms)	11,500 (pan) 4,008 (ms.)	13,824 (pan) 3,000 (msl)
No. CCDs along track	—	4,092	—	7,500 (pan) 2,672 (ms)	7,680 (pan) 2,000 (ms)
Sensor size	6.5 μm	9 μm	6.5 μm	9 μm	12 μm
Radiometric resolution	12 bit	12 bit	12 bit	>12 bit	12 bit

Spectral resolution (nm)	520–760 (pan) 450–510 (blue) 530–576 (green) 642–682 (red) 770–814 (NIR)	RGB Modus: 400–500 (blue) 500–600 (green) 600–680 (red) CIR Modus: 510–600 (green) 600–720 (nor/NIR) 720–920 (NIR)	465–680 (pan) 428–492 (blue) 533–587 (green) 608–662 (red) 703–757 (NIR) or 833–887 (NIR opt.)	390–690 (pan) 390–470 (blue) 420–580 (green) 620–690 (red) 690–900 (NIR)	400–580 (pan) 400–580 (blue) 500–650 (green) 590–675 (red) 675–850 (NIR)
Readout frequency	1640 Lines/s	0.25 Images/s	800 Lines/s	0.75 Images/s	0.5 Images/s
Stabilization	Zeiss T-AS platform	Own platform	LH platform	Own platform	Zeiss T-AS platform
Data recording	High speed recorded	80 GB exchangeable hard disk	MM40 mass storage	RAID disk system	SCU (>1 TB)
Georeferencing	Applanix POS/DG	Applanix POS IMU	Applanix POS IMU	Not specified	POS Z/1 519
Estim. costs incl. pos. system	—	$425,000	$1,200,000	$700,000	$1,600,000

Source: Based on Schiewe, J. and M. Ehlers, *Photogrammetrie-Fernerkundung-Geoinf.* (PFG), 6: 463–474, 2004.

TABLE 17.3
Selected LIDAR Systems with Optional Imaging Sensors

System	FALCON	ALTM 3033, 3070		ALS 50
Company	TopoSys	Optech		Leica geosystems
URL	www.toposys.de	www.optech.on.ca		www.gis.leica-geosystems .com
Recording principle	Glasfiber Array	Rotating mirror		Rotating mirror
Multiple reflections	Max. 2 echoes	Max. 4 echoes		Max. 4 echoes
Image sensor	Line scanner (pixel size 0.5 m)	DSS		ADS 40
(Spectral resolution in nm)	450–490 (blue) 500–580 (green) 580–660 (red) 770–890 (NIR)	RGB modus: 400–500 (blue) 500–600 (green) 600–680 (red)	CIR modus: 510–600 (green) 600–720 (nor/NIR) 720–920 (NIR)	465–680 (pan) 428–492 (blue) 533–587 (green) 608–662 (red) 703–757 (NIR) or 833–887 (NIR opt.)
Pulse frequency	83 kHz	Up to 70 kHz		Up to 83 kHz
Scanning frequency	653 Hz	70 Hz		$412.33 \times FOV^{-0.6548}$ (max. 51°)
Max. flying height	1600 m	3000 m		4000 m
Scan angle (FOV)	±7°	±0 ... 25°		±10 ... 37.5°
Swath width (h = 1000 m)	245 m	930 m		1530 m
Resolution	0.02 m	0.01 m		0.01 m
Vertical accuracy	±0.15 m	±0.15 m (h = 1200 m)		±0.15 m ... ± 0.50 m
Horizontal accuracy	—	±0.50 m (h = 1000 m)		±0.15 m ... ± 0.75 m

Source: After Schiewe, J. and M. Ehlers,   Photogrammetrie-Fernerkundung-Geoinf. (PFG), 6: 463–474, 2004.

sensors are presented in Figure 17.6. The higher density and direct distance measurements of the laser scanners produce more accurate digital surface models with sharper edges. Window-based correlation, on the other hand, tends to create "hills" in areas of distinct discontinuities (i.e., house walls, trees, etc.).

17.4 IMAGE FUSION

In a special issue of the *International Journal of Geographical Information Science* (IJGIS) on data fusion, Edwards and Jeansoulin (2004) state that

Data fusion is a complex process with a wide range of issues that must be addressed. In addition, data fusion exists in different forms in different scientific communities. Hence, for example, the term is used by the image community to embrace the problem of sensor fusion, where images from different sensors are combined. The term is also used by the database community for parts of the interoperability problem. The logic community uses the term for knowledge fusion.

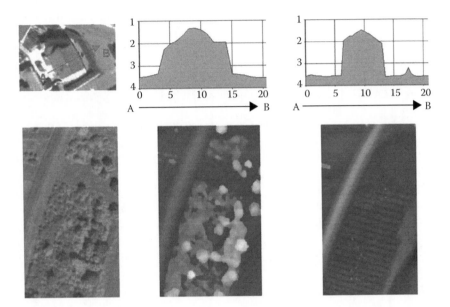

FIGURE 17.6 Comparison of elevation models from ADS image data (left) and Falcon LIDAR data. The LIDAR DSM (right) shows a better edge preservation than image-based correlation for the ADS 40 (center). The other advantage of laser scanners is the ability to use last echoes for retrieving ground elevation even under trees (bottom).

Consequently, it comes as no surprise that several definitions for data fusion can be found in the literature. Pohl and van Genderen (1998) proposed that "image fusion is the combination of two or more different images to form a new image by using a certain algorithm." Mangolini (1994) extended data fusion to information in general and also referred to quality. He defined data fusion as "a set of methods, tools and means using data coming from various sources of different nature, in order to increase the quality (in a broad sense) of the requested information." Wald (1999) defined data fusion as "a formal framework in which are expressed means and tools for the alliance of data originating from different sources. It aims at obtaining information of greater quality; the exact definition of 'greater quality' will depend upon the application."

In the imaging community, fusion techniques are used to merge panchromatic image information of high spatial resolution into multispectral images of lower spatial resolution. These techniques are designed to produce images that present the "best of both worlds" — high spatial resolution combined with high spectral resolution.

17.4.1 FUSION TECHNIQUES

Fusion techniques for remotely sensed data (image fusion) can be classified into three levels: pixel level (ikonic), feature level (symbolic), and knowledge or decision level (Pohl and van Genderen, 1998). Of highest relevance for remote sensing data are techniques for ikonic image fusion to merge panchromatic images of high spatial resolution with multispectral data of lower resolution (see Cliche et al., 1985; Welch and Ehlers, 1987; Zhang, 2002). However, existing techniques hardly satisfy conditions for successful fusion of the new generation of high-resolution satellite images such as IKONOS, Landsat-7, SPOT-5, and QuickBird or ultrahigh resolution airborne data (Zhang, 2002).

All of the new generation satellite and almost all airborne sensors provide high-resolution information only in their panchromatic mode, whereas the multispectral images are of lower spatial resolution. The ratios between high-resolution panchromatic and low-resolution multispectral images vary between 1:2 and 1:8 (Ehlers, 2004b). To produce high-resolution multispectral data sets (as are required for urban remote sensing), the panchromatic information must be merged with the multispectral images. The most significant problem with image fusion techniques is the color distortion of the fused image.

17.4.2 SPECTRAL CHARACTERISTICS PRESERVING IMAGE FUSION (EHLERS FUSION)

The principal idea behind spectral characteristics preserving image fusion is that the high-resolution image must sharpen the multispectral image without adding new information to the spectral components. As a basic image fusion technique, we will make use of the IHS transform. This technique can be extended to include more than the standard three bands (red, green, and blue color transform) from color theory. In addition, filter functions for the multispectral and panchromatic images must be developed. The filters need to be designed so that the effect of color change from the high-resolution component is minimized. The ideal fusion function would add the high resolution spatial components of the panchromatic image (i.e., edges, object changes) but disregard its actual gray values.

For a thorough analysis of information distribution along the spatial frequencies of an image, use is made of Fourier transform (FT) theory (Gonzales and Woods, 2001). An overview flowchart of the method is presented in Figure 17.7 (see Ehlers and Klonus, 2004, for a complete description). The Ehlers fusion was applied to the panchromatic and multispectral Quickbird images (Figure 17.2) and showed excellent color preservation (Figure 17.8). For visual comparison, a subset of the merged image is compared to the results of three other standard fusion methods (Figure 17.9).

The positive results for the new technique have been confirmed for a number of image data sets. Even for multisensoral and multitemporal image fusion, the fast Fourier transform (FFT)-based technique preserves the spectral characteristics of multispectral images while keeping the spatial resolution of panchromatic images. This is reflected by the correlation coefficient for the multispectral bands before and after

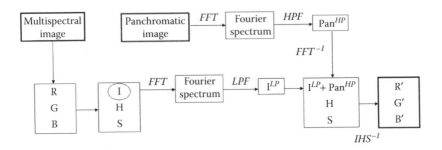

FIGURE 17.7 FFT-based filter fusion using a standard IHS transform. Three selected bands (RGB) of the low-resolution multispectral image are transformed into the IHS domain. Intensity component and high-resolution panchromatic image are transformed into the Fourier domain using a two-dimensional fast Fourier transform (FFT). The power spectrum of both images is used to design the appropriate low-pass (LP) filter for the intensity component and high-pass (HP) filter for the high-resolution panchromatic image. An inverse FFT transforms both components back into the spatial domain. The low-pass filtered intensity (I^{LP}) and the high-pass filtered panchromatic band (P^{HP}) are added and matched to the original intensity histogram. At the end, an inverse IHS transform converts the fused image back into the RGB domain. (After Ehlers, M., in Ehlers, M., F. Posa, H.J. Kaufmann, U. Michel, and G. De Carolis (Eds.), *Remote Sensing for Environmental Monitoring, GIS Applications, and Geology IV, Proc. SPIE*, 5574, Bellingham, WA: 1–13, 2004.)

fusion. The Ehlers fusion achieved a correlation coefficient of 0.994, which is far superior to those of all the other methods (Table 17.4). Because the filter function for the high-resolution image can be adjusted to the size of the geo-objects, the Ehlers fusion technique can also be used for an optimum spatial enhancement of selected geo-objects (e.g., houses, parcels, or field boundaries) (Ehlers, 2005).

FIGURE 17.8 Pansharpened Quickbird image from Figure 17.2 using the Ehlers fusion. Besides the spatial enhancement, colors are also well preserved (see Figure 17.9 for details).

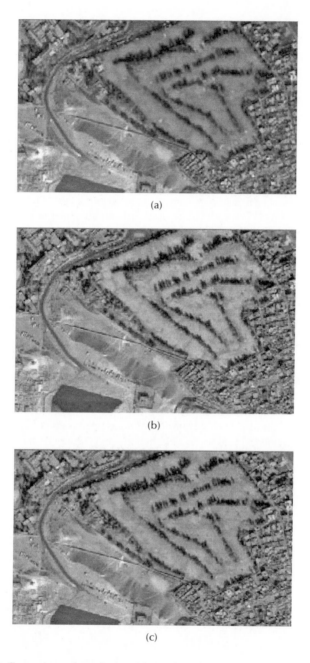

(a)

(b)

(c)

FIGURE 17.9 Comparison of standard pansharpening techniques with the Ehlers fusion. All methods show significant color changes when compared to the original unsharpened image (a). The employed methods were: Brovey transform (b), multiplicative merge (c), and principal component transform (d). The Ehlers fusion (e) shows almost no change in spectral characteristics. Please note that the images are not optimized for resolution enhancement but for color comparison.

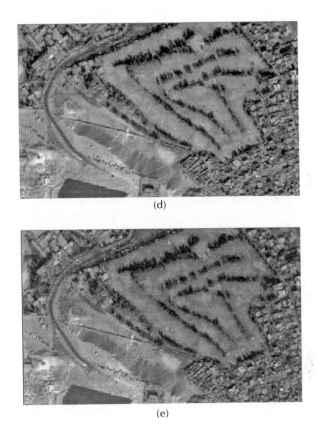

(d)

(e)

FIGURE 17.9 (Continued).

TABLE 17.4
Correlation Coefficients between Multispectral Bands of Original and Pansharpened Images Prove Superiority of the Ehlers Fusion Method

Pansharpening Method	Correlation Coefficient with Original Bands
IHS	0.762
Brovey	0.816
Principal component	0.850
Multiplicative	0.932
Ehlers	0.994

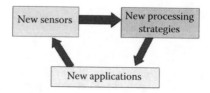

FIGURE 17.10 Idealized causal development in remote sensing.

17.5 CONCLUSION

The development of remote sensing systems capable of very high and ultrahigh resolution has matured over the last few years. The new airborne digital camera systems have the potential to finally end the reign of analog cameras when it comes to image acquisition for large-scale applications. It could be shown that the new very high and ultrahigh resolution sensors offer tremendous potential for urban applications. The three-dimensional capability of the airborne sensors is especially a great advantage for applications such as city modeling. Almost all sensors have the highest resolution only for their panchromatic mode, so image fusion for pansharpening is an essential for the image analysis. FFT-based filtering prior to application of an IHS transform produces fused images of improved spatial resolution without changing the spectral characteristics.

However, emphasis must shift now to development of new automated procedures to deal with the spatial complexity of the new image data adequately (Blaschke and Strobl, 2001; Schiewe, 2005). During a workshop on remote sensing for GIS, emphasis was placed on automated information extraction from very high resolution remote sensing data (see Blaschke, 2002). It is our firm belief that new sensors require new processing techniques that, in turn, offer new application fields. New possibilities for application will eventually lead to new demands on sensor developments so that the cycle of innovations can continue (Figure 17.10).

REFERENCES

Blaschke, T. and J. Strobl, What's wrong with pixels? Some recent developments interfacing remote sensing and GIS. *Geo-Inf.-Syst.*, 6/01: 12–17, 2001.

Blaschke, T. (Ed.), Fernerkundung und GIS: Neue Sensoren — Innovative Methoden, Wichmann Verlag, Heidelberg, 2002.

Cliche, G., Bonn, F., and Teillet, P., Integration of the SPOT pan channel into its multispectral mode for image sharpness enhancement. *Photogrammetric Eng. Remote Sensing*, 51: 311–316, 1985.

Dörstel, C., DMC — practical experiences and photogrammetric system performance, in Fritsch, D. (Ed.), *Photogrammetric Week 03*, Wichmann Verlag: 59–66, 2003.

Edwards, G. and R. Jeansoulin, Data fusion — from a logic perspective with a view to implementation; guest editorial. *Int. J. Geogr. Inf. Sci.*, 18(4): 303–307, 2004.

Ehlers, M., J. Schiewe, and M. Möller, 3D city modeling using high resolution and multisensoral remote sensing. *Geo-Inf.-Syst.* 6(03): 30–37, 2003a.

Ehlers, M., M. Gähler, and R. Janowsky, Automated analysis of ultrahigh resolution remote sensing data for biotope type mapping: new possibilities and challenges. *ISPRS J. Photogrammetry Remote Sensing*, (57): 315–326, 2003b.

Ehlers, M., Remote sensing for GIS applications: new sensors and analysis methods, in Ehlers, M., H.J. Kaufmann, and U. Michel (Eds.), *Remote Sensing for Environmental Monitoring, GIS Applications, and Geology III, Proc. SPIE*, 5239, Bellingham, WA: 1–13, 2004a.

Ehlers, M., Spectral characteristics preserving image fusion based on Fourier domain filtering, in Ehlers, M., F. Posa, H.J. Kaufmann, U. Michel, and G. De Carolis (Eds.), *Remote Sensing for Environmental Monitoring, GIS Applications, and Geology IV, Proc. SPIE*, 5574, Bellingham, WA: 1–13, 2004b.

Ehlers, M. and S. Klonus, Erhalt der spektralen Charakteristika bei der Bildfusion durch FFT basierte Filterung, *Photogrammetrie-Fernerkundung-Geoinformation* (PFG) 6/2004: 495–506, 2004.

Ehlers, M., Urban remote sensing: new developments and trends, in *Proc. 5th In. Symp. Remote Sensing Urban Areas* (URS 2005), Tempe, AZ, (CD proceedings), 6 pp., 2005.

Fricker, P., R. Sandau, U. Tempelmann, and S. Walker, ADS 40 — why LH systems took the three-line road. *GIM Int.*, July: 45–47, 2000.

Gonzales, R.C. and R.E. Woods, *Digital Image Processing*, Prentice Hall, Upper Saddle River, NJ, 2001.

Hoffmann, A. and F. Lehmann, Vom Mars zur Erde — die erste digitale Orthobildkarte Berlin mit Daten der Kamera HRSC-A. *Kartographische Nachrichten* 50(2): 61–71, 2000.

Leberl, F. and M. Gruber, Flying the new large format digital aerial camera UltraCam, in Fritsch, D. (Ed.), *Photogrammetric Week 03*, Wichmann Verlag: 67–76, 2003.

Lemmens, M., Product survey on airborne laserscanners. *GIM Int.*, 5: 45–47, 2004.

Mangolini, M., Apport de la fusion d'images satellitaires multicapteurs au niveau pixel en télédétection et photointerprétation. Thèse de Doctorat, Université Nice — Sophia Antipolis, France, 1994.

Möller, M., Urbanes Umweltmonitoring mit digitalen Flugzeugscannerdaten. Wichmann Verlag, Heidelberg, 2003.

Pohl, C. and J.L. van Genderen, Multisensor image fusion in remote sensing: concepts, methods and applications. *Int. J. Remote Sensing*, 19: 823–854, 1998.

Schiewe, J. and M. Ehlers, Semantisches Potenzial digitaler flugzeuggestützter Fernerkundungssensoren. *Photogrammetrie-Fernerkundung-Geoinformation* (PFG), 6: 463–474, 2004.

Schiewe, J., Status and future perspectives of the application potential of digital airborne sensor systems. *Int. J. Appl. Earth Obs. Geoinf.*, 6: 215–228, 2005.

Spiller, R., DMC — why Z/I imaging preferred the matrix approach. *GIM Int.*, July: 66–68, 2000.

Wald, L., Definitions and terms of references in data fusion. *Int. Arch. Photogrammetry Remote Sensing*, 32/7-4-3 W6, Valladolid, Spain, 1999.

Welch, R. and M. Ehlers, Merging multiresolution SPOT HRV and Landsat TM data. *Photogrammetric Eng. Remote Sensing*, 53(3): 301–303, 1987.

Zhang, Y., Automatic image fusion: a new sharpening technique for IKONOS multispectral images. *GIM Int.* 16(5): 54–57, 2002.

18 Spectral Resolution in the Context of Very High Resolution Urban Remote Sensing

Paolo Gamba and Fabio Dell'Acqua

CONTENTS

18.1 INTRODUCTION

Spatial and spectral resolutions are equally important in urban remote sensing, but they are typically discussed in separate manners (Welch, 1982; Forster, 1985; Herold et al., 2003; Phinn et al., 2002). A limited number of research publications do address spatial and spectral resolutions, typically exploiting data from a single sensor (Mueller et al., 2003; Price, 1997; Roessner et al., 2001). The scope of this chapter is therefore first to provide a very brief discussion of spatial and spectral resolution issues in urban area mapping, with an emphasis on very high resolution (VHR). Then, a more detailed presentation of a new procedure for joint spectral and spatial resolution exploitation will be given in the context of VHR sensors. Finally, mapping results will be used for comparing two airborne hyperspectral and one space-borne multi-spectral data set over the same urban area.

18.2   VHR SPATIAL RESOLUTION AND URBAN AREA MAPPING

Urban remote sensing is usually associated with the finest available spatial resolution of VHR sensors because it is able to characterize urban objects and to obtain sufficiently detailed land-cover/land-use maps. Indeed, VHR images with 1-m resolution reveal a number of important elements of the urban landscape. Undoubtedly, building characterization requires this resolution (Shackelford et al., 2004) because of the complexity of the structures and their extreme geometric variability. Moreover, a fine spatial resolution is necessary to recognize urban land-use classes, often based on a particular land-cover mixture (Mueller et al., 2003). Environmental analysis of urban materials is also made possible using VHR because roofs, road pavements, bridges, and other urban elements are represented in sufficient detail. Thus, a statistical analysis of such objects' spectral response and the recognition of dangerous or degenerated materials are made possible.

Together with the capabilities of VHR data, some problems should be stressed. For instance, VHR urban data are not always recorded with a nadir view and thus may correspond to different viewing angles. The complexity of the three-dimensional structure of the urban landscape adds further problems. Thus, VHR imagery requires more accurate and robust algorithms for co-registration and correction from spatial distortions (Guindon, 1997).

Moreover, the structural limits of the sensor's design are more apparent in a VHR image, and atmospheric effects should be carefully removed, taking into account their spatial variability to remove artifacts and improve spectra discriminability. Thus, new and increased needs for atmospheric correction and geocoding arise from these challenging data.

From the point of view of our mapping application, VHR data call for new procedures and algorithms because they highlight the inadequateness of techniques developed for coarser data sets at recognizing land-use patterns and at analyzing structural features in remotely sensed imagery (Guanaes et al., 2003).

Finally, there are problems involving spectral resolution as well, such as increased variability in the spectral characteristics of built surfaces on a small scale. Material recognition in VHR data requires an improved database of spectral signatures (Herold et al., 2004). Otherwise, local changes in the status of the material, visible at VHR, may cause undue assignments of affected pixels to different mapping classes and therefore result in decreased classification performance.

18.3   JOINT SPATIAL AND SPECTRAL VHR FOR URBAN AREA MAPPING

Urban area mapping requires sufficient spatial resolution to see small urban objects, such as trees in a park or cars on a street. However, sufficient spectral resolution is also required to recognize the "white" of pedestrian crosswalks relative to "white" of a nearby car. Thus, in normal practice VHR data are paired to multispectral data. Generally, a few bands in the visible and near-infrared wavelengths are considered the best compromise between spectral precision, on the one hand, and sensor cost, data size, and complexity of analysis on the other. Multispectral data from VHR satellite sensors have been exploited in urban areas for land-use mapping considering joint analysis of land covers (Tarantino et al., 2003) for land change and cartographic update (Greenhill et al., 2003), for the extraction of environmental indexes (Ridd, 1995), or for characterization of impervious and vegetated areas and their relative abundance in the same location (Marino et al., 2001).

The discussion in the previous section proves, however, that joint availability of spatial and spectral VHR can be of great help in many applications, particularly in the precise mapping of urban areas. The recent availability or development of new airborne sensors with very high spectral resolution may thus allow evaluation of the extent of the advantage arising from joint spectral and spatial VHR data.

Accordingly, the more prolific current research in urban remote sensing is the exploitation of VHR hyperspectral data, in which spectral and spatial VHR may be achieved. VHR hyperspectral data recently have been used for various urban applications. Urban material characterization has been considered by means of the multispectral infrared and visible imaging spectrometer (MIVIS) sensor, allowing not only provision of a precise definition of roof materials, but also building a database of rooftops made of environmentally threatening materials for subsequent substitution and safe disposal (Xiao et al., 1999).

Similarly, vegetation canopies and trees in urban forests provide information that is extremely interesting for energy exchange, air quality analysis (microclimate), and hydrology. Tree distribution, height, stress status, and species already have been derived from low-altitude airborne visible/infrared imaging spectrometer (AVIRIS) measurements (Heiden et al., 2003). As a further example, detection of land cover and sealed sections in an urban area can be accomplished using the multitechnique approach proposed in Gamba (2004), where HyMap data have been analyzed for a better characterization of urban environments.

A more detailed analysis of the spectral properties required by VHR sensors to obtain interesting results in urban areas is given in Herold et al. (2003), where it was shown that AVIRIS fine sampling of the electromagnetic spectrum provides a much better result than (simulated) multispectral Ikonos data because the latter lack information on a significant portion of the infrared wavelength range. However, no discussion of the effect of the spatial resolution is provided in that paper because Ikonos data are simulated at the same resolution as AVIRIS.

18.4 EXAMPLE OF JOINT VERY HIGH RESOLUTION SPECTRAL AND SPATIAL LAND-COVER MAPPING

This section is devoted to examining a practical situation by comparing VHR hyperspectral data with different spectral and spatial resolutions and discussing their ability to provide land-cover maps in urban areas. The focus will be on a subset of the recently released urban remote sensing and data fusion set (Gamba et al., 2003). These data sets consist of two VHR hyperspectral and one VHR multispectral data sets, all of which depict the same urban area — namely, the town of Pavia in northern Italy (see Figure 18.1).

The hyperspectral sensors — namely, the digital airborne imaging spectrometer (DAIS) and reflective optics system imaging spectrometer (ROSIS) — were flown over the test area by the German Space Agency in the framework of the HySens project

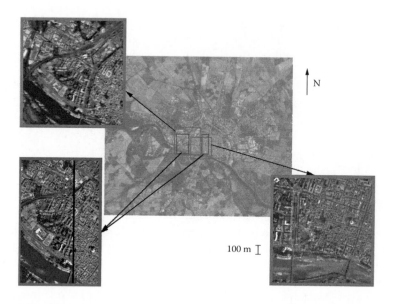

FIGURE 18.1 Samples of the VHR hyperspectral and multispectral data sets against a SPOT panchromatic image of the whole urban area of Pavia. Samples depict a part of the town center, as imaged (false colors) by the DAIS (top left), ROSIS (bottom left), and Quickbird-1 (bottom right).

during the 2002 flight campaign. DAIS is a multiband system with 80 bands in the visible and infrared wavelength range (from 0.4 to 12.6 mm). ROSIS is a multiband sensor focused only on visible and near-infrared frequency bands; it is a spectrometer with 32 frequency bands from 0.45 to 0.85 mm. ROSIS shows higher potential for vegetation and green area mapping than DAIS, which has a broader range of applications because it provides information down to the thermal infrared. The spatial resolution of the hyperspectral data sets is among the finest currently available (2.6- and 1.2-m posting, respectively) and, for ROSIS, almost matches the one by the multispectral VHR data set, a Quickbird-1 pan sharpened multispectral image (0.7 m). However, Quickbird-1 supports a very coarse spectral resolution (four bands: three visible and one near infrared). All the images were acquired in the same time period of July 2002.

To evaluate the effect of different spectral and spatial resolution, the three data sets are not resampled in order to be directly comparable pixel by pixel. Instead, a mapping procedure is applied separately to them and evaluated in two test areas. To this end, mapping results are compared with a ground truth map resampled to each of the different spatial resolutions.

Note that two test areas are used to cover the different urban environments. The largest test area, shown in Figure 18.1, is a part of the town center. The area presents a typical urban land cover for historical Italian towns, dominated by the large number of roofs covered with red tiles. The second test area depicts the Engineering School of the University of Pavia. This area is located in the external and newer city belt, and more recent urban covers are available, such as metallic roofs and gravel. Classification maps for the second test area are shown in Figure 18.3. That the

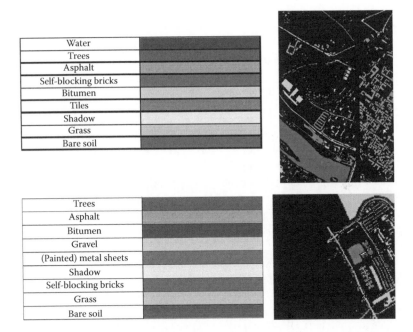

FIGURE 18.2 (Color Figure 18.2 follows page 240.) Test sets for the two validation areas (sampled to match ROSIS spatial resolution), together with corresponding legends.

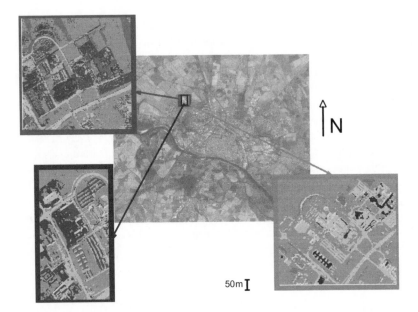

FIGURE 18.3 (Color Figure 18.3 follows page 240.) Second test area (university campus): classification maps obtained using the DAIS (red border), ROSIS (blue border), and Quickbird-1 (green border) data.

ROSIS and DIAS data sets do not cover exactly the same area is recognized and this is due to differences in the swath width. For example, in Figure 18.1 the whole area covered by the DAIS and the Quickbird-1 data sets is split into two nonoverlapping parts in the ROSIS false color image.

18.5 WORKING CHAIN

The classification tool used for the previously mentioned mapping comparison is based on a technique recently developed for multiclassification fusion of spatial and spectral VHR data (Dell'Acqua et al., 2004; Landgrebe, 2003). The first idea for exploiting the information and improving classification is to combine spectral and spatial classifiers in a multiclassification framework. By fusing the classification maps obtained by means of pixel-by-pixel and region-based classifiers, we hope to "take the best from two worlds" and better characterize the test area. A simple scheme of the processing steps implemented in this approach is as follows.

1. A feature extraction procedure is applied to the data to reduce dimensionality.
2. Supervised classifications are performed comparing the results of different spectral and spatial classifiers.
3. All the classification maps are combined using majority voting or linear opinion pools (LOPs) (Gamba and Houshmand, 2001).
4. A further geometrical refinement step is introduced, and spatial information from the neighborhood of each pixel is taken into account to improve the results.

The need for feature extraction (step 1) is clear, at least in principle. As a matter of fact, feature extraction can be viewed as finding a set of vectors that represents an observation while reducing the dimensionality. In pattern recognition, to extract features focused on discriminating between classes is desirable. Although a reduction in dimensionality is required, the information loss due to the reduction in dimension must have the smallest possible impact on the discriminative power of classifiers. In linear feature extraction, the number of input dimensions corresponds to the number of eigenvectors elected. The transformed data are determined by $Y = F^T X$, where F is the transformation matrix composed of the eigenvectors of the feature matrix, X is the data in the original feature space, and Y is the transformed data in the new feature space.

Several feature extraction approaches have been proposed. A recognized method is the discriminant analysis feature extraction (DAFE) technique (Lee and Landgrebe, 1997), which is a method intended to enhance separability. In DAFE, a within-class scatter matrix, S_w, and a between-class scatter matrix, S_B, are defined:

$$S_w = \sum_i P(W_i)S_i \qquad (18.1)$$

$$S_B = \sum_i P(W_i)(M_i - M_0)(M_i - M_0)^T \qquad (18.2)$$

$$M_0 = \sum_i P(w_i)M_i \qquad (18.3)$$

where M_i, S_i, and $P(w_i)$ represent the mean vector, the covariance matrix, and the prior probability of the ith class, respectively.

The index used for optimization may be defined as $J = \text{tr}(S_W^{-1} S_B)$, where "tr( )" denotes the trace of a matrix. New feature vectors are selected to maximize the J index. The necessary transformation from X to Y is found by taking the eigenvalue–eigenvector decomposition of the matrix $S_W^{-1} S_B$ and then taking the transformation matrix as the normalized eigenvectors corresponding to the eigenvalues in decreasing order. This method does have some shortcomings. For example, because discriminant analysis mainly utilizes class mean differences, the feature vectors selected by discriminant analysis are not reliable if mean vectors are near to one another. Because the lumped covariance matrix is used in the criterion, discriminant analysis may lose information contained in class covariance differences. Also, the maximum rank of S_B is $M - 1$ because S_B depends on M_0. Usually, S_W is of full rank and therefore the maximum rank of $S_W^{-1} S_B$ is $M - 1$. This indicates that, at most, $M - 1$ features can be extracted through this approach. Another problem is that the preceding criterion function generally has no direct relationship to the error probability.

In the early 1990s, decision boundary feature extraction (DBFE) was proposed by Lee and Landgrebe (Briem, 2002). They showed that discriminantly informative features and discriminantly redundant features can be extracted from the decision boundary. They also showed that discriminantly informative feature vectors have a component normal to the decision boundary at least at one point on the decision boundary. Furthermore, discriminantly redundant feature vectors are orthogonal to a vector normal to the decision boundary at every point on the decision boundary.

Thus, a decision boundary feature matrix (DBFM) was defined to extract discriminantly informative features and discriminantly redundant features from the decision boundary. It can be shown that the rank of the DBFM is the smallest dimension in which the same classification accuracy can be obtained as in the original feature space. Also, the eigenvectors of the DBFM corresponding to nonzero eigenvalues are the necessary feature vectors to achieve the same classification accuracy as in the original feature space.

Steps 2 and 3 are a consequence of the different behaviors of the proposed classifier on the same data set. It is extremely difficult to find rules for choosing the best classifier in any situation. Past papers (Benediktsson and Kanellopoulos, 1999) have instead proven that the best way to exploit their characteristics is to combine the classification maps into an information fusion framework. The well-known majority voting and opinion polls approaches are discussed here (Gamba and Houshmand, 2001).

Majority voting is a standard procedure that provides a classification map in which each pixel is assigned to the class to which it belongs in the majority of the maps to be jointly considered. The linear opinion pool (LOP) and the logarithmic opinion pool (LOGP) procedures require a training set instead, which inherently defines the output classes to be considered in the final classification map. For any possible pattern of the output classes in the training set, $p(i,j)$ is computed, where $p(i,j)$ represents the probability that the pixels of the ith class of the training set are associated with the jth pattern in the multiple classification set. Finally, the output class is defined by maximizing one of the two following functionals:

$$C_i^{LOP} (\mathbf{X}) = S^n_{j=1} l_j \, p(i,j) \qquad\qquad (18.4)$$

$$C_i^{LOGP} (\mathbf{X}) = S^n_{j=1} \, p(i,j)^\wedge \, l_j \qquad\qquad (18.5)$$

where $\mathbf{X}$ is the input classification pattern, n is the number of classified images, and l_j represents the weight of the jth map (default is $l_j = 1$). Of course, some patterns may not be present in the training set. For these patterns, the output class assignment follows majority voting.

Among the considered classifiers, some are based purely on spectral characteristics (maximum likelihood, Fisher linear discriminant, fuzzy ARTMAP), while others take into account the spatial information as well, such as the fuzzy ARTMAP with spatial reclassification. Just to specify the last spatial classifier, recall that it is the second step of the fuzzy ARTMAP classification chain introduced in Gamba and Dell'Acqua (2003) and Lisini et al. (2005). The output of a pixel-by-pixel classification is refined by a second classification by means of the same fuzzy ARTMAP neural network. This is done using the same training areas of the spectral step but using as input a vector representing the mapping patterns in a window around each pixel. Although very simple, this method seems to be highly effective in a variety of situations.

Step 3 provides as the output a combined classification map, which may be further improved by correcting (again, spatially) the results. This is required due to "salt and pepper" noise still present in the final map, even after combining multiple classifiers. Therefore, a final reclassification step was introduced (step 4 in the preceding list), again by means of the fuzzy ARTMAP classifier, that usually provides more precise results.

Finally, another step (step 5) aimed at geometric refinement of the classification may be introduced. This step is based on a generalization of Dell'Acqua et al. (2005), where a top-down segmentation step is matched with a bottom-up approach based on geometric features and cue grouping. In the top-down step, an initial partitioning of areas of interest within a scene is accomplished using simple supervised classification schemes. A reduced resolution image may be considered, and textural features may be used to complement the original information. In the bottom-up step, each part of the scene and its surroundings are searched for basic patterns of covers, and corrections to misclassifications are made based on the context. Alternatively, geometric features are extracted to recognize objects instead of classification segments, and objects are assigned to classes by majority voting. Details of this procedure may be found in Swayze et al. (2003).

18.6 TEST AREA RESULTS

All the classifiers used in this research work are supervised; thus, in addition to the test sets for validation, training sets are also needed. As already mentioned, these sets are based on a ground truth map drawn after a ground campaign during the HySens flight. This ground truth map was used to choose training sites, well representative of the chosen land-cover classes. Test sites are their nonoverlapping complement with respect to the whole ground truth map; these are shown in Figure 18.2.

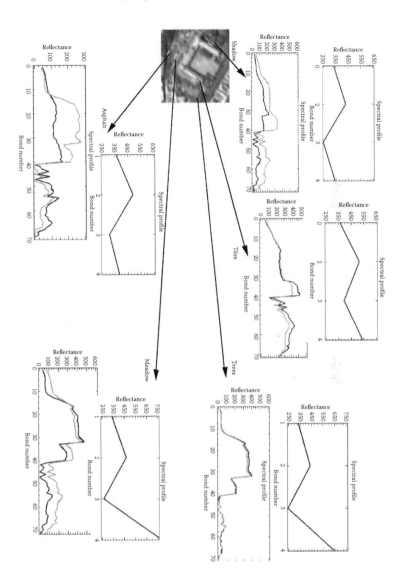

FIGURE 18.4 Spectra of some covers as recorded by DAIS (lower graphs, two data sets) and by Quickbird-1 (upper graphs), sampled in the area around the castle in Pavia. On the horizontal axes the band number is reported, while on the vertical axes the reflectance is reproduced. The double reflectance curve in the DAIS data sets could be drawn, thanks to the overlap between adjacent DAIS swaths, and represent reflectance acquired in the two adjacent flight lines.

The same figure reports the used land-cover legends. Seven classes of urban covers plus water and shadow are considered for the town center. They include natural cover (trees, grass, and bare soil) as well as artificial ones (asphalt, bitumen, self-blocking bricks, and tiles). For the second test area — the Engineering School of the University of Pavia — "tiles" was replaced by (painted) "metallic roof" and a "gravel" class was added.

We should first note that in both cases the legend is questionable because it does not define all the materials consistently. For instance, trees and grass are vegetation life forms, and asphalt, self-blocking bricks, and tiles are land covers specific to this location, but not general land-cover classes; shadow is a spectral class. However, the legends were proposed by the final users of the urban mapping products and thus hold their own (application-oriented) validity. Moreover, the shadow class was added to improve producers' accuracy values for all classes; however, overall accuracy values are computed assuming all shadow pixels as erroneous classification results. Finally, this cover legend does not provide adequate mapping accuracy values (i.e., at least 80%) using Quickbird-1 data. To provide effective mapping (which was the aim of this work), the land-cover legend was reduced (see Table 18.4). Still, lower accuracy values than using hyperspectral data were obtained.

The highest overall accuracy values for the first and the second test areas and for all the three data sets are shown in Table 18.1. The classification algorithm used to obtain these values is also introduced. Speaking only about the algorithms aimed at the multiclassification, results show very clearly that spatial reclassification using the fuzzy ARTMAP neural network is the best option. Please note that the accuracy values are to be interpreted in a comparative sense because they have been obtained on a limited test set. Naturally, complete analysis of the image would probably result in lower values, but this is not relevant to the objective of our comparisons.

For a quantitative comparison of the results, the complete confusion matrices for the first test area referring to DAIS, ROSIS, and Quickbird-1 best classification maps are reported in Table 18.2 to Table 18.4. Each sensor shows a peculiar misclassification behavior. In particular, in DAIS data, "water" is more precisely extracted than in ROSIS data. This is probably due to the largest spectrum, including longer wavelengths, of the DAIS sensor. "Bare soil," "asphalt," and "self-blocking bricks" are misclassified in the DAIS map, and they are similarly misclassified in the ROSIS map. "Bitumen" and "asphalt" are well recognized. Finally, "grass" and "trees" are jointly misclassified almost to the same extent in both data sets.

The map derived from Quickbird-1 data clearly provides the poorest results with the original land-cover class. To achieve a sufficiently large accuracy value, the

---

TABLE 18.1
Comparison of Best Overall Accuracy Values for Land-Cover Maps Obtained from Data by All Sensors and Referring to the Two Test Areas[a]

Overall Accuracy (%)	DAIS	ROSIS	Quickbird-1
Center	96.0	97.4	91.3
Engineering school	87.5	89.2	80.0

Class Algorithm	DAIS	ROSIS	Quickbird-1
Center	ARTMAP (DAFE)	Spatial ARTMAP	Max. likelihood
Engineering school	Majority voting	Spatial ARTMAP	Spatial ARTMAP

[a] See Figure 18.1 and Figure 18.3.

TABLE 18.2
Confusion Matrix for Best Classification Map of First Test Area Using DAIS Data

	Producer's Accuracy (%)	1	2	3	4	5	6	7	8	9	User's Accuracy (%)
1. Water	100.00	4281	0	0	0	0	0	0	0	0	100.00
2. Trees	93.61	0	2269	99	0	2	0	0	0	54	99.04
3. Grass	99.52	0	6	1245	0	0	0	0	0	0	88.11
4. Bare soil	92.61	0	0	69	1366	14	26	0	0	0	98.56
5. Asphalt	98.36	0	0	0	3	1676	18	3	1	3	97.10
6. Self-block	93.03	0	0	0	17	3	267	0	0	0	85.85
7. Tiles	99.55	0	0	0	0	10	0	2227	0	0	99.87
8. Bitumen	97.08	0	0	0	0	20	0	0	665	0	99.85
9. Shadow	92.95	0	16	0	0	1	0	0	0	224	79.72

number of classes was reduced to six (see Table 18.4). Even with this reduced cover legend, misclassifications can be found between "shadow" and "asphalt" or "asphalt" and "water" classes, whose differences in spectra are still not captured by the reduced number of bands of the sensor.

For the second test area (the university campus), the classification maps are presented in Figure 18.3, but the confusion matrices are not reported. This allows visual comparison of the details of the university campus mapped using the three different data sets. It is visually quite clear that both hyperspectral maps, on the left, provide a better spatial characterization with respect to Quickbird-1, on the right. It is more difficult to appreciate the slightly higher overall accuracy of the ROSIS map

TABLE 18.3
Confusion Matrix for Best Classification Map of First Test Area Using ROSIS Data

	Producer's Accuracy (%)	1	2	3	4	5	6	7	8	9	User's Accuracy (%)
1. Water	99.88	65,492	0	0	14	0	0	0	63	0	100.00
2. Trees	95.56	0	7233	336	0	0	0	0	0	0	99.48
3. Grass	98.63	0	38	2950	0	3	0	0	0	0	89.77
4. Bare soil	86.93	0	0	0	2334	349	0	2	0	0	94.61
5. Asphalt	98.80	0	0	0	73	6498	0	6	0	0	94.81
6. Self-block	99.36	0	0	0	20	3	9144	36	0	0	99.41
7. Tiles	98.94	0	0	0	23	0	54	7210	0	0	99.39
8. Bitumen	99.97	0	0	0	3	1	0	0	42,523	7	99.85
9. Shadow	100.00	0	0	0	0	0	0	0	0	2860	99.76

TABLE 18.4
Confusion Matrix for Best Classification Map of First Test Area Using Quickbird-1 Data

	Producer's Accuracy (%)	1	2	3	4	5	6	7	8	9	User's Accuracy (%)
1. Water	100.00	2735	0	0	0	0	0	0	0	0	94.41
2. Trees	100.00	0	46	0	0	0	0	0	0	0	41.82
3. Grass	22.16	0	64	41	0	0	0	0	0	80	60.29
4. Bare soil	0.00	0	0	0	0	0	0	0	0	0	0.00
5. Asphalt	85.30	140	0	0	0	1033	0	0	0	38	98.38
6. Self-block	0.00	0	0	0	0	0	0	0	0	0	0.00
7. Tiles	99.19	0	0	0	0	16	0	2339	0	3	100.00
8. Bitumen	0.00	0	0	0	0	0	0	0	0	0	0.00
9. Shadow	79.51	22	0	27	0	1	0	0	0	194	61.59

(bottom left). The only visible difference between it and the DAIS map (top left) is the reduction in "salt-and-pepper" classification noise, which results in more homogeneously classified regions.

To view results with material spectral characteristics, Figure 18.4 shows Quickbird-1 and DAIS spectra for five land covers sampled in the area of the castle of Pavia. Lower (DAIS) graphs always plot two curves because of the availability of two flight lines for this area. As highlighted by the confusion matrices above, there are spectra virtually indistinguishable in Quickbird-1 data. Roof tiles are instead very well recognized, and this accounts for their high accuracy in Table 18.4.

18.7 DISCUSSION AND CONCLUSIONS

Apparently, the most innovative result of this research work is that, in a number of cases, VHR in the spectral sense is more valuable than in the spatial sense. For our urban site, in fact, maps from DAIS data are more accurate than those from Quickbird-1 data. Thus, our experiments suggest that it is worth investigating to what extent and with what confidence it is more useful for VHR urban mapping to have more bands rather than a spatially more detailed image of the scene.

One reason supporting our preference for spectral resolution in urban areas is the similarity of the spectra for many artificial covers due to their very similar chemical components. Thus, mapping accuracy for these materials improves with spectral resolution, as far as spectrally pure pixels are considered. But once a sufficient spatial resolution is achieved, most pixels are indeed spectrally pure. Therefore, the majority of pixels would benefit from higher spectral resolution, but only border regions between urban objects (a minor set, beyond a certain level of detail) would benefit from higher spatial resolution. Accordingly, when very high spectral and spatial resolutions are not available at the same time, the former is expected to be more valuable than the latter, provided that some minimum requirements are met for both. Problems still remain in border regions between urban objects, where spatial resolution matters, but misclassification may be reduced by exploiting geometric properties of urban objects, e.g., the usual constraints on building shapes.

Finally, we encourage further investigation including many different sites and with many different land-cover legends to further extend the range of cases in which the compromise between resolutions can be optimized. Another valuable finding emerging from our results is that not all the bands in the hyperspectral data are equally useful for urban area mapping. This comment is shared with Herold et al. (2003), (Guindon), and Swayze et al. (2003), where the suggested and/or implemented feature reduction scheme is the selection of a subset of the original bands. The DBFE or DAFE approaches used in this research work differently by somehow "assembling" synthetic bands containing the largest part of the information useful to discriminate the covers under analysis. These approaches are computationally more efficient, but they are less directly connected to the physical properties of the materials.

Finally, the proposed approach suggests jointly exploiting spatial and spectral VHR by classifying a pixel considering its neighborhood. This is done by multiclassification and spatial postclassification refinement. Although Herold et al. (2003) recognized this need, automatic and efficient algorithms for joint spectro–spatial classification are still in their first stages of development and represent a very active research field. This is especially true for VHR hyperspectral data because they need to be coupled with feature reduction schemes. This and related works are a first attempt in this new direction. In summary:

- High spectral resolution is certainly an important feature, but not every band equally concurs with spectral discriminability among land-cover classes. This is shown by the fact that the best results in spectral classification were obtained using a feature reduction scheme (DAFE or DBFE) that greatly reduced the number of useful bands.
- To make a joint spectral and spatial classification for urban area mapping is always useful while performing a feature reduction scheme to ameliorate the efficiency of classification algorithms and reduce computation time.
- Moreover, spectral analysis of the multispectral and hyperspectral data allows understanding that urban material requires a more accurate spectral characterization in the visible and near-infrared bands than that which is currently available from VHR multispectral sensors.

ACKNOWLEDGMENT

The authors thank Francesco Grassi and Alessio Ferrari for performing some of the classification tests.

REFERENCES

Benediktsson, J.A. and Kanellopoulos, I., Classification of multisource and hyperspectral data based on decision fusion, *IEEE Trans. Geosci. Remote Sensing*, 37(3), 1367–1377, 1999.

Briem, G.J., Benediktsson, J.A., and Sveinsson, J.R., Multiple classifiers applied to multi-source remote sensing data, *IEEE Trans. Geosci. Remote Sensing*, 40(10), 2291–2299, Oct. 2002.

Dell'Acqua, F., Gamba, P., Ferrari, A., Palmason, J.A., Benediktsson, J.A., and Arnason, K., Exploiting spectral and spatial information in hyperspectral urban data with high resolution, *IEEE Geosci. Remote Sensing Lett.*, 1(4), 322–326, 2004.

Dell'Acqua, D., Gamba, P., and Lisini, G., Urban land-cover mapping using hyperspectral and multispectral VHR sensors: spatial vs. spectral resolution, *Proc. URBAN2005*, Tempe, AZ, 14–16 Mar. 2005, IAPRS, XXXVI, Part 8/W27.

Forster, B., An examination of some problems and solutions in monitoring urban areas from satellite platforms, *Int. J. Remote Sensing*, 6(1), 139–151, 1985.

Gamba, P. and Houshmand, B., An efficient neural classification chain for optical and SAR urban images, *Int. J. Remote Sensing*, 22(8), 1535–1553, May 2001.

Gamba, P. and Dell'Acqua, F., Improved multiband urban classification using a neuro-fuzzy classifier, *Int. J. Remote Sensing*, 24(4), 827–834, Feb. 2003.

Gamba, P., Dell'Acqua, F., and Ferrari, A., Exploiting spectral and spatial information for classifying hyperspectral data in urban areas, *Proc. IGARSS'03*, July 2003, I, 464–466.

Gamba, P., A collection of data for urban area characterization, *Proc. IGARSS'04*, Anchorage, Alaska, Sept. 2004, I, 69–72.

Greenhill, D., Ripke, L.T., Hitchman, A.P., Jones, G.A., and Wilkinson, G.G., Use of lacunarity index to characterize suburban areas for land-use planning using IKONS-2 multi-spectral imagery, *Proc. 2nd IEEE/ISPRS Joint Workshop Remote Sensing Data Fusion Urban Areas*, Berlin, Germany, 22–23 May 2003, 294–298.

Guanaes Rego, L.F. and Koch, B., Automatic classification of land cover with high resolution data of the Rio de Janeiro city in Brazil, *Proc. 2nd IEEE/ISPRS Joint Workshop Remote Sensing Data Fusion Urban Areas*, Berlin, Germany, 22–23 May 2003, 172–176.

Guindon, B., Computer-based aerial image understanding: a review and assessment of its application to planimetric information extraction from very high resolution satellite images, *Can. J. Remote Sensing*, 23(1), 1997.

Heiden, U., Segl, K., Roessner, S., and Kaufmann, H., Ecological evaluation of urban biotope types using airborne hyperspectral HyMap data, *Proc. 2nd IEEE/ISPRS Joint Workshop Remote Sensing Data Fusion Urban Areas*, Berlin, Germany, 22–23 May 2003, 18–22.

Herold, M., Gardner, M.E., and Roberts, D.A., Spectral resolution requirements for mapping urban areas, *IEEE Trans. Geosc. Remote Sens.*, 41(9), 1907–1919, Sept. 2003.

Herold, M., Roberts, D., Gardner, M., and Dennison, P., Spectrometry for urban area remote sensing — development and analysis of a spectral library from 350 to 2400 nm, *Remote Sensing Environ.*, 91(3–4), 304–319, 2004.

Landgrebe, D.A., *Signal Theory Methods in Multispectral Remote Sensing*, John Wiley & Sons, Hoboken, NJ, 2003.

Lee, C. and Landgrebe, D.A., Decision boundary feature extraction for neural networks, *IEEE Trans. Geosci. Remote Sensing*, 8(1), 75–83, 1997.

Lisini, G., Dell'Acqua, F., Gamba, P., and Thompkinson, W., Image interpretation through problem segmentation for very high resolution data, *Proc. IGARSS'05,* Seoul (Korea), July 2005, 534–5637.

Marino, C.M., Panigada, C., and Busetto, L., Airborne hyperspectral remote sensing applications in urban areas: asbestos concrete sheeting identification and mapping, *Proc. 1st IEEE/ISPRS Joint Workshop Remote Sensing Data Fusion Urban Areas,* Rome, Italy, 8–9 Nov. 2001, 212–216.

Mueller, M., Segl, K., and Kaufmann, H., Discrimination between roofing materials and streets within urban areas based on hyperspectral, shape, and context information, *Proc. 2nd GRSS/ISPRS Joint Workshop Remote Sensing Data Fusion Urban Areas,* Berlin (Germany), 22–23 May 2003, 196–200.

Phinn, S., Stanford, M., Scarth, P., Murray, A.T., and Shy, P.T., Monitoring the composition of urban environments based on the vegetation–impervious surface–soil (VIS) model by subpixel analysis techniques, *Int. J. Remote Sensing,* 23(20), 4131–153, 2002.

Price, J.C., Spectral band selection for visible-near infrared remote sensing: spectral– spatial resolution trade-offs, *IEEE Trans. Geosci. Remote Sens.,* 35(5), 127–128, 1997.

Ridd, M.K., Exploring a V-I-S (vegetation–impervious surface–soil) model for urban ecosystem analysis through remote sensing: comparative anatomy for cities, *Int. J. Remote Sensing,* 16(12), 2165–2185, 1995.

Roessner, S., Segl, K., Heiden, U., and Kaufmann, H., Automated differentiation of urban surfaces based on airborne hyperspectral imagery, *IEEE Trans. Geosci. Remote Sens.,* 39(7), 1525–1532, 2001.

Shackelford, A.K., Davis, C.H., Wang, X., Automated 2-D building footprint extraction from high-resolution satellite multispectral imagery, *Proc. IGARSS04,* 3, 1996–1999, 2004.

Swayze, G.A., Clark, R.N., Goetz, A.F.H., Chrien, T.G., and Gorelick, N.S., Effects of spectrometer band pass, sampling, and signal-to-noise ratio on spectral identification using the tetracorder algorithm, *J. Geophys. Res.,* 108(E9), 5105, 2003.

Tarantino, C., D'Addabbo, A., Pasquariello, G., Blonda, P., Satalino, G., and Castellana, L., Remote sensed high resolution images for cartographic updating, *Proc. 2nd IEEE/ISPRS Joint Workshop Remote Sensing Data Fusion Urban Areas,* Berlin, Germany, 22–23 May 2003, 249–252.

Welch, R., Spatial resolution requirements for urban studies, *Int. J. Remote Sensing,* 3(2), 139–146, 1982.

Xiao, Q., Austin, S.L., McPherson, E.G., and Peper, P.J., Characterization of the structure and species composition of urban trees using high resolution AVIRIS data, *Proc. 1999 AVIRIS Workshop,* Pasadena, CA, 1999, unpaginated CD-ROM.

About the Contributors

Sharolyn Anderson is a geographic information scientist with interests in land-use and land-cover change. She obtained her Ph.D. from Arizona State University, Tempe. She travels the world in search of sustainable solutions to rapid urban growth problems.

Keith C. Clarke is a research cartographer and professor. He holds a B.A. degree with honors from Middlesex Polytechnic, London, England, and M.A. and Ph.D. degrees from the University of Michigan, specializing in analytical cartography. He joined the faculty at the University of California, Santa Barbara, in 1996. Dr. Clarke's most recent research has been on environmental simulation modeling, modeling urban growth using cellular automata, terrain mapping and analysis, and the history of the CORONA remote sensing program. He is the former North American editor of the *International Journal of Geographical Information Systems* and series editor for the Prentice Hall series in geographic information science.

Dr. Clarke is the author of the textbooks, *Analytical and Computer Cartography* (Prentice Hall, 1995) and *Getting Started with GIS* (1997), as well as over a hundred book chapters, journal articles, and papers in the fields of cartography, remote sensing, and geographic information systems. Since 1997, he has been the Santa Barbara director of the National Center for Geographic Information and Analysis. A member of the National Academy of Sciences Mapping Sciences Committee, Dr. Clarke recently chaired National Research Council studies on the National Map and the National Geospatial Intelligence Agency, and he served on the USGS geography discipline long-term science planning team, for which he received the USGS's John Wesley Powell Award in 2005.

Daniel Comarazamy received his M.Sc. degree in mechanical engineering from the University of Puerto Rico-Mayagüez in 200 and is currently a Ph.D. student in mechanical engineering at Santa Clara University.

Fabio Dell'Acqua obtained a first-class honors degree in electronics engineering at the University of Pavia, Italy, in 1996.

Manfred Ehlers is professor for GIS and remote sensing at the University of Osnabrueck, Germany. He is the director of the University's Research Center for Geoinformatics and Remote Sensing and the scientific director of the GiN — Center of Excellence in Geoinformatics, a ten-university consortium in North Germany. Dr. Ehlers teaches courses in geoinformatics, GIS, remote sensing, digital image processing, environmental monitoring, and environmental information systems. He has held academic appointments at various universities and research institutions in Germany, the United States, and the Netherlands. His research interest focuses on the concepts for integrated geographic analysis and modeling techniques, especially the integration of GIS and remote sensing. Other research areas include geoinformatics concepts, data fusion techniques, and advanced techniques for image analysis.

Dr. Ehlers has been the chairman of several international working groups on GIS and remote sensing for the International Society for Photogrammetry and Remote Sensing. He has organized and chaired a number of national and international conferences and workshops including the SPIE (International Society for Optical Engineering) conferences "Remote Sensing for Environmental Monitoring, GIS Applications and Geology," in Toulouse, France (2001); Aghia Pelagia, Crete, Greece (2002); Barcelona, Spain (2003); Maspalomas, Spain (2004); and Bruges, Belgium (2005). Over the last 10 years, Dr. Ehlers has been principal or coprincipal investigator on more than 50 funded research grants totaling over 7.5 million euros, as well as a number of research projects from different funding agencies, including U.S. National Science Foundation (NSF); European Union; German Research Foundation (DFG); German Ministry for Science and Technology (BMBF); German Ministry of the Interior (BMI); NASA; German Agency for Space Applications (DARA); Ministry of Environment, Lower Saxony; Ministry of Science, Lower Saxony; German Environmental Foundation (DBU); Federal Commission for Education and Research (BLK); University of California at Santa Barbara, California; International Institute for Aerospace Survey and Earth Sciences (ITC), The Netherlands; and the German Agency for Technical Cooperation (GTZ).

Dr. Ehlers has published more than 260 papers on GIS, mapping, digital image processing, remote sensing, geoinformatics, digital photogrammetry, and environmental monitoring. He has given over 280 presentations at scientific and professional meetings and conferences including keynote addresses at the 6th International Conference on Systems Research, Informatics and Cybernetics in Baden-Baden, Germany; the International Association of Pattern Recognition (IAPR) Technical Committee 7 (TC7) Workshop in Delft, The Netherlands; the Swiss Academy of Natural Sciences Symposium on GIS and Remote Sensing for Environmental Analysis in Zürich, Switzerland; the GEOINFO 2000 Conference in Santiago, Chile; the AGIT 2001 Symposium in Salzburg, Austria; the 2003 SPIE Conference on Remote Sensing for Environmental Monitoring, GIS Applications and Geology in Barcelona, Spain; and the Asia GIS 2003 Conference in Wuhan, China, as well as numerous invited papers at various geoinformatics, mapping and GIS, remote sensing, and photogrammetry conferences.

Christopher D. Elvidge is a senior scientist at the National Geophysical Data Center of NOAA in Boulder, Colorado. He obtained his Ph.D. in earth science from Stanford University. Chris is an ardent advocate of the utility of nighttime satellite imagery for myriad practical applications.

Paolo Gamba is currently associate professor of telecommunications at the University of Pavia, Italy. He received the laurea degree in electronic engineering *cum laude* from the University of Pavia, Italy, in 1989. He also received his Ph.D. degree in electronic engineering from the University of Pavia in 1993. In 1994 he joined the Department of Electronics of the university, where he is now an associate professor. Dr. Gambia is the recipient (first place) of the 1999 ESRI Award for Best Scientific Paper in Geographic Information Systems. He was chair of Technical Committee 7, "Pattern Recognition in Remote Sensing," of the International Association for Pattern Recognition (IAPR) from October 2002 to October 2004. He is a senior member of IEEE and the present cochair of the Data Fusion Committee of the IEEE Geoscience and Remote Sensing Society.

Dr. Gamba is the organizer and technical chair of the GRSS/ISPRS Joint Workshop on Remote Sensing and Data Fusion over Urban Areas series, whose first meeting was held in Rome in November 2001, the second in Berlin in May 2003, and the third in Tempe, Arizona, in March 2005. He is the guest editor of a special issue of the *ISPRS Journal of Photogrammetry and Remote Sensing* on "algorithm and techniques for multisource data fusion in urban areas"; a special issue of the *IEEE Transactions on Geoscience and Remote Sensing* on "urban remote sensing"; and a special issue of the *International Journal of Information Fusion* on "fusion of urban remotely sensed features." He has published nearly 40 papers in peer-reviewed journals on urban remote sensing and presented more than 100 papers at workshops and conferences.

Jorge E. González holds a B.Sc. and a master's degree from the University of Puerto Rico-Mayagüez (UPRM) and a Ph.D. degree from the Georgia Institute of Technology, all in mechanical engineering. He joined the faculty of the Department of Mechanical Engineering of UPRM after completing the Ph.D. requirements in 1994 and has held numerous appointments, including department chairman for 2000 to 2003. He is currently adjunct faculty at UPRM. He joined the Department of Mechanical Engineering of Santa Clara University in the fall of 2003 with an appointment as professor and David Packard Scholar; he teaches and conducts research in thermodynamics, heat transfer, renewable energy, and urban ecosystems.

Dr. González is actively involved in research related to applications of solar energy, low-energy buildings, sensor development, and atmospheric modeling. His research has been sponsored by the government of Puerto Rico, the U.S. Department of Energy, the National Science Foundation, the National Aeronautics and Space Agency, and several private industries. Dr. González has numerous publications in international refereed journals and in proceedings of international conferences, and holds two patents in solar energy equipment. He was recognized in 1997 by the National Science Foundation as a prominent young researcher with a prestigious CAREERS Award and received the 1999 Outstanding Mechanical Engineering Faculty Award of UPRM. He is a member of the American Society of Mechanical Engineers, the International Solar Energy Society, the International Association of Urban Climatology, and vice-chairmen of the Caribbean Solar Energy Society. In 2000, Dr. González founded the small technology business, Caribbean Thermal Technologies, based in Mayagüez, Puerto Rico. This company is one of only a few small technology businesses in Puerto Rico and has been the recipient of two Small Business Innovation Research Awards from the National Science Foundation.

Jeff Hemphill is a Ph.D. candidate in the University of California, Santa Barbara, Geography Department. His areas of expertise are GIS, aerial photography, and cartography. The research emphasis of his dissertation is theoretical urban growth, the dynamics of urban change. As a graduate student researcher, he has been involved in numerous research projects; the skills he has developed working on various aspects of geospatial data compilation and analysis cover a broad range of geographic research topics.

Initially working for the late Dr. John E. Estes in the Remote Sensing Research Unit, he gained valuable experience; later, under the guidance of Dr. Keith Clarke he developed his interests in urban growth modeling. While working for the late Dr. Estes, he was funded by the California Sea Grant Cooperative Extension Program to compile the California Marine Protected Areas Database (MPA) from 1998 to 1999. He also worked for the Department of Justice, Environmental Division, in 1999 and 2000 doing wetlands mapping. In 2002, he was awarded the Academic Senate Outstanding Teaching Assistant Award and the Geography Department's Excellence in Teaching Award for his efforts in the classroom and for creating the contents of the *Remote Sensing Core Curriculum — Volume 1* (RSCC). This project was sponsored by the International Center for Remote Sensing Education (ICRSEd), International Society of Photogrammetry and Remote Sensing (ISPRS), National Aeronautics and Space Administration (NASA), and UCSB's National Center for Geographic Information and Analysis (NCGIA).

Geoffrey M. Henebry was born in Champaign, Illinois, in 1960. He received a B.A. in liberal arts (great books) from St. John's College (Santa Fe) in 1982 and an M.S. and a Ph.D. in environmental sciences from the University of Texas at Dallas in 1986 and 1989, respectively. He is currently a professor of biology and geography and a senior research scientist at the Geographic Information Science Center of Excellence at South Dakota State University. His research interests are broad and varied, but a recurrent theme is the use of remote sensing to study ecological patterns and processes, including quantitative analysis and modeling of land-surface phenology and land-cover and land-use change. He has worked across a wide spatiotemporal–spectral range of observational data: wavelengths from 10^{-8} to 10^{-2} m, periods from 10^1 to 10^7 s, and areas from 10^{-1} to 10^{12} m².

Martin Herold was born in Leipzig, Germany, in 1975. He received his first graduate degree (diploma in geography) in 2000 from the Friedrich Schiller University of Jena and the BAUHAUS University of Weimar, Germany, and his Ph.D. at the Department of Geography, University of California at Santa Barbara in 2004. Dr. Herold is currently coordinating the ESA GOFC GOLD Land Cover project office at the Friedrich Schiller University, Jena, Germany. In his earlier career, he was interested in multifrequency, polarimetric, and interferometric SAR data analysis for land-surface parameter derivation and modeling. He joined the Remote Sensing Research Unit, University of California at Santa Barbara in 2000, where his research focused on remote sensing of urban areas and the analysis and modeling of urban growth and land-use change processes.

Dr. Herold's most recent interests are in international coordination and cooperation towards operational terrestrial earth observations with specific emphasis on harmonization and validation of land-cover data sets. He is an expert in the field of remote sensing and digital processing and modeling of geographic data, as well as local and regional planning. His publications include more than 20 peer-reviewed journal articles and book chapters. During his university career, he taught several classes in processing and analysis of remote sensing and spatial data, imaging spectrometry, GIS, and modeling of spatial processes.

Patrick Hostert received his first degree in physical geography from Trier University, Germany, in 1994; an M.Sc. in geographical information systems from Edinburgh University, United Kingdom, in 1995; and a Ph.D. in remote sensing from Trier University in 2001. His M.Sc. and Ph.D. degrees with distinction were awarded with dissertation prizes from Edinburgh University and Trier University, respectively. From 2002 until 2005 he held an assistant professorship in urban remote sensing and GIS at Humboldt Universität zu Berlin, Germany. He has recently received a full professorship in geomatics at Humboldt Universität. His research topics center on remote sensing and GIS for environmental monitoring, analysis, and modeling in different environments, with a major focus on urban environmental change. Further research includes transboundary studies and land degradation and desertification issues. Hyperspectral image analysis, VHR satellite data, and time series analysis of satellite data are of special interest.

John A. Kelmelis is senior counselor for earth science at the Department of State. He is on extended detail from the U.S. Geological Survey. At USGS he is senior science advisor for international policy. He has held various other positions including chief scientist for geography and manager of the branch of Research and Applications for the National Mapping Division.

Guiying M. Li was born in Shandong, China, in 1969. She received a B.A. in forestry in 1989 and an M.A. in silviculture in 1992 from Nanjing Forestry University, and an M.A. in geography from Indiana State University in 2003. Currently, she is a Ph.D. candidate in that department. Her research interests include integration of remote sensing and GIS in urban social study, such as population estimation, quality of life, and land-use and land-cover change. She was the recipient of the Outstanding Research Award and the Benjamin Moulton Geography Award by her department and has worked on the Terre Haute Urban Mapping Project and Indiana Department of Natural Resources Habitat Mapping Project during her study at Indiana State.

Weiguo Liu received his Ph.D. degree from Boston University in 2001 and is an assistant professor in the Department of Geography and Planning of The University of Toledo. His research interests range from theory to applications of GIS, remote sensing, and artificial neural networks (ANNs). His current research efforts include integrating spatial data mining (geographical knowledge discovery) and GIS to resolve real-world geographic problems related to environmental health and land-use change (especially urbanization monitoring, simulation, and prediction).

Xiaohang Liu is an assistant professor at San Francisco State University. She received her Ph.D. in geography from the University of California, Santa Barbara, in 2003, and M.Sc. degrees in geography and computer science from Rutgers—the State University of New Jersey in 1999. Her research interests are in using GIS, remote sensing, and spatial statistics to examine urban environmental issues. Dr. Liu has been involved in research on urban growth modeling, population estimation using remote sensing, community-based GIS, and health disparity in the San Francisco Bay area.

Dengsheng Lu is currently an assistant research scientist at the Center for the Study of Institutions, Population, and Environmental Change at Indiana University. He is a forester and biogeographer by training and received his Ph.D. in geography from Indiana State University. Before that, he received a B.A. in forestry from Zhejiang Forestry College in 1986, and an M.A. in forestry from Beijing Forestry University in 1989. Dr. Lu specializes in remote sensing and its application to the study of forests, land use, land cover, and biomass estimation.

Jeffrey C. Luvall is currently employed by NASA as a senior research scientist at Marshall Space Flight Center. He holds a B.S. (1974, forestry) and an M.S. (1976, forest ecology) from Southern Illinois University, Carbondale, and a Ph.D. (1984, tropical forest ecology) from the University of Georgia, Athens. His current research involves the modeling of forest canopy thermal response using airborne thermal scanners on a landscape scale. He is also investigating the relationships of forest canopy temperatures and the evapotranspiration process. He has used remotely sensed surface temperatures to develop evapotranspiration estimates for eastern deciduous and tropical rain forests. These investigations have resulted in the development of a thermal response number (TRN), which quantifies a land surface's energy response in terms of $kJ\ m^{-2}\ C^{o-1}$, which can be used to classify land surfaces in regional surface budget modeling by their energy use. A logical outgrowth of characterizing surface energy budgets of forests is the application of thermal remote sensing to quantify the urban heat island effect. One important breakthrough is the ability to quantify the importance of trees in keeping the city cool. His current research involves alternate mitigation strategies to reduce ozone production through the use of high-albedo surfaces for roofs and pavements and increasing tree cover in urban areas to cool cities.

Dr. Luvall's recent work on urban heat islands was the focus of several CNN, CBN, CBS Evening News, NBC, and ABC Discovery News programs during 1998. It was also featured in a November 23, 1998, *Newsweek* article "Blue Skies Ahead: Hot Ways to Cool Down Our Cities." He worked closely with Salt Lake City's 2002 Olympic Organizing Committee in revitalizing the city by planting greenways and high-albedo surface materials. Dr. Luvall was invited by the USSR Academy of Sciences and the United Nations Environment Program to speak at the International Symposium on the State of the Art of Remote Sensing Technology for Biosphere Studies in Moscow (September 1989). He was invited to participate in and coauthored a paper at the Space Conference of the Americas in San Jose, Costa Rica, in 1990 and was an invited delegate in August 1991. Dr. Luval was a member of the steering committee for organizing a symposium, "Thermal Remote Sensing of the Energy and Water Balance over Vegetation in Conjunction with Other Sensors," held in La Londe Les Maures, France in 1993. He organized a symposium at the Intecol 1994 meeting at Manchester, England: "A Thermodynamic Perspective of Ecosystem Development" (with J. Kay and E. Schneider). Dr. Luval was appointed to serve from 1994 to 1997 on the La Selva advisory committee by the Organization for Tropical Studies (OTS). La Selva is a biological research field station in Costa Rica funded by the National Science Foundation. OTS is a consortium of 50 U.S. and international universities that manage several field stations and courses in Costa Rica.

Soe W. Myint received his doctorate in geography from Louisiana State University. Before attending LSU for his Ph.D., he spent 4 years in the United Nations Environment Program — Environment Assessment Program for Asia and the Pacific as a research specialist. Prior to joining ASU Geography, he was an assistant professor of GIScience in the Department of Geography at the University of Oklahoma for 4 years. He received four best student paper awards and two second-place paper awards at different professional meetings during his Ph.D. study. He also received a USGS Scholar Award at the first international conference on GIScience in 2000 and an Intergraph Young Scholar Award at the UCGIS 2002 meeting.

Dr. Myint has been selected for inclusion in the 60th diamond edition of *Who's Who in America* (October 2005) and the 7th edition of *Who's Who in American Education* (December 2005) by Marquis Who's Who in America. He received a research grant from the NASA EPSCoR program to examine and develop fractal- and lacunarity-based approaches. He has been awarded a grant by NASA through the Institute for Advanced Education in Geospatial Sciences to develop a model curriculum in geospatial sciences highlighting the application of GIScience technologies for community growth. Recently, Dr. Myint's research efforts have focused on frequency-based, multiscale, multidecomposition techniques in comparison to other advanced geospatial approaches to identify land-use and land-cover classes effectively for his single PI NSF project.

Janet Elizabeth Nichol has a background in physical geography, specializing in biogeography and remote sensing. She obtained her B.Sc. at London University, her M.A. at the University of Colorado, and her Ph.D. at the University of Aston in Birmingham. She has subsequently worked in the United Kingdom, Nigeria, Singapore, and the Republic of Ireland as a university lecturer. Since 2001, Dr. Nichol has been an associate professor at the Department of Land Surveying and GeoInformatics of The Hong Kong Polytechnic University. Her research interests are in the application of remote sensing techniques to environmental assessment and monitoring, including the urban heat island, urban environmental quality, vegetation mapping, landslide hazard assessment, air quality monitoring, and aspects of image processing. She has published widely on these topics and is a reviewer for journals specializing in remote sensing, planning, and environmental issues.

Ana J. Picón received her M.Sc. degree in electrical engineering from the University of Puerto Rico-Mayagüez in 2005 and is currently a Ph.D. student in computer sciences and engineering at the same institution. She is a fellow of the NASA Harriett G. Jenkins Predoctoral Fellowship Program.

Douglas L. Rickman is employed by NASA at the Marshall Space Flight Center. He received his doctorate at the University of Missouri, Rolla, in 1981. His training and experience cover a broad range: mineral exploration, image processing, remote sensing, software development, medical applications, and computer systems integration. His current work includes public health applications of satellite data, leading the applied sciences effort at MSFC/NSSTC, and land process and airborne remote sensing research. In 1994 he was inducted into the Space Technology Hall of Fame for his contributions to image processing of MRI images of the human head.

Dar A. Roberts was born in Torrance, California, in 1960. He received a B.A. with a double major in environmental biology and geology from the University of California, Santa Barbara, in 1982; an M.A. in applied earth sciences from Stanford University in 1986; and a Ph.D. in geological sciences from the University of Washington in 1991. He is currently a professor of geography at the University of California, Santa Barbara, where he has taught since 1994. He has published over 60 articles in refereed journals, contributed 14 book chapters, and published over 100 nonrefereed papers and proceedings. His primary research interests are in spectroscopy, land-use and land-cover change, fire danger assessment, and vegetation analysis, primarily using remote sensing.

Dr. Roberts has worked with a large variety of sensors, including hyperspectral thermal (SEBASS), several hyperspectral VNIR sensors (AIS, HYDICE, Hyperion, HYMAP, AVIRIS), active sensors (SAR, LIDAR, IFSAR), and broadband data (MSS, ETM+, TM, IKONOS, MODIS). Research sites include a diversity of sites in North America, all of North Africa, Madagascar, and the Brazilian Amazon. He has been a major participant in several large campaigns, including DOE-sponsored research at the Wind River Canopy Crane site in south-central Washington, LBA in Brazil, and, most recently, the North American Carbon Program. Recently, he has worked in urban environments, studying the spectral properties of urban materials and evaluating methods for mapping urban infrastructure, including road quality. He teaches advanced courses in optical and microwave remote sensing.

Aparajithan Sampath received his bachelor's degree in geoinformatics engineering from Anna University, India, and M.Sc degree in geomatics engineering from Purdue University, Layfayette, Indiana. Currently, he is a Ph.D candidate at Purdue University. His research interests are LIDAR mapping and spatial data handling.

Sebastian Schiefer was born in Düsseldorf, Germany, in 1976. He has studied at the University of Trier, Germany, and the University of Edinburgh, Scotland, and received a master's degree in applied environmental sciences in 2002. He is currently working at the Institute of Geography of the Humboldt Universität zu Berlin and the Center for Remote Sensing of Land Surfaces, Bonn, within the scholarship program of the German Environmental Foundation (DBU). His research focuses on urban remote sensing and the processing of imaging spectrometry data. In this and other contexts, he has worked intensively with the normalization of bidirectional effects as well as transformation and hierarchical classification of hyperspectral data.

Karen C. Seto is a center fellow with the Freeman Spogli Institute for International Studies at Stanford University and an assistant professor in the Department of Geological and Environmental Sciences. Her research focuses on monitoring urban growth trajectories, understanding the causes of land-use change, and evaluating social and ecological impacts of land-use dynamics.

Jie Shan received his Ph.D degree in photogrammetry and remote sensing from Wuhan University, China. Since then he has worked as a faculty member at universities in

China, Sweden, and the United States and has been a research fellow and visiting professor in Germany. Currently, he is an associate professor in geomatics engineering at the School of Civil Engineering of Purdue University, Layfayette, Indiana, where he teaches subjects in photogrammetry and remote sensing, and geographic information science. His research interests are digital mapping, geospatial databases, and urban modeling.

Uwe Soergel was born in Zell (Mosel), Germany, in 1969. In 1997 he received a diploma degree (Dipl. Ing.) in electrical engineering from the University of Erlangen, Nuremberg, Germany. From fall 1997 until the end of 2005 he was a research assistant at the Institute of Optronics and Pattern Recognition (FOM) of FGAN, a German research establishment for defense-related studies. He dealt mainly with pattern recognition of man-made objects from remote sensing imagery, with an emphasis on SAR data. In 2003 he received a doctoral degree (equivalent of a Ph.D.) from the faculty of electrical engineering and information processing of the University of Hanover, Germany. Since January 2006 he has been an assistant professor for radar remote sensing at the Institute of Photogrammetry and GeoInformation on the faculty of civil engineering and geodesy there.

Conghe Song was born in Anhui province, China in December 1965. He received his bachelor's degree in forestry from Anhui Agricultural University in 1988, a master's degree in forest ecology from Beijing Forestry University in 1991, and a Ph. D. degree in geography from Boston University in 2001. He is currently an assistant professor in the Department of Geography at the University of North Carolina at Chapel Hill. He has been a principal investigator with support from the National Science Foundation, the National Aeronautics and Space Administration, and the USDA Forest Service. He was a recipient of the Charles Bullard Fellowship Award from Harvard University in 2005. His primary research interests include remote sensing of vegetation, ecological modeling of terrestrial ecosystem energy, and water and carbon budgets in the context of land-cover and land-use change on a regional scale.

Uwe Stilla was born in Cologne, Germany, in 1957. In 1980 he received a diploma degree (Dipl. Ing.) in electrical engineering from Gesamthochschule Paderborn, Germany, and in 1987 he received an additional diploma degree (Dipl.-Ing.) in biomedical engineering from the University of Karlsruhe, Germany. From 1990 until 2004 he was with the Institute of Optronics and Pattern Recognition (FGAN-FOM), a German research establishment for defense-related studies. In 1993 he received his Ph.D. in engineering from the University of Karlsruhe with work in the field of pattern recognition. Since 2004 Dr. Stilla has been head of the Department of Photogrammetry and Remote Sensing, and he is currently director of the Institute of Photogrammetry and Cartography at the Technical University of Munich. He is student dean for the diploma, bachelor's, and master program "Geodesy and Geoinformation" in the faculty of civil engineering and geodesy. Additionally, he is involved in the new master's course "Earth-Oriented Space Science and Technology." His research focuses on image analysis in the field of photogrammetry and remote sensing.

Zhanli Sun is a postdoctoral research associate in the Department of Urban and Regional Planning, University of Illinois at Urbana–Champaign. He received his Ph.D. in geographic information systems and cartography from the Institute of

Geography, Chinese Academy of Sciences, 1999. He has broad research interests in spatial information theory and technology, including spatial modeling, spatial data handling, and system development. He has recently focused his research interests upon decision support systems (DSSs) for urban planning by employing system dynamic (DS) concepts and cellular automata.

Paul C. Sutton is a population geographer with interests in sustainability and human–environment interactions. He obtained his Ph.D. from the University of California at Santa Barbara. He lives in Conifer, Colorado, with his wife and son.

Matthew J. Taylor is a cultural geographer who specializes in human–environment interaction in Latin America. He obtained his Ph.D. from Arizona State University, Tempe. He lives in an East Bulgarian neighborhood of Denver, Colorado, with his wife and two daughters.

Yong Tian is an assistant professor in the Department of Environmental, Earth, and Ocean Sciences at the University of Massachusetts, Boston. He is also the director of the Certificate Program in Geographic Information Technology. His research has been in the area of mathematical modeling and computer simulation of ecosystem dynamics for environmental and sustainable production management.

Qihao Weng was born in Fuzhou, China, in 1964. He received a B.A. degree in geography from Minjiang University in 1984, an M.Sc. degree in physical geography from South China Normal University in 1990, an M.A. in geography from the University of Arizona in 1996, and a Ph.D. in geography from the University of Georgia in 1999. He is currently an associate professor of geography, and director of the Center for Urban and Environmental Change at Indiana State University. His research focuses on remote sensing and GIS analysis of urban ecological and environmental systems, land-use and land-cover change, urbanization impacts, and human–environment interactions. He is the author of more than 35 peer-reviewed journal articles and numerous book chapters, and was the recipient of the Robert E. Altenhofen Memorial Scholarship Award from the American Society for Photogrammetry and Remote Sensing (1999) and the Best Student Authored Paper Award from the International Geographic Information Foundation (1998). He has worked extensively with optical and thermal remote sensing data, primarily for urban heat island study, land-cover and impervious surface mapping, urban growth detection, spectral mixture analysis, and socioeconomic characteristics derivation.

Man Sing Wong was born in Hong Kong in 1980. He received his first degree, bachelor of surveying and geo-informatics, from the department of Land Surveying and Geo-Informatics of The Hong Kong Polytechnic University in 2003, and his M.Phil. degree in remote sensing and GIS in the same department in 2005. He is currently a Ph.D. candidate research fellow and Fulbright scholar, visiting the University of Maryland for research on air-quality monitoring using remote sensing. He is the official site manager for the AERONET station in Hong Kong, which forms a part of a collaborative agreement with NASA. His main research interests are in the use of remote sensing to study urban heat islands, urban environmental quality, landslides, vegetation and ecosystems, spectral mixture analysis, and aerosol mapping and monitoring.

George Xian is currently working for Science Applications International Corporation (SAIC) at the U.S. Geological Survey Center for Earth Resources Observation and Science (EROS). He received a B.Sc. from Yunnan University in Kunming, China; an M.Sc. from Colorado State University; and a Ph.D. from the University of Nevada, Reno. He has worked for several projects related to satellite remote sensing information systems, urban environments, and urban dynamic research conducted by USGS, NASA, and the U.S. EPA since 1997 at EROS. Dr. Xian has conducted urban growth and environmental influence research using remote sensing information for several metropolitan areas, including Chicago, Illinois; Tampa, Florida; Las Vegas, Nevada; Atlanta/Columbus, Georgia; and Detroit, Michigan. His research interests include urban land-use and land-cover change detection and impacts on regional climate change and air quality. He has published several articles on these topics.

Guoqing Zhou received his Ph.D. degree in remote sensing and information science from the Wuhan University in 1995. He then worked in the Department of Computer Science and Technology of Tsinghua University and Information Science Institute of Beijing University as a visiting scholar and postdoctoral researcher. After he had worked at the Technical University of Berlin, Germany, for more than a year, he worked as a postdoctoral researcher at the Ohio State University from 1998 through 2000. Since August 2000, he has worked at Old Dominion University as an associate professor.

Dr. Zhou has had a 20-year career in teaching and researching remote sensing and information engineering. He has published three books and 114 publications and worked on 45 projects as principal or coprincipal investigator. He serves as chair, editorial board member, editorial advisor, or editor-in-chief of several journals, and serves as chair and cochair of working groups of several international societies and organizations. Dr. Zhou has won five prestigious national or international awards: Alexander von Humboldt Award (Germany), the Talbert Abrams Award (U.S.), the NASA–NIAC Fellow Award, the Outstanding Young Author Best Paper Award (Japan), and Outstanding Innovative Technology Teaching Award (U.S.).

Index